Springer Series in Synergetics   Editor: Hermann Haken

Synergetics, an interdisciplinary field of research, is concerned with the cooperation of individual parts of a system that produces macroscopic spatial, temporal or functional structures. It deals with deterministic as well as stochastic processes.

Volume 1 **Synergetics** An Introduction   2nd Edition
By H. Haken

Volume 2 **Synergetics** A Workshop
Editor: H. Haken

Volume 3 **Synergetics** Far from Equilibrium
Editors: A. Pacault and C. Vidal

Volume 4 **Structural Stability in Physics**
Editors: W. Güttinger and H. Eikemeier

Volume 5 **Pattern Formation** by Dynamic Systems and **Pattern Recognition**
Editor: H. Haken

Volume 6 Dynamics of **Synergetic Systems**
Editor: H. Haken

Volume 7 Problems of **Biological Physics**
By L. A. Blumenfeld

Volume 8 **Stochastic Nonlinear Systems** in Physics, Chemistry, and Biology
Editors: L. Arnold and R. Lefever

Volume 9 Numerical Methods in the **Study of Critical Phenomena**
Editors: J. Della Dora, J. Demongeot, and B. Lacolle

Volume 10 **The Kinetic Theory of Electromagnetic Processes**
By Yu. L. Klimontovich

Volume 11 **Chaos and Order in Nature**
Editor: H. Haken

Volume 12 **Nonlinear Phenomena in Chemical Dynamics**
Editors: C. Vidal and A. Pacault

Volume 13 **Handbook of Stochastic Methods** for Physics, Chemistry and the Natural Sciences
By C. W. Gardiner

Volume 14 **Concepts and Models of a Quantitative Sociology**
The Dynamics of Interacting Populations
By W. Weidlich and G. Haag

Volume 15 **Noise Induced Transitions.** Theory and Applications in Physics, Chemistry, and Biology
By W. Horsthemke and R. Lefever

Volume 16 **Physics of Bioenergetic Processes**
By L. A. Blumenfeld

Volume 17 **Evolution of Order and Chaos** in Physics, Chemistry, and Biology
Editor: H. Haken

Volume 18 **The Fokker-Planck-Equation** By H. Risken

Yu. L. Klimontovich

The Kinetic Theory of Electromagnetic Processes

Translated by A. Dobroslavsky

Springer-Verlag Berlin Heidelberg New York 1983

Professor Yuri L. Klimontovich, Ph.D.

Department of Physics, Moscow State University, Moscow 117234, USSR

Translator:
A. Dobroslavsky

115551 Moscow, Kashirskoe shosse, 102-2-358 USSR

Series Editor
Professor Dr. Hermann Haken

Institut für Theoretische Physik der Universität Stuttgart, Pfaffenwaldring 57/IV
D-7000 Stuttgart 80, Fed. Rep. of Germany

Title of the original Russian edition:
Kineticheskaya teoriya elektromagnitnykh protsessov
© by "Nauka" Publishing House, Moscow 1980

ISBN-13:978-3-642-81824-0   e-ISBN-13:978-3-642-81822-6
DOI: 10.1007/978-3-642-81822-6

Library of Congress Cataloging in Publication Data. Klimontovich, ÎU. L. (ÎUriĭ L'vovich). The kinetic theory of electromagnetic processes. (Springer series in synergetics ; v. 10). Translation of: Kineticheskaĭa teoriĭa elektromagnitnykh protsessov. Includes bibliographical references and index. 1. Many-body problem. 2. Electromagnetic theory. 3. Quantum field theory. 4. Matter, Kinetic theory of. I. Title. II. Series. QC174.17.P7K5513 1982 530.1'41 82-10586

Typesetting: K + V Fotosatz, Beerfelden

2153/3130-543210

Preface

The best developed of today's kinetic theories are those for gases and completely ionized plasmas. In recent years, however, kinetic theories of more complicated systems — consisting of free particles as well as those bound in atoms, and an electromagnetic field — have played an increasingly important role. An example of such a system is a partially ionized plasma of gas discharges or in semiconductors. The main purpose of this book is the further development of the kinetic theory of systems of this kind.

Naturally, it would be impossible to encompass at once all the problems concerning the kinetic theory of these extremely complicated systems. This book is mainly concerned with processes dominated by weak but collective interactions of charged particles and atoms, as well as processes determined by the interaction with an electromagnetic field. These topics determined the method adopted here for constructing the kinetic equations of the distribution functions for free and bound charged particles. The results of contemporary scattering theory make it possible to take strong interactions, which are interpreted as collisions, into account without any basic difficulties. More complicated, however, is the task of taking both strong interactions at small distances and weak but collective interactions into account simultaneously. The solution of this problem would open an approach to a number of fundamental questions, one of which is the construction of a kinetic theory of nonideal chemically reacting systems of charged particles.

The present book is divided into two parts, which deal with classical and quantum theories, respectively. In the classical approach the starting point is a closed set of microscopic equations for the phase density of charged particles and microscopic field strengths. In quantum theory the starting point is a corresponding set of operator equations, which represents the Heisenberg approach in quantum electrodynamics. The method of constructing a kinetic theory adopted here can be considered a further development of the method employed earlier by the author in obtaining the kinetic equations for gases and for completely ionized plasmas. The present study is in this respect a sequel to earlier ones by the author.

The scope of problems discussed in this book is very wide. It includes the statistical foundation of Maxwell equations, the kinetic equations for the distribution functions of free and bound charged particles, the kinetic theory of fluctuations in a chemically reacting system (partially ionized plasma), the kinetic theory of spectral line broadening of atoms' radiation, the influence of the correlations of atoms' positions on the processes of spontaneous and induced emission of radiation, light absorption in nonideal systems, the fluctuation dissipation theorem for nonequilibrium states, fluctuations in classical and quantum self-oscillatory systems, and equilibrium and induced phase transitions in a system of atoms and a field. Naturally, it would be impossible to give an equally exhaustive analysis of each problem, so some of our models may perhaps seem overly simplified. This drawback is hopefully exculpated by the fact that all the problems mentioned are treated according to a single common approach, allowing consideration of the subject in its entirety. But, of course, it is up to the reader to judge to what extent the author has succeeded in this.

This book is based on work the author has performed at different times; the earliest was described in a volume written in cooperation with Vassily S. Fursov, under whose guidance the author prepared his graduation thesis. I would like to take this occasion to express my gratitude for his benevolent attitude, good advice, and help.

Certain questions were discussed at one time or another with F. V. Bunkin, D. N. Zubarev, L. V. Keldysh, V. S. Lisitsa, L. P. Pitayevsky, R. L. Stratonovich, and many others. I am sorry not to be able to name everyone here to whom my thanks are due for their friendly concern and assistance.

Certain sections of this book employ results from the collaboration of my students and colleagues, from whom I also received advice and help. I am grateful to E. A. Asmaryan, V. V. Belyi, V. I. Yemelianov, A. S. Kovalyov, Yu. A. Kukharenko, S. N. Luzgin, M. A. Osipov, and S. A. Sukhin.

Moscow, September, 1982   *Yu. L. Klimontovich*

Contents

1. Introduction 1

Part I Classical Theory

2. Free Charged Particles and a Field 8
  2.1 The Equations of Particle Motion. Local Field 8
  2.2 The Equations for Microscopic Field Strengths (Lorentz Equations) 9
  2.3 A Coulomb Plasma 13
  2.4 The Complete Set of Microscopic Equations for a Plasma 14
  2.5 The Equation of Motion of Free Charged Particles and
      Field Oscillators 14
  2.6 Lagrange Function 17
  2.7 Hamiltonian Function 18
  2.8 The Equations of Motion for Phase Densities of Particles and
      Field Oscillators 19
  2.9 Distribution Functions of Particles and Field Oscillators 21
  2.10 Chain of Equations for Distribution Functions of Particles and
      Field Oscillators 23
  2.11 Equations for Moments 25
  2.12 The Relation Between Moments and Distribution Functions 27

3. Atoms and Field 30
  3.1 The Equations of Motion of Pairs of Free Charged Particles
      and Atoms 30
  3.2 The Equations for Microscopic Phase Density of Atoms 33
  3.3 Microscopic Field Equations 36
  3.4 Lagrange Function 42
  3.5 Hamiltonian Function 45
  3.6 The Closed Equation for Phase Density of Atoms 48
  3.7 The Interaction of Atoms 49
  3.8 Atoms and Field Oscillators 50
  3.9 The Method of Distribution Functions for a System of Atoms and
      Field Oscillators 54

4. The Kinetic Equations for a System of Free Charged Particles
  and a Field 56
  4.1 The Principal Parameters (Free Charged Particles) 56

4.2   Principal Parameters for a System of Atoms 58
      4.2.1   Interaction Parameters 58
      4.2.2   The Relaxation Processes Parameters for Atoms 59
      4.2.3   Comparison of the Density Parameters ε_d and ε_{em} 61
4.3   First Moments Approximation 62
4.4   The Kinetic Equations for a Coulomb Plasma 65
4.5   Electromagnetic Interaction in the Kinetic Equations of a Plasma   70
4.6   The Polarization Approximation for the System of Charged
      Particles and Field Oscillators 73
4.7   The Equilibrium Fluctuations of an Electromagnetic Field 78
4.8   The Kinetic Theory of Fluctuations 81
4.9   Nonequilibrium Fluctuations of the Field 89

5. Brownian Motion 93
5.1   The Langevin Equations 93
5.2   The Fokker – Planck Equation 94
5.3   Diffusion of Brownian Particles 97
5.4   The Brownian Motion of a Harmonic Oscillator.
      The Nyquist Formula 99
5.5   Nondissipative Nonlinearity. Brownian Motion at Phase
      Transitions 100
      5.5.1   Fokker – Planck and Einstein – Smoluchowsky Equations   100
      5.5.2   The Equilibrium Distribution 101
      5.5.3   Energy Fluctuations 103
      5.5.4   Self-Consistent Approximation with Respect to Energy ... 104
      5.5.5   The Order Parameter 105
      5.5.6   Fluctuations of the Order Parameter 106
5.6   Spectral Distribution of the Mean Energy 107
5.7   The Response of the System to External Factors 108
      5.7.1   Dynamic Response. Spectral of Fluctuations 110
      5.7.2   The Critical Region 111
5.8   Phase Transition in a Distributed System 111
      5.8.1   A Langevin Source in the Ginzburg – Landau Equation .. 111
      5.8.2   The Landau Theory 113
      5.8.3   Fluctuations of the Order Parameter 113
      5.8.4   Spatial Correlations 114
      5.8.5   Extrapolation of Landau Theory into the Critical Region . 116
      5.8.6   Critical Indices 117
      5.8.7   Coordination of Limit Transitions 118
      5.8.8   The Critical Point 120
      5.8.9   The Region of Scale Invariance 120
      5.8.10 The Transition to the Results of Landau Theory 121
      5.8.11 Correlation Times and Spectrum Widths at a Phase
              Transition 122
5.9   Dissipative Nonlinearity 125
5.10 The Langevin Equations for a Self-Oscillatory System.
      The Fokker – Planck Equation 127

5.11 The Stationary Distribution of the Energy of Oscillations 131
5.12 Fluctuations of Amplitude. Diffusion of a Phase 133
5.13 Spectral Distribution of the Energy of Autooscillations 135
5.14 The Response to Resonant Force 139
5.15 Kinetic Theory of Fluctuations in Brownian Motion 143

6. **Kinetic Equations for an Atom – Field System** 147
   6.1 Electromagnetic Fluctuations in a Gas 147
   6.2 The Kinetic Equation. The Collision Integral 152
   6.3 The Equation for the Polarization Vector 157
   6.4 The Effective Lorentz Field 161
   6.5 Dissipative Processes Due to Close Correlations 163
   6.6 The Equation for the Polarization Vector 165
   6.7 The Dielectric Permittivity. The Lorentz – Lorenz Formula.
        The Equation of Dispersion 167
   6.8 Fluctuations of Polarization and Field at Above-Critical
        Temperatures 170
   6.9 Phase Transition in an Atom – Field System 173
        6.9.1 Initial Equations 173
        6.9.2 Fluctuations of the "Source" δP^{source} 175
        6.9.3 Induced Fluctuations of the Polarization Vector 176

Part II **Quantum Theory**

7. **Microscopic Equations** 180
   7.1 A System of Free Charged Particles with Coulomb Interaction ... 180
   7.2 Partially Ionized Plasma 184
   7.3 The Hamiltonian with Electromagnetic Interaction (Extreme Cases) 188
   7.4 The Equations for Operators of Field and Particles 189
   7.5 Operator Equations for an Atom – Field System in Dipole
        Approximation 191

8. **The Kinetic Equations for Partially Ionized Plasma.**
   The Coulomb Approximation 194
   8.1 The Polarization Approximation 194
   8.2 The Correlation of the Source Fluctuations 196
   8.3 Dielectric Permittivity 199
   8.4 The Spectral Density of the Electric Field Fluctuations 205
   8.5 The Collision Integral 207
   8.6 The Structure of Collision Integrals 209
   8.7 The Equations for Concentrations 213
   8.8 The Kinetic Theory of Fluctuations in Partially Ionized Plasma .. 215

9. **Kinetic Equations for Partially Ionized Plasma. The Processes**
   Conditioned by a Transverse Electromagnetic Field 220
   9.1 Dielectric Permittivity 220

9.2   The Spectral Density of Transverse Field Fluctuations 222
9.3   The Collision Integral 223
9.4   The Structure of the Collision Integrals for the Transparency
      Region 225
9.5   The Evolution of the Distribution Function of Atoms 226
9.6   The Equations for Concentrations of Free Charged Particles
      and Atoms. The Contribution from the Interaction of Particles
      and Waves 230
9.7   Cooling and Heating of Atoms by Resonant Field.
      Classical Theory 232
9.8   Cooling and Heating of Atoms by a Resonant Field.
      Quantum Theory 235

10. Spectral Emission Line Broadening of Atoms
    in Partially Ionized Plasma 238
10.1 The Foundations of the Kinetic Theory of Spectral Line
      Broadening 238
10.2 The Dissipative Matrix. The Frequency Shift 243
10.3 The Influence of the Source Fluctuations on Linewidth and
      Frequency Shift 245
10.4 The Probabilities of the Transition. The Broadening at
      Spontaneous and Induced Processes 247
10.5 Spectral Line Broadening by a Plasma's Electrons 250
10.6 Resonant Broadening of Spectral Lines Due to Atoms'
      Collisions 253
10.7 Spectral Line Broadening upon Elastic Collisions of Atoms ... 255
10.8 Radiation Capture (Inprisonment of Radiation) 259
10.9 The Influence of the Static Electric Field upon the Atomic
      Emission Spectrum 263
10.10 The Distribution of Microfields Created by Ions. The Holzmark
      Formula 265
10.11 The Atomic Emission Spectrum with the Ion Field Distribution
      Taken into Account 267
10.12 The Influence of an Electron Field on the Intensity Distribution
      at the Wings of the Spectral Line 270
10.13 Taking Strong Short-Range Interactions and Collective
      Long-Range Interactions into Account Simultaneously 271
10.14 Some Problems of the Kinetic Theory of Spectral Line
      Broadening 272

11. Fluctuations and Kinetic Processes in Systems Composed
    of Strongly Interacting Particles 275
11.1 The Influence of the Correlations of Atoms' Positions on Their
      Spontaneous Radiation 275
11.2 The Effective Lorentz Field 277
11.3 The Influence of the Correlations of Atoms' Positions
      on the Coefficient of Spontaneous Emission 278
11.4 The Kubo and the Callen – Welton Equations 280

11.5   Fluctuations of the Distribution Function of the Density
         Matrix. The Random Source in the Liouville Equation   286
11.6   Fluctuations in an Extended System. The Polarization
         Approximation   288
11.7   The Fluctuations of Polarization and Field.
         The Callen – Welton Equation for Nonequilibrium States   290
11.8   The Kinetic Equation Giving the Distribution Function for the
         States of a System of Interacting Atoms   294
11.9   The Transition to the Kinetic Equation for One-Particle
         Distribution Functions of Atoms   298
11.10   The Transparency Region. Probabilities of Transition   300
11.11   The Distribution Function of a System of Atoms and Mean
         Field. The First-Moments Approximation   301
11.12   The Influence of the Correlations of Atoms' Positions on the
         Absorption and Scattering of Electromagnetic Waves   302

12. Fluctuations in Quantum Self-Oscillatory Systems   305
12.1   A System Composed of Two-Level Atoms and a Field   305
12.2   Stationary Generation Regime, Without Taking Fluctuations
         into Account   308
12.3   Sources of Fluctuations in a Quantum Generator   310
12.4   Field Equations with Fluctuations Taken into Account   313
12.5   Fluctuations of Radiation in a Quantum Optical Generator   315
12.6   Spatial and Temporal Correlations of a Field Below the
         Generation Threshold   318
12.7   Spatial and Temporal Correlations of the Fluctuations of Laser
         Radiation   320

13. Phase Transitions in a System Composed of Atoms and a Field   324
13.1   A Phase Transition in a System Composed of Two-Level Atoms
         and a Field   324
13.2   Fluctuations in the Polarization and the Field at Above-Critical
         Temperatures   327
13.3   Fluctuations in the Polarization and the Field in the Critical
         Region   329
         13.3.1 The Equation for the Polarization Vector   329
         13.3.2 Induced Fluctuations of Polarization   331
         13.3.3 Field Fluctuations   332
13.4   Laser-Radiation-Induced Phase Transitions in a System
         of Two-Level Atoms   333
13.5   The Influence of a Phase Transition on Generation   338
         13.5.1 In Ferroelectrics   338
         13.5.2 In Liquid Crystals   342

14. Conclusion   347

References   349

Subject Index   361

1. Introduction

The title of this volume is very general and therefore demands explanation. Today the best-developed kinetic theories are those for a gas of structureless atoms and for a completely ionized plasma [1.1 – 14]. The kinetic theory of chemically reacting gases and partially ionized plasmas is immensely more complicated and, in spite of the vast amount of research done in this field, many fundamental problems remain unsolved.

This book is principally concerned with a system of charged particles – both free as well as bound in atoms – and an electromagnetic field. For the sake of brevity such a system will be called a 'partially ionized plasma'.

From the great variety of processes taking place in such a complex system we shall not choose those which are dominated by strong interactions at small distances (pair collisions), but those dominated by weak collective interactions of free and bound charged particles. Then even in a Coulomb plasma approximation the field cannot be totally excluded. We shall deal, consequently, with a kinetic theory in which the field is at least as important as the particles. This justifies the title of this study.

There is a large body of literature on various aspects of the kinetic theory of electromagnetic processes. Monographs providing detailed analyses of concrete systems are available [1.15 – 24], just as textbooks are (e. g. [1.25 – 27]). Nevertheless, there is a growing demand for books which give a general statistical presentation of a wide scope of problems in the kinetic theory of electromagnetic processes and employ a unified method to describe simple models.

Since the subject of this book is partially ionized plasma, systems consisting of at least four components have to be considered. These are two kinds of free charged particles (i. e., electrons and singly charged ions), atoms and, finally, an electromagnetic field. An extreme case of such a system is a totally ionized plasma (the concentration of neutral atoms is zero). Another extreme case is a gas of structureless atoms.

For rarefied partially ionized plasma we shall often employ the conditions of smallness of density and plasma parameters. On a model level, however, we shall discuss dense systems for which these parameters are not small.

The method adopted in the present book is a generalization of the method which was developed earlier to describe processes in a completely ionized plasma [1.5, 11] and in the kinetic theory of gases [1.11]. In the former case the starting point was a closed set of equations for the microscopic phase densities of electrons ($a = e$) and ions ($a = i$) in a plasma

$$N_a(r,p,t) = \sum_{1 \leqslant i \leqslant N_a} \delta(r - r_{ia}(t)) \delta(p - p_{ia}(t)) \tag{1.1}$$

and for the microscopic strengths of electric and magnetic fields

$$E^m(r, t), \quad B^m(r, t).$$ (1.2)

The set of equations for these functions can be written in the form

$$\frac{\partial N_a(r, p, t)}{\partial t} + v \frac{\partial N_a}{\partial r} + e_a \left\{ E^m(r, t) + \frac{1}{c} [v B^m(r, t)] \right\} \frac{\partial N_a}{\partial p} = 0,$$

$$\mathrm{rot}\, B^m = \frac{1}{c} \frac{E^m}{\partial t} + \frac{4\pi}{c} j^m(r, t),$$

$$j^m = \sum_a e_a \int v N_a(r, p, t) dp$$

$$\mathrm{rot}\, E^m = - \frac{1}{c} \frac{\partial B^m}{\partial t},$$

$$\mathrm{div}\, B^m = 0,$$

$$\mathrm{div}\, E^m = 4\pi q^m(r, t), \quad q^m = \sum_a e_a \int N_a(r, p, t) dp.$$ (1.3)

A similar set of equations can also be employed in the kinetic theory of a gas of structureless particles [Ref. 1.11, Sect. 5]. The state of the system can then be defined by giving the phase density

$$N(r, p, t) = \sum_{1 \leqslant i \leqslant N} \delta(r - r_i(t)) \, \delta(p - p_i(t)),$$ (1.4)

and the initial equations can be written in the form of a set of equations for phase density and microscopic force

$$\frac{\partial N(r, p, t)}{\partial t} + v \frac{\partial N}{\partial r} + F^m(r, t) \frac{\partial N}{\partial p} = 0,$$ (1.5)

where

$$F^m(r, t) = - \mathrm{grad}_r \int \Phi(|r - r'|) N(r', p', t) dr' dp'.$$

Based on the microscopic equations (1.3 – 5) one can get a chain of equations for the moments of initial random functions (N_a, E^m, B^m for a plasma and N, F^m for a gas), which is equivalent to the hierarchy of equations for a succession of distribution functions, introduced in the works of *Bogolyubov* [1.2], *Born* and *Green* [1.28], *Kirkwood* [1.29] and *Yvon* [1.30].

In a model of partially ionized plasma the charged particles may be free as well as bound in atoms. We assume that charged particles are bound, if ever, in atoms by pairs. The state of the system is then conveniently described by the phase density in twelve-dimensional space of pairs of oppositely charged particles

$$N_{ab}(x', x'', t) = \sum_{1 \leqslant i \leqslant N} \delta(x' - x_{ia}(t)) \, \delta(x'' - x_{ib}(t)) \, ,$$

$$a \neq b, \quad x = (r, p), \quad \int N_{ab}(x', x'') dx' dx'' = N \tag{1.6}$$

where N is the total number of pairs of electrons and ions.

The equation for the phase density of pairs of charged particles can be written in the following form

$$\left[\frac{\partial}{\partial t} + v' \frac{\partial}{\partial r'} + v'' \frac{\partial}{\partial r''} + F_a^m(x', t) \frac{\partial}{\partial p'} + F_a^m(x'', t) \frac{\partial}{\partial p''} \right] N_{ab} = 0 \tag{1.7}$$

where F_a^m and F_b^m are the Lorentz forces at points x' and x'', i. e.,

$$F_a^m(x, t) = e_a \left[E^m(r, t) + \frac{1}{c} B^m(r, t) \right], \quad (a = e, i) \, . \tag{1.8}$$

The microscopic strengths of charge and current can be expressed through the phase density of pairs of charged particles

$$q^m(r, t) = \int [e_a \delta(r - r') + e_b \delta(r - r'')] N_{ab}(x', x'', t) \, dx' dx'',$$

$$j^m(r, t) = \int [e_a v' \delta(r - r') + e_b v'' \delta(r - r'')] N_{ab}(x', x'', t) \, dx' dx'' \, . \tag{1.9}$$

For the initial set of equations, therefore, one can take a closed set for the phase density of pairs of charged particles and microscopic field strengths, i.e., for functions

$$N_{ab}(x', x'', t), \quad E^m(r, t), \quad B^m(r, t) \, . \tag{1.10}$$

Naturally, for the case of completely ionized plasma, when all charged particles are free, one can go from equations for functions (1.10) to equations for phase densities of free charged particles and a field (1.3).

In the other extreme case of a gas of neutral particles (where all charged particles are bound in pairs), it is more convenient to go from (1.7) to the equation for phase density in the space R, P (space of atoms' centers of mass) and r, p (internal variables). The initial set of equations will be for random functions

$$N_{ab}(r, p, R, P, t), \quad E^m(q, t), \quad B^m(q, t), \tag{1.11}$$

which, as a matter of fact, is more general than the set of equations for a gas of structureless atoms (1.5). Indeed, on the basis of these equations for random functions (1.11) one can build a kinetic theory of processes conditioned by the inner structure of atoms; for example, one can obtain a kinetic equation for the polarization vector accounting for atoms' interactions, or describe a phase transition in a system of atoms and a field (Chap. 6).

The book is divided into two parts, dealing respectively with classical and quantum theories of electromagnetic processes. Classical theory is based on the equations for random functions (1.10, 11), and quantum theory on corresponding operator equations.

What then are the main problems dealt with in this book? One of the tasks is to discuss different ways of presenting initial microscopic equations, which is necessary because of the complexity of the studied system. For instance, in some cases the equations for random functions (1.10, 11) are conveniently replaced by the set of equations for the phase densities of particles and field oscillators. Choosing the right presentation of the initial equations facilitates the construction of the kinetic equations in each case. We also discuss in detail different presentations of microscopic field equations. This provides an opportunity to choose the most practical form of transition to the equations for mean fields – Maxwell equations.

Since all dissipative characteristics, and the collision integrals in particular, are expressed through correlations of the fluctuations of phase densities (or operator matrices of density) and field, one of the main tasks lies in the calculation of nonequilibrium fluctuations. The general scheme of these calculations is similar to that employed in an earlier study [1.11]. Here the fluctuations are also calculated in two steps:

1) the calculation of correlations of small-scale (fast) fluctuations, which determine the collision integrals in the kinetic equations for distribution functions of electrons, ions, and atoms; and

2) the calculation of large-scale fluctuations, determined by the kinetic equations themselves.

Here, however, the calculation of fluctuations brings forth fundamentally new problems due to the great complexity of the system in question. For example, one encounters the problem of calculating the fluctuations in chemically reacting systems. To calculate the kinetic fluctuations in simple gases by Langevin's method, it was sufficient just to determine the statistical characteristics of the Langevin source in the kinetic equation for the distribution function of particles; now the same problem has to be solved for the kinetic equations for both particles and field. As a result a corresponding set of Langevin equations appears. The existence of random sources is now due to two causes: the atomic structure of the substance and the structure of the field as a system of electromagnetic oscillators. Such equations are the starting point, e. g., for calculating the fluctuations of radiation in lasers (Chap. 12).

Small-scale fluctuations play a much more varied role here than in the theory of gases or completely ionized plasma. Here, for example, the problem arises of calculating spectral line broadening in partially ionized plasma. In such a complex system the causes of this broadening are very numerous.

Yet another task of this book is to show that the theory of broadening may be considered part of the general kinetic theory of plasmas. This allows considering problems of spectral line broadening in nonequilibrium plasma (Chap. 10).

If the density of atoms becomes high enough, one has to take into account the difference between the mean field acting upon the atoms and the mean Maxwell field, which is due to the correlations of atoms' positions. Taking these correlations into account leads to significant alteration of the equation for the polarization vector (Chap. 6).

One such change is in the renormalization of the coefficient of extinction of radiation. This renormalized coefficient of extinction is now proportional to the relative fluctuation of density and hence to the coefficient of isothermal compressibility. This leads to an abrupt increase of absorption and scattering of light upon approaching the critical point. This enables one, starting from microscopic equations, to obtain the results of Einstein's theory for critical opalescence. The corresponding calculations for a quantum system are carried out in Chap. 11.

Another principal change in the equation for the polarization vector, also due to correlation of atoms' positions, is the introduction of the so-called Lorentz correction, which brings about a frequency shift proportional to atoms' density. At a sufficiently strong concentration this leads to the appearance of a soft mode in a system made up of atoms and a field, which in turn may lead to a phase transition resulting in spontaneous polarization of the system.

We have already noted that small-scale fluctuations determine the collision integrals in the kinetic equations. The kinetic equations themselves (taking of relevant Langevin sources) describe, in turn, large-scale fluctuations. An example of equations of this kind is given by a set of equations for polarization vector and field. Kinetic (large-scale) fluctuations may play an important role in both equilibrium and nonequilibrium states. Their role in equilibrium states is especially important at phase transitions.

The investigation of fluctuations in the critical region is one of the most important problems in the modern theory of phase transitions. The pattern of fluctuations depends, of course, of the character of the interactions of particles in the system. In the present book we shall discuss two extreme cases.

1) In Chap. 5 we shall discuss fluctuations in a classical system of oscillating atoms, connected through spatial diffusion. The equations for the "polarization vector" can then be written in the form of a Ginzburg-Landau equation with a corresponding Langevin source. In the vicinity of the critical point such a system has a so-called region of scale invariance, where the critical indices which characterize the dependence of the parameter of order, correlation radius, and other parameters on the temperature difference $|T - T_c|$, differ markedly from the corresponding critical indices of the Landau theory.

In that chapter some results of a qualitative theory of fluctuations in the critical region will be given based on the condition of consistency of two limit transition, the thermodynamic limit transition and the limit transition to the critical point $T \rightarrow T_c$.

2) Another extreme situation is − either in a classical system of oscillating atoms (Chap. 6), or in a corresponding quantum system of two-level atoms (Chap. 13) − when atoms only interact through the electromagnetic field. The character of the fluctuations in the critical region is here different, and there are no reasons for singling out the region of scale invariance. The Landau theory in the thermodynamic limit remains valid up to the critical point.

Nonequilibrium situations in which large-scale fluctuations play a significant role are very numerous, turbulent states providing a vivid example. Recently a large class of systems has emerged, where so-called nonequilibrium phase transitions occur in open systems. Here we discuss examples of nonequilibrium phase transitions both for classical (Chap. 6) and quantum (Chap. 12) systems. They are represented in the first case by crossing the generation threshold in a classical self-oscillatory system, and in the second case by exceeding the generation threshold in a quantum generator (laser).

Fluctuations at both equilibrium and nonequilibrium phase transitions can be considered as examples of Brownian motion in nonlinear distributed systems. Then an analogy is revealed between crossing the generation threshold (nonequilibrium phase transition) and the phase transition of the second kind. This analogy was recently the subject of discussion in a number of works [1.31]. When turning to this analogy one should bear in mind that the physical nature of these processes is profoundly different.

Indeed, in a quantum generator, for example, the coherent radiation is emitted at frequencies which are close to the frequency of the atom transition; the process of attaining the coherent state is then governed by dissipative nonlinearity. In contrast, in the second kind of phase transition the coherent state is established for the range of frequencies close to zero. The principal role in this type of phase transition is played by nondissipative nonlinearity. Due to this difference, the generation threshold can only be exceeded in an open system; such a transition is therefore necessarily a nonequilibrium process. The dissimilarities in the nature of the fluctuations near the generation threshold and in the critical region at phase transitions of the second kind are also very important.

Today we know substances which simultaneously display sufficiently great dissipative and nondissipative nonlinearities. Examples are ferroelectric laser crystals and liquid crystals with dissolved optically active dope.

Naturally, it is very interesting to study the mutual influence of equilibrium and nonequilibrium phase transitions in such systems: the influence of the second type of phase transition in the "solvent" on the process of generation, and, reciprocally, the influence of the laser field on the characteristics of the second kind of phase transition. Two examples of such phenomena are discussed in Chap. 13. One of them is an induced phase transition, i.e., a transition occurring under the influence of laser radiation. The converse phenomenon is illustrated by the influence of fluctuations at phase transitions of the second kind in piezoelectric and liquid crystals on the generation threshold and on the wavelength of emitted radiation.

The kinetic equations for one-particle distribution functions can be employed while studying kinetic processes and obtaining the equations for macroscopic characteristics in sufficiently rarefied systems. For a partially ionized plasma this is a set of kinetic equations for the distribution functions of electrons, ions, and atoms, and the equations for the field. To investigate kinetic processes in profoundly nonideal systems one will, of course, need equations for more general distribution functions.

For that purpose we shall discuss in Chap. 11 a kinetic equation for the distribution function of the states of a macroscopic system of strongly interacting particles. The dissipation is then determined by an additional weak interaction of the same particles through a fluctuating electromagnetic field. This equation enables one to follow the relaxation towards the Gibbs distribution and the establishment of the equilibrium distribution of the fluctuations' spectra in accordance with the fluctuation dissipation theorem, extrapolating thus the Callen – Welton and the Kubo formulae on nonequilibrium states.

Even such a brief introduction certainly must provide some information about the diversity and complexity of the problems confronting the kinetic theory of electromagnetic processes. Naturally, all the problems cannot be given an equally detailed treatment, and some of them are presented in quite a schematic manner. Also, some of the results may be unripe or require further revision. A few problems beyond our scope will be mentioned in Chap. 14.

2. Free Charged Particles and a Field

2.1 The Equations of Particle Motion. Local Field

In discussing completely ionized electron-ion plasma, i.e., one which is neutral as a whole system of charged particles (electrons and ions) and a field, we shall use the following notation. The subscript a denotes the kind of component (it can takes on one of two forms: e for electrons and i for ions); $x_{ai} = (r_{ai}, p_{ai})$ gives the coordinates and momenta of the ith particle of kind a; e_a is the charge of the particle. The charge of the ion is here assumed equal to that of the electron,

$$e_i = |e_e| \equiv e . \tag{2.1.1}$$

The total number of particles of kind a is given as N_a; because of (2.1.1)

$$N_i = N_e = N . \tag{2.1.2}$$

The system as a whole is electrically neutral, so

$$\sum_a e_a N_a = 0 . \tag{2.1.3}$$

Here $\Phi_{ab} = e_a e_b / r$ is the Coulomb interaction potential, and $\Phi(|r|)$ an arbitrary potential of interaction.

We shall distinguish between microscopic field strengths $E^m(r, t)$ and $B^m(r, t)$ induced at point r by all charged particles, and local (acting) fields. Local fields at point r are induced by all particles except the one chosen. We shall denote them $E_l^m(r, t)$ and $B_l^m(r, t)$. They naturally appear in the equations of motion of charged particles,

$$\frac{dr_{ia}}{dt} = \frac{p_{ia}}{m_a} \equiv v_{ia} ,$$

$$\frac{dp_{ia}}{dt} = e_a \left\{ E_l^m(r_{ia}, t) + \frac{1}{c} [v_{ia} B_l^m(r_{ia}, t)] \right\} . \tag{2.1.4}$$

The force acting on the charged particle is determined by the local field strengths.

For the sake of convenience, instead of using the equations of motion (2.1.4), in statistical theory one can use related equations for the microscopic phase densities of charged particles in the six-dimensional space of coordinates and momenta. By definition

$$N_a(r, p, t) = \sum_{1 \leqslant i \leqslant N_a} \delta(r - r_{ia}(t)) \, \delta(p - p_{ia}(t)) , \qquad (2.1.5\,a)$$

which can be written in more compact form,

$$N_a(x, t) = \sum_{1 \leqslant i \leqslant N_a} \delta(x - x_{ia}(t)) . \qquad (2.1.5\,b)$$

From these definitions it follows that $N_a(x, t) \, dx$ equals the number of type a particles whose coordinates and momenta at time t fall within the domain $dx = dr \, dp$ in the neighborhood of $x = (r, p)$. The integral over the whole phase space

$$\int N_a(x, t) \, dx = N_a \qquad (2.1.6)$$

equals the total number of type a particles. The evolution of $N_a(x, t)$ with time is described by the equation

$$\frac{\partial N_a}{\partial t} + v \frac{\partial N_a}{\partial r} + e_a \left\{ E_l^m(r, t) + \frac{1}{c} [v \times B_l^m(r, t)] \right\} \frac{\partial N_a}{\partial p} = 0 , \qquad (2.1.7)$$

reflecting the conservation of the total number of type a particles.

The equations of motion (2.1.4) as well as those of phase density are not self-contained because they include local-field strengths, which in turn are themselves dependent on the motion of the particles.

2.2 The Equations for Microscopic Field Strengths (Lorentz Equations)

The equations for microscopic field strengths were first derived by Lorentz. In the accepted notation they are as follows:

$$\mathrm{rot}\, B^m = \frac{1}{c} \frac{\partial E^m}{\partial t} + \frac{4\pi}{c} j^m(r, t) , \quad \mathrm{div}\, B^m = 0 ,$$

$$\mathrm{rot}\, E^m = -\frac{1}{c} \frac{\partial B^m}{\partial t} , \quad \mathrm{div}\, E^m = 4\pi q^m(r, t) \qquad (2.2.1)$$

where q^m and j^m are microscopic charge and current densities, determined as

$$q^m(r, t) = \sum_a e_a \sum_i \delta(r - r_{ia}(t)) ,$$

$$j^m(r, t) = \sum_a e_a \sum_i v_{ia} \delta(r - r_{ia}(t)) .\tag{2.2.2}$$

Using the definition of $N_a(x, t)$, one can express the densities of charge and current through phase densities,

$$q^m(r, t) = \sum_a e_a \int N_a(r, p, t) dp ,$$

$$j^m(r, t) = \sum_a e_a \int v N_a(r, p, t) dp .\tag{2.2.3}$$

The equations of motion (2.1.4) together with the field equations (2.2.1, 2) are not yet self-contained because the equations of motion include local-field strengths. We shall deal with this question in the next section.

Along with (2.2.1) we shall use the equations for vector potential A and scalar potential φ that are related to the field strengths,

$$E = -\frac{1}{c}\frac{\partial A}{\partial t} - \mathrm{grad}\,\varphi , \quad B = \mathrm{rot}\,A .\tag{2.2.4}$$

The definition of potentials is ambiguous, and there is room for additional constraints. We shall use the Coulomb gauge, which demands that

$$\mathrm{div}\,A = 0 .\tag{2.2.5}$$

With this condition (2.2.5) taken into account, the equations for potentials become

$$\Delta A - \frac{1}{c^2}\frac{\partial^2 A}{\partial t^2} = \frac{4\pi}{c}j^{\perp}, \quad \Delta \varphi = -4\pi q .\tag{2.2.6}$$

The superscript $\perp$ denotes the rotational component of a vector. Any given vector A can be decomposed into two components – rotational and potential – in the following manner:

$$A^{\perp}(r) = A(r) + \frac{1}{4\pi}\int \frac{\mathrm{grad}_{r'}\,\mathrm{div}_{r'}A(r')}{|r - r'|} dr' \equiv A - A^{\parallel},\tag{2.2.7}$$

where $A^{\parallel}$ is the potential component of A. From (2.2.7) it follows that

$$\mathrm{div}\,A^{\perp} = 0 , \quad \mathrm{rot}\,A^{\parallel} = 0 .$$

In order to prove, for example, the former of these equations one should bear in mind that

$$\frac{1}{4\pi}\mathrm{div}_r\int \frac{\mathrm{grad}_{r'}\,\mathrm{div}_{r'}A(r')}{|r - r'|} dr' = -\mathrm{div}\,A$$

because

$$\frac{1}{4\pi} \Delta_r \frac{1}{|r-r'|} = -\delta(r-r') \, .$$

(2.2.8)

Yet another form of (2.2.7) can be helpful,

$$A^{\perp}(r) = -\frac{1}{4\pi}(\Delta_r - \text{grad}_r \, \text{div}_r) \int \frac{A(r')}{|r-r'|} dr' \, .$$

(2.2.9a)

Using the vector identity

$$\text{rot rot } A = \text{grad div } A - \Delta A \, ,$$

(2.2.10)

we can rewrite (2.2.9a) as

$$A^{\perp}(r) = \frac{1}{4\pi} \text{rot}_r \, \text{rot}_r \int \frac{A(r')}{|r-r'|} dr' \, .$$

(2.2.9b)

Let us now consider the solutions of equations for potentials. For the scalar potential the solution is given by

$$\varphi^{m}(r, t) = \int \frac{q^{m}(r, t)}{|r-r'|} dr' = \sum_{a} \sum_{i} \frac{e_a}{|r-r_{ia}(t)|}$$

(2.2.11)

from the definition of charge density (2.2.2).

It is clear from (2.2.11) that the scalar potential is determined solely by the distribution of charged particles. The solution of the equation for the vector potential is totally different in nature,

$$A(r, t) = A_0(r, t) + \frac{1}{c} \int_{|r-r'| \leq ct} \frac{[j^{\perp}(r', t)]_{t-|r-r'|/c}}{|r-r'|} dr' \, ,$$

(2.2.12)

where $A_0(r, t)$ is the distribution of the vector potential at a time t, which is determined by the initial ($t = 0$) distribution of the field. In the second term on the right-hand side of the equation, which represents the solution of a nonuniform equation for the vector potential at zero initial conditions, the range of integration is constrained,

$$|r-r'| \leq ct \, .$$

(2.2.13)

This constraint reflects the fact that at a time t only those charges can contribute to the field at point r (starting from $t = 0$) which are within the region (2.2.13). The influence of charges outside this region does not reach point r by time t. At $t = 0$ the second term in the right-hand side of (2.2.13) vanishes.

Now let us turn to the equations for the rotational and potential components of the electric field strength. According to the first equation in (2.2.4),

$$E = E^{\perp} + E^{\parallel}, \quad E^{\perp} = -\frac{1}{c}\frac{\partial A}{\partial t}, \quad E^{\parallel} = -\operatorname{grad}\varphi. \tag{2.2.14}$$

The set of equations for a potential field is

$$\operatorname{rot}E = 0, \quad \operatorname{div}E = 4\pi q. \tag{2.2.15}$$

Using (2.2.11) we can write the solution of these equations as

$$[E^{\mathrm{m}}(r,\,t)]^{\parallel} = -\operatorname{grad}_r\int\frac{q^{\mathrm{m}}(r',\,t)}{|r-r'|}dr' = -\sum_{a,\,i}\operatorname{grad}_r\frac{e_a}{|r-r_{ia}(t)|}. \tag{2.2.16}$$

The equations for a rotational electric field follow from (2.2.6, 14),

$$\Delta E - \frac{1}{c^2}\frac{\partial^2 E}{\partial t^2} = \frac{4\pi}{c^2}\frac{\partial j^{\perp}}{\partial t}, \quad \operatorname{div}E = 0. \tag{2.2.17}$$

According to (2.2.7) the rotational component of the vector of electric current is given by

$$j^{\perp}(r,\,t) = j + \frac{1}{4\pi}\int\frac{\operatorname{grad}_{r'}\operatorname{div}_{r'}j(r',\,t)}{|r-r'|}dr'. \tag{2.2.18}$$

In future yet another form of the first equation in (2.2.17) will be helpful. In order to obtain it we shall employ the continuity equation for charge density,

$$\frac{\partial q(r,\,t)}{\partial t} + \operatorname{div}j = 0. \tag{2.2.19}$$

From (2.2.18, 19) and taking the solution of the equation for the scalar potential into account, we find

$$j^{\perp} = j - \frac{1}{4\pi}\frac{\partial}{\partial t}\operatorname{grad}\varphi. \tag{2.2.20}$$

Thus the equations for a rotational field take on the following form:

$$\Delta E - \frac{1}{c^2}\frac{\partial^2 E}{\partial t^2} = \frac{4\pi}{c^2}\frac{\partial}{\partial t}\left(j - \frac{1}{4\pi}\frac{\partial}{\partial t}\operatorname{grad}\varphi\right), \quad \operatorname{div}E = 0. \tag{2.2.21}$$

The solution of (2.2.17) can be presented as [cf. (2.2.12)]

$$E(r, t) = E_0(r, t) - \frac{1}{c^2} \frac{\partial}{\partial t} \int \frac{j^{\perp}(r, t - |r-r'|/c)}{|r-r'|} dr' . \tag{2.2.22}$$

As in (2.2.12), the range of integration is constrained by the condition (2.2.13).

2.3 A Coulomb Plasma

In the equation for the phase density (2.1.7) the term with B^m is absent for a Coulomb plasma, and the electric field strength E^m satisfies the set of equations (2.2.15).

The equation for the phase density of a Coulomb plasma (no distinction being made between E^m and E_l^m) is therefore

$$\frac{\partial N_a}{\partial t} + v \frac{\partial N_a}{\partial r} + e_a E^m(r, t) \frac{\partial N_a}{\partial p} = 0 . \tag{2.3.1}$$

The dissimilarity between (2.1.7) and (2.3.1) is determined by the difference between two terms,

$$E^m(r, t) N_a(x, t) - E_l^m(r, t) N_a(x, t) . \tag{2.3.2}$$

The field strength E^m is given by (2.2.16), and the local field strength by

$$E_l^m(r_i, t) = - \sum_a \sum_{j \neq i} \mathrm{grad}_{r_i} \frac{e_a}{|r_i - r_j|} . \tag{2.3.3}$$

Using (2.2.16, 3.3) for field strengths we transform (2.3.2) into

$$\int \mathrm{grad} \left(\frac{1}{|r-r'|} \right) \delta(r - r') dr' N_a(x, t) = 0 . \tag{2.3.4}$$

This integral equals zero because the substitution of $(r-r')$ for $-(r-r')$ leads to the change of the integrand sign.

Thus equations (2.1.7) and (2.3.1) for a Coulomb plasma are identical. Due to this fact, (2.3.1), together with the equations of a potential field

$$\mathrm{rot}\, E^m = 0 , \quad \mathrm{div}\, E^m = 4\pi q^m = 4\pi \sum_a e_a \int N_a dp , \tag{2.3.5}$$

form a self-contained set of equations for N_a and E^m. It is self-evident that the substitution of the local field for E^m amounts to the inclusion of an eigenfield into the Hamiltonian.

Instead of the set of equations (2.3.1, 5) for a Coulomb plasma one can use the equivalent equations involving only $N_a(x, t)$. Take the solution of the set (2.3.5),

$$E^m(r, t) = - \text{grad}_r \sum_b e_b \int \frac{N_b(x', t)}{|r-r'|} dr' dp' \qquad (2.3.6)$$

and substitute it into (2.3.1), obtaining as a result

$$\frac{\partial N_a}{\partial t} + v \frac{\partial N_a}{\partial r} - \frac{\partial}{\partial r} \sum_b \int \frac{e_a e_b}{|r-r'|} N_b(x', t) dx' \frac{\partial N_a}{\partial p} = 0 . \qquad (2.3.7)$$

The set of equations (2.3.1, 5) is commonly accepted as the basic set of equations in the kinetic theory of completely ionized plasma [2.1 – 4].

2.4 The Complete Set of Microscopic Equations for a Plasma

Let us now attend to the fact that the integral in (2.3.4) is equal to zero because in a Coulomb plasma the potential which determines the microscopic force only depends on the absolute value of distance. If the interactions are noncentral, we come to a different result. Below, however, we shall make no distinction between total and local fields in our basic microscopic equations for phase density.

This should by no means be interpreted as a total lack of difference between these fields. On the contrary, in Chaps. 6, 11, 13 we shall see that the difference between the mean field and the averaged local field becomes quite significant in the vicinity of critical points in phase transitions of the first and the second kind. It is just a matter of convenience that the effects of this difference can be more readily seen in the averaged equations. Accordingly let us substitute total fields in lieu of local in (2.1.7) and obtain

$$\left(\frac{\partial}{\partial t} + v \frac{\partial}{\partial r} + e_a \left\{ E^m(r, t) + \frac{1}{c} [v \times B^m(r, t)] \right\} \frac{\partial}{\partial p} \right) N_a(x, t) = 0 , \qquad (2.4.1)$$

which together with (2.2.1, 2) serves as a basic microscopic equation in the statistical approach to the processes in a completely ionized electron-ion plasma.

2.5 The Equations of Motion of Free Charged Particles and Field Oscillators

The electromagnetic field has so far been described by the microscopic electric and magnetic field strengths for every point in space. In many applications especially in solving both classical and quantum radiation theory problems, another way of describing an electromagnetic field can be helpful [2.5 – 7].

Let us expand a vector potential in a Fourier series

$$A(r, t) = \sqrt{\frac{4\pi c^2}{V}} \sum_k e_k A_k(t) e^{ikr} \tag{2.5.1}$$

where e_k is a unit vector. From the condition of calibration (2.2.5) it follows that

$$k e_k = 0. \tag{2.5.2}$$

Let us further assume that V is the volume of a cube with a side L. The summation in the expansion (2.5.1) is carried out over all admissible values of integral vector k, whose components take on the values

$$k_\beta = \frac{2\pi}{L} n_\beta, \quad \beta = 1, 2, 3, \tag{2.5.3}$$

where n_β are positive and negative integers.

The inverse Fourier transform can be written as

$$A_k(t) = \frac{1}{\sqrt{4\pi c^2 V}} \int e_k A(r, t) e^{-ikr} dr. \tag{2.5.4}$$

Instead of complex quantities A_k we shall use the corresponding real quantities, which are defined as

$$A_k = Q_k^2 - i Q_k^1, \ Q_k^2 = \frac{1}{2}(A_k + A_k^*) = Q_{-k}^2$$

$$A_k^* = Q_k^2 + i Q_k^1, \ Q_k^1 = \frac{i}{2}(A_k - A_k^*) = -Q_{-k}^1. \tag{2.5.5}$$

The expansion (2.5.1) then becomes

$$A(r, t) \equiv \sum_{k,\alpha} e_k A_{k,\alpha} = \sqrt{\frac{4\pi c^2}{V}} \sum_{k,\alpha} e_k Q_k^\alpha(t) \begin{Bmatrix} \sin kr \\ \cos kr \end{Bmatrix}, \tag{2.5.6}$$

where $\sin kr$ corresponds to $\alpha = 1$, and $-\cos kr$ to $\alpha = 2$. The appropriate inverse transform is written as

$$Q_k^\alpha = \frac{1}{\sqrt{4\pi c^2 V}} \int e_k A(r, t) \begin{Bmatrix} \sin kr \\ \cos kr \end{Bmatrix} dr. \tag{2.5.7}$$

The factor $\sqrt{4\pi c^2/V}$ in (2.5.1, 6) is chosen to make it possible to express the electromagnetic field energy as a sum of the energies of harmonic field oscillators (2.5.12).

As we shall see in Sect. 2.7, the generalized momentum of the field for a system of free charged particles and field $\Pi(r, t)$ is proportional to the rotational component of the electric field strength

$$\Pi = - \frac{1}{4\pi c} E^\perp . \tag{2.5.8}$$

We shall write the expansion of electric field strength in the form

$$E^\perp(r, t) = - \frac{1}{c} \frac{\partial A}{\partial t} = - \sqrt{\frac{4\pi}{V}} \sum_{k,\alpha} e_k P_k^\alpha \begin{Bmatrix} \sin kr \\ \cos kr \end{Bmatrix} \equiv \sum_{k,\alpha} e_k E_k^\alpha . \tag{2.5.9}$$

From a comparison of (2.5.1, 9) it follows that

$$P_k^\alpha = \dot{Q}_k^\alpha . \tag{2.5.10}$$

The Fourier expansion for the magnetic field strength can be written as

$$B = \operatorname{rot} A = \sqrt{\frac{4\pi c^2}{V}} \sum_{k,\alpha} [k \times e_k] \begin{Bmatrix} -Q_k^2 \sin kr \\ Q_k^1 \cos kr \end{Bmatrix} . \tag{2.5.11}$$

Using (2.5.9, 11) one can express the electromagnetic field energy H_f as a sum of the energies of electromagnetic oscillators with eigenfrequencies $\omega_k = ck$,

$$H_f = \frac{1}{8\pi} \int (E^2 + B^2) dr = \frac{1}{2} \sum_{k,\alpha} (P_{k,\alpha}^2 + c^2 k^2 Q_{k,\alpha}^2) . \tag{2.5.12}$$

Finding a set of equations of motion for charged particles and field oscillators, we substitute the local-field strengths for E^m, B^m in the equations of motion of particles (see the end of Sect. 2.4). The equations of motion then become

$$\frac{dr_{ia}}{dt} = \frac{p_{ia}}{m_a} = v_{i,a} ,$$

$$\frac{dp_{i,a}}{dt} = e_a \left\{ E^m(r_{ia}, t) + \frac{1}{c} [v_{ia} \times B^m(r_{ia}, t)] \right\} . \tag{2.5.13}$$

Into these equations we substitute the Fourier transforms (2.5.9, 11) for field strengths $E^\perp$ and B, and (2.2.16) for $E^\parallel$.

In order to obtain the equations of motion for field oscillators, we must substitute the Fourier expansion (2.5.6) of a vector potential into the equation for A (2.2.6). By dint of (2.5.10) the set of equations for Fourier components of the field can be written as follows:

$$\dot{Q}_k^{\alpha} = P_k^{\alpha},$$

$$\dot{P}_k^{\alpha} + c^2 k^2 Q_k^{\alpha} = \sqrt{\frac{4\pi}{V}} \int e_k j(r, t) \begin{Bmatrix} \sin kr \\ \cos kr \end{Bmatrix} dr$$

$$= \sqrt{\frac{4\pi}{V}} \sum_{a, i} e_a e_k v_{ia} \begin{Bmatrix} \sin kr_{ia} \\ \cos kr_{ia} \end{Bmatrix}. \qquad (2.5.14)$$

The equations of motion of particles and a field can also be obtained using Lagrangian and Hamiltonian functions. We shall discuss this next since knowledge of the Hamiltonian is essential for obtaining quantum equations of motion.

2.6 Lagrange Function

The Lagrange function for a system of free charged particles and field can be written

$$L = \sum_a \sum_i \left[\frac{m_a v_{ia}^2}{2} + \frac{e_a}{c} v_{ia} \cdot A^m(r_{ia}, t) \right] - \frac{1}{2} \sum_{a,b} \sum_{i,j} \Phi_{ab}(|r_i - r_j|)$$

$$+ \frac{1}{8\pi} \int \left[\left(-\frac{1}{c} \frac{\partial A^m}{\partial t} \right)^2 - (\text{rot } A^m)^2 \right] dr. \qquad (2.6.1)$$

When using this equation one should bear in mind that the vector potential $A = A^{\perp}$ (2.2.7); thus

$$\text{div } A = 0. \qquad (2.6.2)$$

Lagrange's equations for particles

$$\frac{d}{dt} \frac{\partial L}{\partial v_{ia}} - \frac{\partial L}{\partial r_{ia}} = 0 \qquad (2.6.3)$$

give the equations of motion (2.5.13) where

$$E^m(r, t) = -\frac{1}{c} \frac{\partial A^m}{\partial t} - \text{grad}_r \sum_b \sum_j \frac{e_b}{|r - r_{ib}|},$$

$$B^m = \text{rot } A^m. \qquad (2.6.4)$$

Lagrange's equations for a field

$$\frac{\partial}{\partial t} \frac{\delta L}{\delta A^m} - \frac{\delta L}{\delta A^m} = 0 \qquad (2.6.5)$$

lead to the equation for vector potential (2.2.6). When varying the interaction term in (2.6.1) with respect to A, one should note that A is a rotational vector. By virtue of this only the rotational component of j^m appears on the right-hand side of (2.2.6).

We must point out that in the Lagrangian no distinction is made between total and local fields. Therefore a Lorentz force, accounting for self-influence of the particles, enters the equations of motion for particles. One should be aware of this in carrying out practical calculations.

Making use of the definition of phase density (2.1.5), we can rewrite the mathematical expression for the Lagrangian (2.6.1) in the form

$$L = \sum_a \int \left[\frac{m_a v^2}{2} + \frac{e_a}{c} v \cdot A^m(r, t) \right] N_a(x, t) dx$$

$$- \frac{1}{2} \sum_{a,b} \int \Phi_{ab} N_a(x, t) N_b(x', t) dx\, dx'$$

$$+ \frac{1}{8\pi} \int \left[\left(-\frac{1}{c} \frac{\partial A^m}{\partial t} \right)^2 - (\text{rot}\, A^m) \right] dr . \qquad (2.6.6)$$

Instead of the set of Lagrange's equations (2.6.3) we now have just one equation

$$\left(\frac{d}{dt} \frac{\partial}{\partial v} - \frac{\partial}{\partial r} \right) \frac{\delta L}{\delta N_a} = 0 , \qquad (2.6.7)$$

from which the characteristic equation

$$\frac{dp}{dt} = e_a \left(-\frac{1}{c} \frac{\partial A^m}{\partial t} - \text{grad}\, \varphi^m + \frac{1}{c} [v \times B^m] \right),$$

where $\quad \varphi^m = \sum_b \int \frac{N_b(x', t)}{|r - r'|} dx' , \qquad (2.6.8)$

follows for the equation for phase density (2.4.1).

Lagrange's equations for the field remain intact, and in the first equation of (2.2.6)

$$[j^m]^\perp = \left[\sum_a e_a \int v N_a(x, t) dp \right]^\perp .$$

2.7 Hamiltonian Function

The generalized momentum of a type a particle is defined as

$$P_{ia} = \frac{\partial L}{\partial v_{ia}} = P_{ia} + \frac{e_a}{c} A^m(r_{ia}, t) , \qquad (2.7.1)$$

and the generalized momentum of the field

$$\Pi^m(r,\, t) = \frac{\delta L}{\delta A^m} = \frac{1}{4\pi c^2}\frac{\partial A^m}{\partial t} = -\frac{1}{4\pi c}(E^m)^\perp \tag{2.7.2}$$

is proportional to the rotational component of the electric field strength.

Using these two equations, we proceed in a conventional way to find the expression for the Hamiltonian function from the Lagrangian. Using the latter two equivalent presentations of the Lagrangian (2.6.1, 6),

$$H = \sum_a \int \frac{1}{2m_a}\left[P - \frac{e_a}{c}A^m(r,\, t)\right]^2 N_a(x,\, t)\, dx$$

$$+ \frac{1}{2}\sum_{a,b}\int \Phi_{ab}N_a(x,\, t)N_b(x',\, t)\, dx\, dx' + \frac{1}{8\pi}\int[(E^m)^2 + (\mathrm{rot}\, A^m)^2]\, dx \quad . \tag{2.7.3}$$

The corresponding Hamilton's equation for particles leads to (2.6.8) which is the characteristic equation for the equation for phase density $N_a(x,\, t)$ (2.4.1). Hamilton's field equations lead to (2.2.6) and in turn to (2.2.1).

In the Hamiltonian, as in case of the Lagrangian, no distinction is made between local-field strength and field strength in particle-free points. This corresponds to the inclusion of self-influence in the Hamiltonian function.

From (2.7.3) and using (2.5.6, 12), we can obtain the Hamiltonian function for a system of charged particles and field oscillators,

$$H = \sum_a \int \frac{1}{2m_a}\left(P - e_a\sqrt{\frac{4\pi}{V}}\sum_{k,\alpha}e_k Q_k^\alpha \begin{Bmatrix}\sin kr\\ \cos kr\end{Bmatrix}\right)^2 N_a(x,\, t)\, dx$$

$$+ \frac{1}{2}\sum_{a,b}\int \Phi_{ab}N_a(x,\, t)N_b(x',\, t)\, dx\, dx' + \frac{1}{2}\sum_{k,\alpha}(P_{k,\alpha}^2 + e^2 k^2 Q_{k,\alpha}^2). \tag{2.7.4}$$

Hamilton's equations for particles bring us back to the equations of motion (2.6.8) with the vector potential now determined by (2.5.6). Hamilton's equations for field oscillators lead to (2.5.14).

2.8 The Equations of Motion for Phase Densities of Particles and Field Oscillators

So far we have presented an electromagnetic field either by giving the microscopic field strengths for every point in space for a given time, or by giving all the coordinates and momenta of field oscillators. There is also a third way, similar to the method of phase densities for particles.

Let us postulate one of the possible pairs of values for k, α, by letting Q_k^α, P_k^α be the coordinates of a point in two-dimensional phase space of oscillator k, α.

We introduce the notation

$$\mathcal{N}(Q_k^\alpha, P_k^\alpha, t) = \delta(Q_k^\alpha - Q_k^\alpha(t))\,\delta(P_k^\alpha - P_k^\alpha(t)) \tag{2.8.1a}$$

or, in a more compact form,

$$\mathcal{N}(X_k^\alpha, t) \equiv \mathcal{N}_{k,\alpha}(X, t) = \delta(X_k^\alpha - X_k^\alpha(t)) . \tag{2.8.1b}$$

The value of $\mathcal{N}_{k,\alpha}dX_k^\alpha$ is equal to one unit if at a time t the values of Q_k^α and P_k^α fall within $dQ_k^\alpha dP_k^\alpha$. The values of k and α can be considered component numbers.

With the help of (2.8.1a), the expression for the Hamiltonian function (2.7.4) can be written in another form,

$$H = \sum_a \int \frac{1}{2m_a}\left(P - e_a\sqrt{\frac{4\pi}{V}}\sum_{k,\alpha}\int e_k Q_k^\alpha \begin{Bmatrix}\sin kr\\ \cos kr\end{Bmatrix}\mathcal{N}_{k,\alpha}dX_k^\alpha\right)^2$$

$$\times N_a(x, t)\,dx + \frac{1}{2}\sum_{a,b}\int \Phi_{ab}N_a(x, t)N_b(x', t)\,dx\,dx'$$

$$+ \frac{1}{2}\sum_{k,\alpha}\int (P_{k,\alpha}^2 + c^2 k^2 Q_{k,\alpha}^2)\,\mathcal{N}_{k,\alpha}dX_k^\alpha . \tag{2.8.2}$$

The equations for phase densities $N_a(x, t)$ and $\mathcal{N}_{k,\alpha}(X, t)$ are

$$\left[\frac{\partial}{\partial t} + v\frac{\partial}{\partial r} - \frac{\partial}{\partial r}\sum_b\int \Phi_{ab}N_b(x', t)\,dx'\frac{\partial}{\partial p}\right]N_a(x, t)$$

$$+ \frac{e_a}{e}\sum_{k,\alpha}\int\left[\frac{1}{c}\frac{\partial A_{k,\alpha}}{\partial t} + [v\,\mathrm{rot}\,A_{k,\alpha}]\right]\mathcal{N}_{k,\alpha}(X, t)\,dX_k^\alpha\frac{\partial N_a}{\partial p} = 0 \tag{2.8.3}$$

and

$$\left(\frac{\partial}{\partial t} + P_k^\alpha\frac{\partial}{\partial Q_k^\alpha} - c^2 k^2 Q_k^\alpha\frac{\partial}{\partial P_k^\alpha}\right)\mathcal{N}_{k,\alpha}(X, t)$$

$$+ \sum_a e_a\sqrt{\frac{4\pi}{V}}\int e_k v\begin{Bmatrix}\sin kr\\ \cos kr\end{Bmatrix}N_a(x, t)\,dx\frac{\partial \mathcal{N}_{k,\alpha}}{\partial P_k^\alpha} = 0 . \tag{2.8.4}$$

In (2.8.3) we use the notation

$$A_{k,\alpha} = \sqrt{\frac{4\pi c^2}{V}}\,e_k Q_k^\alpha\begin{Bmatrix}\sin kr\\ \cos kr\end{Bmatrix}. \tag{2.8.5}$$

The self-contained set of equations (2.8.3, 4) is equivalent to the set of equations of the motion of charged particles and field oscillators considered previously. It is natural that on the microscopic level all the discussed methods of describing the field and the motion of particles are equivalent. The advantages of one or the other become clear when we seek approximate solutions of averaged equations. We shall take the sets of microscopic equations (2.2.1, 4.1) and (2.8.3, 4) for basic when applying the method of microscopic phase densities (Sects. 2.11, 12).

In the future we must present the Fourier expansions (2.5.6, 9, 11) through $\mathcal{N}_{k,\alpha}$. Making use of definition (2.8.1), we obtain

$$A^m(r, t) \equiv \sum_{k,\alpha} A^m_{k,\alpha} = \sqrt{\frac{4\pi c^2}{V}} \sum_{k,\alpha} \int e_k Q^\alpha_k \begin{Bmatrix} \sin kr \\ \cos kr \end{Bmatrix} \mathcal{N}_{k,\alpha}(X, t) dX^\alpha_k, \qquad (2.8.6)$$

$$[E^m(r, t)]^\perp \equiv \sum_{k,\alpha} (E^m_{k,\alpha})^\perp = -\sqrt{\frac{4\pi}{V}} \sum_{k,\alpha} \int e_k P^\alpha_k \begin{Bmatrix} \sin kr \\ \cos kr \end{Bmatrix} \mathcal{N}_{k,\alpha}(X, t) dX^\alpha_k, \qquad (2.8.7)$$

and

$$B^m(r, t) \equiv \sum_{k,\alpha} B^m_{k,\alpha} = \sqrt{\frac{4\pi e^2}{V}} \times$$

$$\times \sum_{k,\alpha} \int [k \times e_k] \begin{Bmatrix} -Q^2_k \sin kr \\ Q^1_k \cos kr \end{Bmatrix} \mathcal{N}_{k,\alpha}(X, t) dX^\alpha_k. \qquad (2.8.8)$$

We also remember that

$$[E^m(r, t)]^\parallel = -\text{grad}_r \sum_b \int \frac{e_b N_b(x', t)}{|r - r'|} dx'. \qquad (2.8.9)$$

Drawing an analogy with hydrodynamics, we could name the approach used to present the processes in a system of charged particles and field oscillators in (2.8.3, 4) the Euler method. The presentation used in (2.2.1, 3, 4.1) can then be called "hybrid" since we describe particles according to the Euler method, and field oscillators according to the Lagrangian. Using definition (2.8.1) we can relate the Lagrangian variables for field oscillators to the function $\mathcal{N}_{k,\alpha}$,

$$Q^\alpha_k(t) = \int Q^\alpha_k \mathcal{N}_{k,\alpha}(X, t) dX^\alpha_k, \quad P^\alpha_k(t) = \int P^\alpha_k \mathcal{N}_{k,\alpha}(X, t) dX^\alpha_k. \qquad (2.8.10)$$

In the future we shall use these equations quite often.

2.9 Distribution Functions of Particles and Field Oscillators

The method of moments for phase densities N_a and $\mathcal{N}_{k,\alpha}$ will be used in close connection with that of the distribution functions.

We denote by

$$f(\ldots, x_{ia}, \ldots, X_k^\alpha, \ldots, t) \tag{2.9.1}$$

the distribution function of coordinates and momenta of charged particles and field oscillators. The distribution function (2.9.1) is normalized as follows:

$$\int f \prod_{i,a} dx_{i,a} \prod_{k,\alpha} dX_k^\alpha = 1, \quad i = 1, 2, \ldots, N_a, \quad a = e, i, \quad \alpha = 1, 2. \tag{2.9.2}$$

Using the distribution function (2.9.1) one can define simpler distribution functions, the simplest of which are so-called one-particle distribution functions for charged particles and field oscillators. They can be defined as

$$f_{a'}(x', t) \equiv \bar{N}_a / n_a = V \int f \delta(x' - x_{1,a'}) \prod_{i,a} dx_{i,a} \prod_{k,\alpha} dX_k^\alpha \tag{2.9.3}$$

(distribution function of particles of a' component for a point in six-dimensional phase space x'), and

$$\mathscr{F}_{k',\alpha'}(X' t) \equiv \bar{\mathscr{N}}_{k',\alpha'}(X', t) = \int f \delta(X_{k'}^{\alpha'} - X_k^\alpha) \prod_{i,a} dx_{i,a} \prod_{k,\alpha} dX_k^\alpha \tag{2.9.4}$$

(distribution function of field oscillators of "component" k' and α' for a point in phase space X').

In a similar way one can define three "two-particle" distribution functions. One of them is a function of the distribution of two charged particles

$$f_{a'a''}(x', x'', t) \equiv f_{a'}(x', t) f_{a''}(x'', t) + g_{a'a''}(x', x'', t) \tag{2.9.5}$$

where $g_{a'a''}$ denotes the appropriate correlation function. Let us indicate by

$$\mathscr{F}_{k_1\alpha_1, k_2\alpha_2}(X', X'', t) = \mathscr{F}_{k_1, \alpha_1} \mathscr{F}_{k_2, \alpha_2} + G_{k_1\alpha_1, k_2\alpha_2}(X', X'', t) \tag{2.9.6}$$

the distribution function of the coordinates and momenta of two field oscillators; $G_{k_1\alpha_1, k_2\alpha_2}$ is here the appropriate correlation function. The combined two-particle distribution function (one of the variables being a particle, and the other a field oscillator) is denoted by

$$\Phi_{a,k,\alpha}(x, X, t) = f_a \mathscr{F}_{k,\alpha} + H_{a,k,\alpha}(x, X, t), \tag{2.9.7}$$

where $H_{a,k,\alpha}$ is a combined correlation function.

Let us now write the Liouville equation – the equation for the most universal distribution function (2.9.1) –

$$\frac{\partial f}{\partial t} + \sum_{i,a} \left\{ v_{ia} \frac{\partial}{\partial r_{ia}} - \frac{\partial}{\partial r_{ia}} \sum_{j,b} \Phi_{ab}(|r_i - r_j|) \frac{\partial}{\partial p_{ia}} \right.$$

$$+ \frac{e_a}{c} \sum_{k,\alpha} \left(- \frac{\partial A_{k,\alpha}}{\partial t} + [v_{ia} \operatorname{rot} A_{k,\alpha}] \right) \frac{\partial}{\partial p_{ia}} \right\} f + \sum_{k,\alpha} \left[P_k^\alpha \frac{\partial}{\partial Q_k^\alpha} \right.$$

$$- c^2 k^2 Q_k^\alpha \frac{\partial}{\partial P_k^\alpha} + \sqrt{\frac{4\pi}{V}} \sum_{i,a} e_a(e_k v_{ia}) \begin{Bmatrix} \sin kr_i \\ \cos kr_i \end{Bmatrix} \frac{\partial}{\partial P_k^\alpha} \right] f = 0 . \qquad (2.9.8)$$

Here we make use of Fourier expansions for the vector potential and current, and of (2.8.5) for $A_{k,\alpha}$.

The Liouville equation is nonsymmetrical in the sense that the terms corresponding to charged particles and field oscillators bear little similarity. For instance, particles are capable of direct interaction, while field oscillators interact only through particles. This is a natural consequence of the nonsymmetry of the basic equations of motion of charged particles and field oscillators.

2.10 Chain of Equations for Distribution Functions of Particles and Field Oscillators

It is possible, based on the Liouville equation, to build a chain of equations for the distribution function of the variables of particles and field oscillators, similar to the set of equations for a many-particle system after *Bogolyubov* [1.2], *Born* and *Green* [1.28], *Kirkwood* [1.29], and *Yvon* [1.30]. It turns out that there are two independent chains of equations for a system of charged particles and a field. The first begins with the equations for the one-particle distribution functions for particles and field oscillators (2.9.3, 4). The second chain starts with the equation for the two-particle distribution function of coordinates and momenta of field oscillator, i. e., with the equation for the distribution function (2.9.6).

We are not going to list these equations here since all the necessary equations will be obtained below through the method of microscopic phase densities. We confine ourselves to citing the references [2.8 – 11] where these chains of equations have been investigated.

Let us introduce some notation which will be used in the next sections. An operator is written

$$\hat{L}_{a_1,\dots,a_n} = \frac{\partial}{\partial t} + \sum_{1 \leqslant i \leqslant n} \left(v_i \frac{\partial}{\partial r_i} + e_{a_i} \left\{ E(r, t) + \frac{1}{c}[v_i \times B] \right\} \frac{\partial}{\partial p_i} \right). \qquad (2.10.1)$$

In particular,

$$\hat{L}_a = \frac{\partial}{\partial t} + v \frac{\partial}{\partial r} + e_a \left\{ E(r, t) + \frac{1}{c}[v \times B(r, t)] \right\} \frac{\partial}{\partial p} , \qquad (2.10.2)$$

where E and B are mean strengths of electric and magnetic fields. They are determined via one-particle distribution functions f_a and $\mathscr{F}_{k,\alpha}$ according to

$$E(r, t) = E^\| + E^\perp = - \text{grad} \sum_b n_b \int \frac{e_b}{|r-r'|} f_b(x', t)\,dx'$$

$$- \sqrt{\frac{4\pi}{V}} \sum_{k,\alpha} \int e_k P_k^\alpha \begin{Bmatrix} \sin kr \\ \cos kr \end{Bmatrix} \mathscr{F}_{k,\alpha}(X, t)\,dX, \qquad (2.10.3)$$

$$B(r, t) = \sqrt{\frac{4\pi c^2}{V}} \sum_{k,\alpha} \int [k \times e_k] \begin{Bmatrix} -Q_k^2 \sin kr \\ Q_k^1 \cos kr \end{Bmatrix} \mathscr{F}_{k,\alpha}(X, t)\,dX. \qquad (2.10.4)$$

Here we use (2.8.7, 8) and the relation of one-particle distribution functions f_a and $\mathscr{F}_{k,\alpha}$ with the mean values of random functions N_a and $\mathscr{N}_{k,\alpha}$ (2.9.3, 4).

The operators acting upon functions of coordinates and momenta of field oscillators are

$$\hat{L}_{k_1\alpha_1,\ldots,k_n\alpha_n} = \frac{\partial}{\partial t} + \sum_{1 \le i \le n} \left[\left(P_{k_i}^{\alpha_i} \frac{\partial}{\partial Q_{k_i}^{\alpha_i}} - c^2 k_i^2 Q_{k_i}^{\alpha_i} \frac{\partial}{\partial P_{k_i}^{\alpha_i}} \right) \right.$$

$$\left. + 4\pi e_{k_i} j_{k_i}^{\alpha_i} \frac{\partial}{\partial P_{k_i}^{\alpha_i}} \right]. \qquad (2.10.5)$$

Here we employ the notation for the mean value of Fourier component of mean current [cf. (2.5.14)]

$$j_k^\alpha = \sum_a e_a n_a \int v \begin{Bmatrix} \sin kr \\ \cos kr \end{Bmatrix} f_a(x, t)\,dx. \qquad (2.10.6)$$

The simplest form of (2.10.5) is the operator $\hat{L}_{k,\alpha}$, which acts upon the functions of coordinates and momenta of one single field oscillator.

We shall also need the simplest "combined" operator

$$\hat{L}_{a,k,\alpha} = \frac{\partial}{\partial t} + v\frac{\partial}{\partial r} + P_k^\alpha \frac{\partial}{\partial Q_k^\alpha} - c^2 k^2 Q_k^\alpha \frac{\partial}{\partial P_k^\alpha}$$

$$+ e_a \left\{ E^\|(r, t) + E^\perp(r, t) + \frac{1}{c}[v \times B(r, t)] \right\} \frac{\partial}{\partial p} + 4\pi e_k j_k^\alpha \frac{\partial}{\partial P_k^\alpha}. \qquad (2.10.7)$$

This acts upon functions which depend on the coordinates and momenta of one particle and on those of a field oscillator.

2.11 Equations for Moments

In obtaining equations for moments we shall employ two methods. The first is based on the set of equations for random functions E^m, B^m and $N_a(x, t)$ (2.2.1, 3 and 2.4.1). This method of getting the equations for moments is now common practice in the statistical theory of completely ionized plasma [2.1 – 4].

In another approach, (2.8.3, 4) are considered basic for the random functions $N_a(x, t)$ and $\mathcal{N}_{k, a}(X, t)$. This second method is far less developed [2.9 – 11].

Let us average (2.2.1, 3, 4.1) for functions N_a and microscopic field strengths, and use the relation between the first moments of the function N_a and one-particle distribution functions for charged particles f_a,

$$\bar{N}_a(x, t) = n_a f_a(x, t), \quad n_a = N_a/V,$$ (2.11.1)

with the identities

$$\overline{E^m N_a} = E n_a f_a + \overline{\delta E \, \delta N_a}, \quad \overline{B^m N_a} = B n_a f_a + \overline{\delta B \, \delta N_a}.$$ (2.11.2)

As a result, we get the following equations for distribution functions f_a and mean field strengths

$$E = \overline{E^m}, \quad B = \overline{B^m}.$$ (2.11.3)

The equations for functions f_a are

$$\left\{ \frac{\partial}{\partial t} + v \frac{\partial}{\partial r} + e_a \left(E + \frac{1}{c} [v \times B] \right) \frac{\partial}{\partial p} \right\} f_a(x, t) \equiv \hat{L} f_a = I_a,$$ (2.11.4)

where

$$I_a(x, t) = \frac{-1}{n_a} \frac{\partial}{\partial p} \left(e_a \overline{\delta E \, \delta N_a} + \frac{e_a}{c} \overline{[v \, \delta \times BN_a]} \right) \equiv \frac{-1}{n_a} \frac{\overline{\partial \delta F_a \delta N_a}}{\partial p}$$ (2.11.5)

is "the collision integral," determined by the fluctuations of the distribution functions of particles and field.

The equations for mean field strengths can be written

$$\operatorname{rot} B = \frac{1}{c} \frac{\partial E}{\partial t} + \frac{4\pi}{c} j, \quad j = \sum_a e_a n_a \int v f_a dp,$$

$$\operatorname{rot} E = -\frac{1}{c} \frac{\partial B}{\partial t},$$

$$\operatorname{div} B = 0,$$

$$\operatorname{div} E = 4\pi q, \quad q = \sum_a e_a n_a \int f_a dp.$$ (2.11.6)

The set of equations (2.11.4, 6) is not closed since the collision integral is determined by the second moments of the fluctuations of field and phase density. In order to build our chain of equations further, we must first write the equations for the deviations δE, δB, and δN_a, which arise from (2.2.1, 3, 4.1), and also from (2.11.4, 6),

$$\hat{L}_a \delta N_a + \delta F_a \frac{\partial n_a f_a}{\partial p} = - \frac{\partial}{\partial p} (\overline{\delta F_a \delta N_a} - \overline{\delta F_a \delta N_a}) , \tag{2.11.7}$$

$$\text{rot } \delta B = \frac{1}{c} \frac{\partial \delta E}{\partial t} + \frac{4\pi}{c} \sum_a e_a \int v \delta N_a dp , \quad \text{div } \delta B = 0 ,$$

$$\text{rot } \delta E = - \frac{1}{c} \frac{\partial B}{\partial t} , \quad \text{div } \delta E = 4\pi \sum_a e_a \int \delta N_a dp . \tag{2.11.8}$$

The Lorentz force fluctuation is denoted

$$\delta F_a = e_a \left(\delta E + \frac{1}{c} [v \times \delta B] \right) . \tag{2.11.9}$$

The operator $\hat{L}_a$ is determined by (2.10.2).

Equation (2.11.7) is nonlinear with respect to fluctuations. For this reason the equations for second moments derived from it will include third moments. In this way a chain of equations for the moments is formed, which we shall further investigate in Sect. 3.4.5.

Now we shall turn to another chain of equations which begins by averaging the equations for random functions N_a and $\mathcal{N}_{k,a}$ (2.8.3, 4). We use the relation between the moments of these functions and one-particle distribution functions f_a and $\mathcal{F}_{k,a}$,

$$\bar{N}_a(x, t) = n_a f_a(x, t) , \quad \bar{\mathcal{N}}_{k,a}(X, t) = \mathcal{F}_{k,a}(X, t) , \tag{2.11.10}$$

and the identities

$$\overline{N_a(x, t) \mathcal{N}_{k,a}(X, t)} = n_a f_a \mathcal{F}_{k,a} + \overline{\delta N_a \delta \mathcal{N}_{k,a}} ,$$

$$\overline{N_a(x, t) N_b(x', t)} = n_a n_b f_a f_b + \overline{\delta N_a \delta N_b} . \tag{2.11.11}$$

Averaging (2.8.3, 4) we get the equations for the first moments of random functions N_a and $\mathcal{N}_{k,a}$. Let us first write for the function f_a

$$\hat{L}_a f_a = I_a(x, t) . \tag{2.11.12}$$

This differs from (2.11.4) only in the collision integral, which is now expressed in terms of the second moments of the fluctuations δN_a and $\delta \mathcal{N}_k^a$,

$$I_a(x, t) = \sum_b \int \frac{\partial \Phi_{ab}}{\partial r} \frac{\partial \overline{(\delta N_a \delta N_b)}_{x, x', t}}{\partial p} dx'$$

$$- \frac{e_a}{c} \sum_{k, \alpha} \int \left(- \frac{\partial A_{k, \alpha}}{\partial t} + [v \operatorname{rot} A_{k, \alpha}] \right) \frac{\partial \overline{(\delta N_a \delta \mathcal{N}_{k, \alpha})}_{x, X, t}}{\partial p} dX .$$

$$(2.11.13)$$

The meaning of $A_{k, \alpha}$ is as in (2.8.5).

The equation for the distribution function of coordinates and moments of field oscillators is

$$\hat{L}_{k, \alpha} \mathcal{F}_{k, \alpha} = I_{k, \alpha}(X, t) .$$

$$(2.11.14)$$

Here we employ the expression (2.10.5) for $\hat{L}_{k, \alpha}$, and introduce another collision integral

$$I_{k, \alpha}(X, t) = - \sum_a e_a n_a \sqrt{\frac{4\pi}{V}} \int (e_n \cdot v) \begin{Bmatrix} \sin kr \\ \cos kr \end{Bmatrix} \frac{\partial \overline{\delta \mathcal{N}_{k, \alpha} \delta N_a}}{\partial P_k^\alpha} dx$$

$$\equiv - 4\pi \frac{\partial}{\partial P_k^\alpha} \overline{\delta j_k^\alpha \delta \mathcal{N}_{k, \alpha}} .$$

$$(2.11.15)$$

Compare these equations with (2.11.6) for the first moments of the field.

If we use the Fourier expansions (2.8.7, 8) for mean field strengths and field fluctuations, we can derive from (2.11.15) the equations for the mean rotational electromagnetic field (2.10.3, 4). As a matter of fact, these equations coincide with the rotational part of (2.11.6). When carrying out this transition, one should bear in mind that

$$\int \delta \mathcal{N}_{k, \alpha}(X, t) dX = 0$$

by definition, and thus the contribution of the collision integral (2.11.15) to the equations for the mean field is equal to zero.

Equations (2.11.14, 15) are, however, more universal than (2.11.6) since the former make it possible to derive the equations for the higher moments of Fourier components $Q_k^\alpha(t)$ and $P_k^\alpha(t)$ from them. In particular, one can obtain the equation for mean energy of the field (2.5.12), which can be presented as a sum of mean energies of field oscillators. The contribution of the collision integral (2.11.5) to the equation for mean energy of the field in this case will, naturally, no longer be equal to zero.

2.12 The Relation Between Moments and Distribution Functions

As we already know, the first moment of phase density N_a is related to the distribution function f_a according to

$$\bar{N}_a(x,\,t) = n_a f_a(x,\,t)\,.\tag{2.12.1}$$

The second moment is related to one- and two-particle distribution functions according to [Ref. 2.2, Sects. 5, 26]

$$\overline{N_a(x,\,t)\,N_b(x',\,t)} = \frac{N_a N_b - \delta_{ab} N_a}{V^2} f_{ab}(x,\,x',\,t) + n_a \delta_{ab} \delta(x-x') f_a\,.\tag{2.12.2}$$

Hence, using the identity

$$\overline{N_a(x,\,t)\,N_b(x',\,t)} = \bar{N}_a \bar{N}_b + (\overline{\delta N_a\,\delta N_b})_{x,\,x',\,t}\tag{2.12.3}$$

and the definition of a two-particle correlation function

$$f_{ab}(x,\,x',\,t) = f_a(x,\,t) f_b(x',\,t) + g_{ab}(x,\,x',\,t)\,,\tag{2.12.4}$$

we can find the relationship of the second central moment with functions g_{ab} and f_a,

$$(\overline{\delta N_a\,\delta N_b})_{x,\,x',\,t} = n_a n_b g_{ab} + n_a \delta_{ab}[\delta(x-x') f_a - \frac{1}{V} f_a(x,\,t) f_b(x',\,t)]\,.\tag{2.12.5}$$

In the extreme case, when $N \to \infty$ and $V \to \infty$ but $N/V = $ const, this expression is simplified,

$$(\overline{\delta N_a\,\delta N_b})_{x,\,x',\,t} = n_a n_b g_{ab} + n_a \delta_{ab} \delta(x-x') f_a\,.\tag{2.12.6}$$

Employing the definitions of phase density and distribution functions, one can find the relationship between higher moments of phase density of charged particles and distribution functions [2.1, 2, 9, 10].

Now we shall turn to the relationship between the moments of the phase density of electromagnetic field oscillators and the corresponding distribution functions. Recalling (2.11.10) we can write

$$\bar{\mathcal{N}}_{k,\,\alpha}(X,\,t) = \mathcal{F}_{k,\,\alpha}(X,\,t)\,,\tag{2.12.7}$$

which relates the first moment to the one-particle distribution function of field oscillators. We can also write the equation for the second moment, analogous to (2.12.2),

$$\overline{\mathcal{N}_{k,\,\alpha}(X,\,t)\,\mathcal{N}_{k',\,\alpha'}(X',\,t)} = \mathcal{F}_{k,\,\alpha;k',\,\alpha'}(X,\,X',\,t) + \delta_{k,\,k'} \delta_{\alpha,\,\alpha'} \delta(X-X') \mathcal{F}_{k,\,\alpha}(X,\,t)\,.\tag{2.12.8}$$

Using the identity

$$\overline{\mathcal{N}_{k,\,\alpha}\,\mathcal{N}_{k',\,\alpha'}} = \bar{\mathcal{N}}_{k,\,\alpha} \bar{\mathcal{N}}_{k',\,\alpha'} + \overline{\delta\mathcal{N}_{k,\,\alpha}\,\delta\mathcal{N}_{k',\,\alpha'}}\tag{2.12.9}$$

and the definition of the field's correlation function

$$\mathscr{F}_{k,\alpha;k',\alpha'}(X, X', t) = \mathscr{F}_{k,\alpha}\mathscr{F}_{k',\alpha'} + G_{k,\alpha;k',\alpha'} ,\tag{2.12.10}$$

we find the relationship between the second central moment and the two-particle correlation function,

$$\overline{\delta\mathscr{N}_{k,\alpha}\delta\mathscr{N}_{k',\alpha'}} = G_{k,\alpha;k',\alpha'}(X, X', t)$$
$$+ \delta_{k,k'}\delta_{\alpha,\alpha'}[\delta(X - X')\,\mathscr{F}_{k,\alpha} - \mathscr{F}_{k,\alpha}(X, t)\,\mathscr{F}_{k',\alpha'}(X', t)] .\tag{2.12.11}$$

Concerning the appropriate formulae for "hybrid" functions of particles and field oscillators, the second combined moment is related to the two-particle distribution function according to

$$\overline{N_a(x, t)\,\mathscr{N}_{k,\alpha}(X, t)} = n_a\,\Phi_{a,k,\alpha}(x, X, t) .\tag{2.12.12}$$

Using the identity

$$\overline{N_a(x, t)\,\mathscr{N}_{k,\alpha}(X, t)} = \bar{N}_a\,\mathscr{F}_{k,\alpha} + \overline{\delta N_a\delta\mathscr{N}_{k,\alpha}}\tag{2.12.13}$$

and the defintion of the correlation function

$$\Phi_{a,k,\alpha}(x, X, t) = f_a\,\mathscr{F}_{k,\alpha} + H_{a,k,\alpha} ,\tag{2.12.14}$$

we obtain

$$\overline{\delta N_a\delta\mathscr{N}_{k,\alpha}} = n_a H_{a,k,\alpha}(x, X, t) .\tag{2.12.15}$$

We shall need these equations in Chap. 4 to obtain the kinetic equations.

3. Atoms and Field

The results obtained in Chap. 2 are generalized for the case of a system of atoms (charged particles bound in pairs) and a microscopic electromagnetic field.

3.1 The Equations of Motion of Pairs of Free Charged Particles and Atoms

Let us now return to the equatons of motion of free charged particles in electron-ion plasma (2.1.4). Consistant with what was said in Sects. 2.3, 4, we shall make no distinction between local fields and microscopic field strengths. Therefore we write the equations of motion in the form

$$\frac{d\boldsymbol{r}_{ia}}{dt} = \frac{\boldsymbol{p}_{ia}}{m_a} = \boldsymbol{v}_{ia}\,, \tag{3.1.1}$$

$$\frac{d\boldsymbol{p}_{ia}}{dt} = e_a \left\{ \boldsymbol{E}^{\mathrm{m}}(\boldsymbol{r}_i,\, t) + \frac{1}{c}\, [\boldsymbol{v}_i \times \boldsymbol{B}^{\mathrm{m}}(\boldsymbol{r}_i,\, t)] \right\}. \tag{3.1.2}$$

In the right-hand side of (3.1.2) we single out the Coulomb interaction between the chosen particle and those surrounding it, and obtain

$$\frac{d\boldsymbol{p}_{ia}}{dt} = -\sum_b \sum_j \frac{\partial \Phi_{ab}}{\partial \boldsymbol{r}_i} + e_a \left\{ (\boldsymbol{E}^{\mathrm{m}})^{\perp} + \frac{1}{c}\, [\boldsymbol{v}_i \times \boldsymbol{B}^{\mathrm{m}}] \right\}. \tag{3.1.3}$$

A plasma, consisting of free charged particles, can be considered as an extreme case of a partially ionized gas. Another extreme case is a gas of neutral particles, where all charged particles are bound in pairs and ionization is equal to zero. To make it possible to describe charged particles in bound states, we must transform the set of equatons of motion (3.1.1, 2).

We introduce the variables $\boldsymbol{r}_i$ and $\boldsymbol{p}_i$ to describe the relative motion of pairs of charged particles, and variables $\boldsymbol{R}_i$ and $\boldsymbol{P}_i$ to describe the center of mass motion. They are determined by the following equations (subscript i omitted):

$$V = \dot{R} = \frac{m_a v_a + m_b v_b}{m_a + m_b}, \quad v = v_a - v_b,$$

$$P = (m_a + m_b) V = p_a + p_b,$$

$$p = \frac{m_a m_b}{m_a + m_b} v, \quad R = \frac{m_a r_a + m_b r_b}{m_a + m_b}, \quad r = r_a - r_b \equiv r_+ - r_- . \tag{3.1.4}$$

In these expressions $a \neq b$, $a = i$, $b = e$ and the subscript ab at R, P, V, r, v, p is omitted. From now on for pairs of bound particles $a = i$ and $b = e$. We shall also use the following symbols for the total mass of a pair and for equivalent mass:

$$m_a + m_b = M ,$$

$$\frac{m_a m_b}{m_a + m_b} = \mu . \tag{3.1.5}$$

From (3.1.4) one may obtain inverse relations

$$r_{a,b} = R \pm \frac{m_{b,a}}{M} r, \quad v_{c,b} = V \pm \frac{m_{b,a}}{M} v, \quad p_{a,b} = \frac{m_{a,b}}{M} P \pm p . \tag{3.1.6}$$

Using (3.1.1, 2) the equations for the center of mass motion of pairs of charged particles are

$$\frac{dR_i}{dt} = \frac{P_i}{M} = V_i, \quad i = 1, 2, \ldots, N ,$$

$$\frac{dp_i}{dt} = e_a E^m \left(R_i + \frac{m_b}{M} r_i, t \right) + e_b E^m \left(R_i - \frac{m_a}{M} r_i t \right)$$

$$+ \frac{e_a}{c} \left[\left(V_i + \frac{m_b}{M} v_i \right) B^m \left(R_i + \frac{m_b}{M} r_i, t \right) \right]$$

$$+ \frac{e_b}{c} \left[\left(V_i - \frac{m_a}{M} v_i \right) B^m \left(R_i - \frac{m_a}{M} r_i, t \right) \right] . \tag{3.1.7}$$

The relevant equations, describing the relative motion of particles in a pair, have the form

$$\frac{\partial p_i}{\partial t} = \frac{p_i}{\mu} = v_i ,$$

$$\frac{dp_i}{dt} = \frac{e_a m_b}{M} E^{\mathrm{m}} \left(R_i + \frac{m_b}{M} r_i,\, t \right) - \frac{e_b m_a}{M} \left(R_i - \frac{m_a}{M} r_i,\, t \right)$$

$$+ \frac{e_a m_b}{cM} \left[\left(V_i + \frac{m_b}{M} r_i \right) B^{\mathrm{m}} \left(R_i + \frac{m_b}{M} r_i,\, t \right) \right]$$

$$- \frac{e_b m_a}{cM} \left[\left(V_i - \frac{m_a}{M} v_i \right) B^{\mathrm{m}} \left(R_i - \frac{m_a}{M} v_i,\, t \right) \right]. \tag{3.1.8}$$

Naturally, (3.1.7, 8) are equivalent to the initial equations, but they are more convenient for executing the transition to the equations of motion of atoms – bound pairs of particles. These equations, describing bound states (atoms), can be substantially simplified provided that the characteristic scale of field changes – e.g., radiation wavelength – greatly exceeds the size of atom r_0.

Let us introduce a small parameter μ_{at}, which will be called the "atom parameter",

$$\mu_{\mathrm{at}} \sim r_0/\lambda \sim v_0/c, \tag{3.1.9}$$

where r_0 is the radius of the atom, $v_0 \sim r_0 \omega_0$ is the velocity of electrons in the atom, and ω_0 is the eigenfrequency of the oscillating atom.

Now we expand (3.1.7, 8) in powers of μ_{at}, and retain only first-order terms. Dropping the subscript i, we get

$$\frac{dP}{dt} = e_a (r\,\mathrm{grad}_R) E^{\mathrm{m}}(R,\, t) + \frac{e_a}{c} [V (r\,\mathrm{grad}_R) B^{\mathrm{m}}(R,\, t)] + \frac{e_a}{c} [v \times B^{\mathrm{m}}(R,\, t)], \tag{3.1.10}$$

$$\frac{dp}{dt} = -\frac{\partial \Phi}{\partial r} + e_a E(R,\, t) + \frac{e_a(m_b - m_a)}{M} (r\,\mathrm{grad}_R) E^{\mathrm{m}}(R,\, t)$$

$$+ \frac{e_a}{c} \frac{m_b - m_a}{M} [v \times B^{\mathrm{m}}(R,\, t)]$$

$$+ \frac{e_a}{c} \left[V, B^{\mathrm{m}} + \frac{m_b - m_a}{M} (r\,\mathrm{grad}_R) B^{\mathrm{m}} \right] \tag{3.1.11}$$

(here we take into account that $e_a + e_b = 0$).

In (3.1.11) the interaction of particles within a pair is singled out, and thus the field only determines the influence upon a given atom which comes from charged particles in other atoms.

If the thermal velocity of atoms is much less than the speed of light, there exists another small parameter

$$\mu_{\mathrm{T}} \sim V_{\mathrm{T}}/c. \tag{3.1.12}$$

In the linear approximation with respect to two parameters μ_T and μ_{at}, the equations of motion (3.1.10, 11) can be further simplified and reduced to

$$\frac{dP}{dt} = e_a (r \, \text{grad}_R) \, E^m (R, t) + \frac{e_a}{c} [v \times B^m (R, t)] , \qquad (3.1.13)$$

$$\frac{dp}{dt} = -\frac{\partial \Phi}{\partial r} + e_a \left\{ E^m (R, t) + \frac{1}{c} [V \times B^m (R, t)] \right\}$$

$$+ \frac{e_a (m_b - m_a)}{M} (r \, \text{grad}_R) \, E^m (R, t) + \frac{e_a}{c} \frac{m_b - m_a}{M} \times [v \times B^m] . \quad (3.1.14)$$

Let us consider two other special cases which we shall frequently need in future. The first of them is the zero approximation with respect to the atom parameter μ_{at}. In this case

$$\frac{dP}{dt} = 0 , \qquad (3.1.15a)$$

$$\frac{dp}{dt} = -\frac{\partial \Phi}{\partial r} + e_a \left\{ E^m (R, t) + \frac{1}{c} [V \times B^m (R, t)] \right\}. \qquad (3.1.15b)$$

In this approximation the atom as a whole is moving freely, but its inner motion is modified by the Lorentz force.

The second case is the zero-order approximation in respect to two parameters μ_{at} and μ_T. In this case the equations of motion take the simplest form

$$\frac{dP}{dt} = 0 , \quad \frac{dp}{dt} = -\frac{\partial \Phi}{\partial r} + e_a E^m (R, t) . \qquad (3.1.16)$$

Both these approximations will be called "dipole approximations."

Let us emphasize one more important point. In the equations of motion of atoms (3.1.10) all terms are of first order with respect to μ_{at}. Yet in equation (3.1.11) for the internal motion of particles within atoms the first two terms are of zeroth order, and the rest of first order with respect to μ_{at}. Consequently, there exist approximations in which the first nondisappearing terms are retained in the right-hand sides of the equations after expansion in powers of the small parameter μ_{at}. In such an approximation (3.1.10) remains intact, and (3.1.11) has a simpler appearance (3.1.15b).

3.2 The Equations for Microscopic Phase Density of Atoms

In Sect. 2.1 we defined the phase densities $N_a(x, t)$ of free charged particles of each plasma component. As we have shown, a set of equations for functions

$N_a(x, t)$ and the microscopic field strengths can be taken for an initial set of equations describing processes in completely ionized plasma.

In order to obtain similar equations for a system composed of atoms and a field, we shall introduce the phase density of pairs of oppositely charged particles,

$$N_{ab}(x', x'', t) = \sum_{1 \leq i \leq N} \delta(x' - x_{ia}(t)) \delta(x'' - x_{ib}(t)) ,$$

$$a \neq b , \quad x = (r, p) , \tag{3.2.1}$$

where N is the total number of electron-ion pairs, $\int N_{ab} dx' dx'' = N$. The quantity $N_{ab} dx' dx''$ determines the number of pairs of charged particles such that one of the particles in a pair (of kind a) is confined within the neighborhood dx' of point x', and the other particle of a pair (of kind b) is confined in the neighborhood dx'' of point x'. The functions $N_a(x, t)$ are linked to N_{ab} according to the formula

$$N_a(x, t) = \int N_{ab}(x, x'', t) dx'' \tag{3.2.2}$$

which ensues from definitions (2.1.5, 3.2.1).

The equation for phase density of pairs of particles can be written in the form

$$\left[\frac{\partial}{\partial t} + v' \frac{\partial}{\partial r'} + v'' \frac{\partial}{\partial r''} + F_a^m(x', t) \frac{\partial}{\partial p'} + F_b^m(x'', t) \frac{\partial}{\partial p''} \right] N_{ab} = 0 , \tag{3.2.3}$$

where F_a^m, F_b^m are Lorentz forces in points x' and x''.

Using (3.2.2), one can derive from (3.2.3) the equation for phase density (2.4.1), and in this sense (3.2.3) is equivalent to (2.4.1). Instead of using the function $N_{ab}(x', x'', t)$ to describe the bound states of charged particles, it is more convenient to use an appropriate microscopic phase density in the space of variables r, p, R and P, which are linked to $x' = (r', p')$ and $x'' = (r'', p'')$ (3.1.4, 6).

In order to pass from (3.2.3) to the equation for function $N_{ab}(r, p, R, P, t)$, one must make use of the operator equation

$$v' \frac{\partial}{\partial r'} + v'' \frac{\partial}{\partial r''} = V \frac{\partial}{\partial R} + v \frac{\partial}{\partial r} \tag{3.2.4}$$

and the appropriate equation for force terms. Then for the phase density

$$N_{ab}(r, p, R, P, t) = \sum_{1 \leq i \leq N} \delta(r - r_i(t)) \delta(p - p_i(t)) \delta(R - R_i(t)) \delta(P - P_i(t)) , \tag{3.2.5}$$

we get the equation

$$\left\{ \frac{\partial}{\partial t} + V\frac{\partial}{\partial R} + v\frac{\partial}{\partial r} + \left[F_a^m \left(R + \frac{m_b}{M}r, V + \frac{m_b}{M}v, t \right) \right. \right.$$

$$\left. + F_b^m \left(R - \frac{m_a}{M}r, V - \frac{m_a}{M}v, t \right) \right] \frac{\partial}{\partial p}$$

$$+ \left[\frac{m_b}{M} F_a^m \left(R + \frac{m_b}{M}r, V + \frac{m_b}{M}v, t \right) \right.$$

$$\left. \left. - \frac{m_a}{M} F_b^m \left(R - \frac{m_a}{M}r, V - \frac{m_a}{M}v, t \right) \right] \frac{\partial}{\partial p} \right\} N_{ab} = 0 . \qquad (3.2.6)$$

In the first nondisappearing approximation with respect to the atom parameter μ_{at} this equation can be substantially simplified,

$$\left\{ \frac{\partial}{\partial t} + V\frac{\partial}{\partial R} + \frac{\partial}{\partial r} - \frac{\partial \Phi}{\partial r}\frac{\partial}{\partial p} + \left\{ e_a (r \, \mathrm{grad}_R) \, E^m(R, t) \right. \right.$$

$$\left. + \frac{e_a}{c} [v \times B^m(R, t)] + \frac{e_a}{c} [V (r \, \mathrm{grad}_R) B^m] \right\} \frac{\partial}{\partial p}$$

$$+ e_a \left\{ E^m(R, t) + \frac{1}{c}[V \times B^m(R, t)] \right\} \frac{\partial}{\partial p} \right\} N_{ab} = 0 . \qquad (3.2.7)$$

Here, as was done in (3.1.11), the interaction of particles within a pair (an atom) is singled out. The characteristic equations of (3.2.7) coincide, no doubt, with (3.1.10, 15 b).

In the dipole approximation (3.2.7) takes the form

$$\left(\frac{\partial}{\partial t} + V\frac{\partial}{\partial R} + v\frac{\partial}{\partial r} - \frac{\partial \Phi}{\partial r}\frac{\partial}{\partial p} \right.$$

$$\left. + e_a \left\{ E^m(R, t) + \frac{1}{c}[V \times B^m(R, t)] \right\} \frac{\partial}{\partial p} \right) N_{ab} = 0 . \qquad (3.2.8)$$

Naturally, with the equation for function N_{ab} the question of distinguishing between local field strengths and microscopic field strengths still remains.

3.3 Microscopic Field Equations

Charge and current densities in field equations (2.2.1) for a system of free charged particles were defined as

$$q^m(r, t) = \sum_a e_a \int N_a(x, t)\,dp \; ,$$

$$j^m(r, t) = \sum_a e_a \int v N_a(x, t)\,dp \; . \tag{3.3.1}$$

When we come to use (3.2.3) for the phase density of pairs of particles, the charge and current densities ought to be expressed by way of function N_{ab},

$$q^m(r, t) = \int [e_a \delta(r - r') + e_b \delta(r - r'')]\, N_{ab}(x', x'', t)\,dx'\,dx'' \tag{3.3.2}$$

$$j^m(r, t) = \int [e_a v' \delta(r - r') + e_b v'' \delta(r - r'')]\, N_{ab}(x', x'', t)\,dx'\,dx'' \; . \tag{3.3.3}$$

From definitions of phase densities N_a and N_{ab} and formula (3.2.2) it follows that (3.3.1 – 3) are equivalent.

Equations (3.2.3) and (2.2.1) with (3.3.2, 3) constitute a self-contained set of equations for the functions $N_{ab}(x', x'', t)$, $E^m(r, t)$, and $B^m(r, t)$.

When one goes from the equation for phase density $N_{ab}(x', x'', t)$ to the equation for function $N_{ab}(r, p, R, P, t)$, one similarly has to transform the expressions for charge and current densities.

Taking into account (3.1.4, 6) we shall use

$$q^m(q, t) = \int \left[e_a \delta\left(q - R - \frac{m_b}{M} r \right) + e_b \delta\left(q - R + \frac{m_a}{M} r \right) \right] N_{ab}(X, t)\,dx \; ,$$

$$X = (r, p, R, P) \; , \quad dX = dr\,dp\,dR\,dP \; , \tag{3.3.4a}$$

$$j^m(q, t) = \int \left[e_a \left(V + \frac{m_b}{M} v \right) \delta\left(q - R - \frac{m_b}{M} r \right) \right.$$

$$\left. + e_b \left(V - \frac{m_a}{M} v \right) \delta\left(q - R + \frac{m_a}{M} r \right) \right] N_{ab}(X, t)\,dX \tag{3.3.4b}$$

instead of (3.3.2, 3).

Let us expand the right-hand sides of these equations in series with respect to $r \dfrac{\partial}{\partial R}$, which amounts to expansion in powers of parameter μ_{at} – "expansion in multifields" [3.1 – 3]. Starting with the expression for charge density, we replace q by R and obtain

$$q^{\mathrm{m}}(R, t) = \int \sum_{n=0}^{\infty} \frac{1}{n!} \left[(-1)^n e_a \left(\frac{m_b}{M} r \, \mathrm{grad}_R \right)^n \right.$$

$$\left. + e_b \left(\frac{m_a}{M} r \, \mathrm{grad}_R \right)^n \right] N_{ab}(X, t) \, dr \, dp \, dP \, . \qquad (3.3.5)$$

The term with $n = 0$ gives a zero contribution since $e_a + e_b = 0$; therefore (3.3.5) can be rewritten in the form

$$q^{\mathrm{m}}(R, t) = - \, \mathrm{div}_R \, P^{\mathrm{m}}(R, t) \, . \qquad (3.3.6)$$

Here we have introduced a microscopic polarization vector (in the summand n becomes $n + 1$)

$$P^{\mathrm{m}}(R, t) = \int \sum_{n=0}^{\infty} \frac{1}{(n+1)!} \left[(-1)^n \frac{e_a m_b}{M} r \left(\frac{m_b}{M} r \, \mathrm{grad}_R \right)^n \right.$$

$$\left. - \frac{e_b m_a}{M} r \left(\frac{m_a}{M} r \, \mathrm{grad}_R \right)^n \right] N_{ab}(X, t) \, dr \, dp \, dP \equiv \int P^{\mathrm{m}}(R, P, t) dP \, .$$
$$(3.3.7)$$

This expression can be written in a more concise form with the help of the designations

$$r_l = \frac{m_b}{M} r \quad \text{if} \quad l = a \, , \quad r_l = - \frac{m_a}{M} r \quad \text{if} \quad l = b \, , \qquad (3.3.8)$$

$$P^{\mathrm{m}}(R, t) = \int \sum_{l=a,b} \sum_{n=0}^{\infty} \frac{(-1)^n}{(n+1)!} \, e_l r_l (r_l \, \mathrm{grad}_R)^n N_{ab}(X, t) \, dr \, dp \, dP \, . \qquad (3.3.9)$$

Now we apply the same treatment to the right-hand side of (3.3.4b) for current density. Designating the current due to relative motion within pairs of particles as

$$J^{\mathrm{m}}(R, t) = \sum_{l=a,b} \int e_l v_l N_{ab}(X, t) \, dr \, dp \, dP \, , \quad v_l = \dot{r}_l \, , \qquad (3.3.10)$$

we can rewrite (3.3.4b) in the form

$$j^{\mathrm{m}}(R, t) = - \int V \, \mathrm{div}_R \, P^{\mathrm{m}} dP + J^{\mathrm{m}}(R, t)$$

$$+ \int \sum_{l=a,b} e_l v_l \sum_{n=0}^{\infty} \frac{(-1)^n}{n!} (r_l \, \mathrm{grad}_R)^n N_{ab}(X, t) \, dr \, dp \, dR \, . \qquad (3.3.11)$$

This expression can be treated in the following manner. We multiply the equation for phase density $N_{ab}(X, t)$ (3.2.6) by

$$\frac{(-1)^n}{(n+1)!} e_l r_l (r_l \operatorname{grad}_R)^n N_{ab} ,$$

carry out summation with respect to l and n, and integrate over r, p and P. The contributions of only the first three terms of (3.2.6) are nonzero. Integrating by parts over r and using the designations in (3.3.7, 10) we get the equation

$$\frac{\partial P^m(R, t)}{\partial t} + \int (V \operatorname{grad}_R) P^m(R, P, t) \, dP = J^m(R, t)$$

$$+ \int \sum_{l=a,b} \sum_{n=0}^{\infty} \frac{(-1)^n}{(n+1)!} e_l [v_l (r_l \operatorname{grad}_R) + n r_l (v_l \operatorname{grad}_R)]$$

$$\times (r_l \operatorname{grad}_R)^{n-1} N_{ab}(X, t) \, dr \, dp \, dP .$$

This equation will help us to exclude J^m from (3.3.11). After reduction of similar terms we obtain the expression for microscopic current density,

$$j^m(R, t) = \frac{\partial P^m}{\partial t} + \int [(V \operatorname{grad}_R) P^m(R, P, t)$$

$$- V \operatorname{div} P^m(R, P, t)] \, dP + \int \sum_{l=a,b} e_l \sum_{n=1}^{\infty} \frac{(-1)^n}{(n+1)!}$$

$$\times n [v_l (r_l \operatorname{grad}_R) - r_l (v_l \operatorname{grad}_R)] (r_l \operatorname{grad}_R)^{n-1} N_{ab}(X, t) \, dr \, dp \, dP .$$

$$(3.3.12)$$

From a textbook vector identity

$$\operatorname{rot} [a \times b] = (b \operatorname{grad}) a - (a \operatorname{grad}) b + a \operatorname{div} b - b \operatorname{div} a , \qquad (3.3.13)$$

we obtain equations

$$(V \operatorname{grad}_R) P - V \operatorname{div} P = \operatorname{rot}_R [P \times V] ,$$

$$- [v_l (r_l \operatorname{grad}_R) - r_l (v_l \operatorname{grad}_R)] A = (v_l \operatorname{grad}_R)(r_l A) - v_l \operatorname{div} (r_l A)$$

$$= \operatorname{rot}_R [r_l A \times v_l] . \qquad (3.3.14)$$

With the help of these equations together with the definition of microscopic magnetization vector

$$M^m(R, t) \equiv \int M^m(R, P, t) \, dP = \frac{1}{c} \int [P^m(R, P, t) \, V] \, dP$$

$$+ \frac{1}{c} \int \sum_{l=a,b} \sum_{n=1}^{\infty} \frac{(-1)^{n-1}}{(n+1)!} n \, [r_l \times v_l] (r_l \operatorname{grad}_R)^{n-1} N_{ab}(X, t) \, dr \, dp \, dP \qquad (3.3.15)$$

the expression for current density (3.3.12) can be written in the form

$$j^m(R, t) = \frac{\partial P^m(R, t)}{\partial t} + c \operatorname{rot} M^m(R, t). \qquad (3.3.16)$$

It turns out that on the microscopic level the current density j^m can be presented as a sum of the polarization and magnetization currents. Vectors P^m and M^m are linked to phase density $N_{ab}(X, t)$ through equations (3.3.9, 15).

The equation for phase density $N_{ab}(X, t)$ (3.2.6) of pairs of charged particles together with Lorentz equations (2.2.1), in which charge and current densities are defined by (3.3.6, 16) and polarization and magnetization vectors by (3.3.9, 15), comprise a closed set of exact microscopic equations. Naturally, for a completely ionized plasma these equations are equivalent to the set (2.2.1, 3, 4.1) for the functions E^m, B^m, and $N_a(x, t)$.

Let us write down in "pairs presentation" microscopic equations for a Coulomb plasma (approximation $c \to \infty$). From (3.2.1, 6) we find

$$\left\{ \frac{\partial}{\partial t} + V \frac{\partial}{\partial R} + v \frac{\partial}{\partial r} + \left[e_a E^m \left(R + \frac{m_b}{M} r, t \right) \right. \right.$$

$$\left. + e_b E^m \left(R - \frac{m_a}{M} r, t \right) \right] \frac{\partial}{\partial P} + \left[\frac{m_b}{M} e_a E^m \left(R + \frac{m_b}{M} r, t \right) \right.$$

$$\left. - \frac{m_a}{M} e_b E^m \left(R - \frac{m_a}{M} r, t \right) \right] \frac{\partial}{\partial p} \right\} N_{ab}(X, t) = 0,$$

$$\operatorname{rot} E^m = 0, \quad \frac{\partial E^m}{\partial t} + 4\pi j^m = 0, \quad \operatorname{div} E^m = 4\pi q^m. \qquad (3.3.17)$$

Field and current strengths are defined by (3.3.6, 16) and polarization and magnetization vectors are related to N_{ab} through (3.3.9, 15).

It is important to understand that "pairs presentation" and expansion into multifields also hold for a system of free charged particles, i.e., completely ionized plasma. The ratio of Debye radius (the effective interaction radius) to the wavelength here acts as the parameter of expansion. In the dipole approximation the equation for (3.3.17) takes the form

$$\left[\frac{\partial}{\partial t} + V\frac{\partial}{\partial R} + v\frac{\partial}{\partial r} - \frac{\partial \Phi_{ab}}{\partial r}\frac{\partial}{\partial p} + e_a E^m (R, t)\frac{\partial}{\partial p}\right] N_{ab}(X, t) = 0 \, . \quad (3.3.18)$$

The interaction of charged particles within a pair here, as in (2.2.8), is singled out. In averaging (correlations taken into account), the Coulomb potential will be replaced by the Debye potential (see [Ref. 3.1, Sect. 56] and Sect. 4.4).

Let us consider another form of equations for microscopic field strengths, which represents a direct microscopic analogy to Maxwell equations. To this effect we rewrite the Lorentz equations in a new form, using (3.3.6, 16) for charge and current densities. We introduce two new microscopic functions $D^m(r, t)$ and $H^m(r, t)$ which correspond to Maxwell's electric induction vector D and magnetic field strength vector H. Then in place of the Lorentz equations (2.2.1) we get

$$\text{rot}\,H^m = \frac{\partial D^m}{\partial t}, \quad \text{rot}\,E^m = -\frac{1}{c}\frac{\partial B^m}{\partial t},$$

$$\text{div}\,B^m = 0, \quad \text{div}\,D^m = 0, \quad\quad\quad\quad\quad (3.3.19)$$

to which we supplement

$$D^m = E^m + 4\pi P^m, \quad B^m = H^m + 4\pi M^m. \quad\quad (3.3.20)$$

After averaging, (3.3.19) are directly converted into Maxwell equations.

The microscopic Maxwell equations (3.3.19) together with formulae (3.3.20), the definitions of vectors P^m and M^m (3.3.9, 15), and the for phase density N_{ab} (3.2.6) also comprise a closed set of equations. This succession of different forms of microscopic equations enables us to describe any state between two extremes, from completely ionized plasma (all charged particles are free) to a gas (all charged particles are paired in atoms).

Let us consider one more presentation of microscopic field equations, which happens to be the most convenient for both classical and quantum radiation theories.

Making use of microscopic Maxwell equations (3.3.19) we shall write down the equations for electric and magnetic induction vectors. As a result we get two sets of equations,

$$\frac{\partial^2 D^m}{\partial t^2} - c^2 \Delta D^m = 4\pi c^2 \,\text{rot}\,\text{rot}\,P^m - 4\pi c\frac{\partial \,\text{rot}\,M^m}{\partial t},$$

$$\text{div}\,D^m = 0, \quad\quad\quad\quad\quad\quad\quad\quad (3.3.21)$$

$$\frac{\partial^2 B^m}{\partial t^2} - c^2 \Delta B^m = 4\pi c^2 \,\text{rot}\,\text{rot}\,M^m + 4\pi c\frac{\partial \,\text{rot}\,P^m}{\partial t},$$

$$\text{div}\,B^m = 0. \quad\quad\quad\quad\quad\quad\quad\quad (3.3.22)$$

Both vectors in these equations are rotational. Since the equation for phase density (3.3.6) includes vectors E^m and B^m, we must concede that

$$E = E^\perp + E^\parallel, \tag{3.3.23}$$

where − taking into account (3.3.20) −

$$E^\perp = D - 4\pi P^\perp, \quad E^\parallel = -4\pi P^\parallel. \tag{3.3.24}$$

In this way we have obtained a closed set of equations for vectors D^m and B^m and phase density N_{ab}. Vectors P^m and M^m are, as before, defined by formulae (3.3.7, 15).

Let us make one more final remark. The most convenient presentation of the Lorentz force when using microscopic Maxwell equations in the form (3.3.21, 22) is

$$F(R, t) = -e_a 4\pi P + e_a D + \frac{e_a}{c}[v \times B]. \tag{3.3.25}$$

The first term in the right-hand side of this equation is determined by polarization vector at point R at time t. Through formula (3.3.7), or (3.3.9), it can be expressed through function N_{ab} at the same point and at the same time. The remaining two terms in the right-hand side of (3.3.25) can only be expressed by solving the nonuniform equations (3.3.21, 22), and are thus determined by values of N_{ab} for earlier times and at points other than R.

Returning to expressions for polarization and magnetization vectors, within the framework of our model, formulae (3.3.7, 15) are exact. In the first approximation with respect to parameters μ_{at} and μ_T they are significantly simplified and become

$$P^m(R, t) = \sum_{l=a,b} e_l \int r_l N_{ab}(X, t) \, dr \, dp \, dP$$

$$- \sum_{l=a,b} \frac{1}{2} e_l \int r_l (r_l \, \mathrm{grad}_R) N_{ab}(X, t) \, dr \, dp \, dP, \tag{3.3.26}$$

$$M^m(R, t) = \frac{1}{c} \int [P^m(R, P, t) V] \, dP + \frac{1}{2c} \sum_{l=a,b} e_l \int [r_l v_l] N_{ab}(X, t) \, dr \, dp \, dP. \tag{3.3.27}$$

We recall that r_l and v_l are defined by (3.3.8). In the zero approximation with respect to μ_{at} we get

$$P^m(R, t) = \sum_{l=a,b} e_l \int r_l N_{ab}(X, t) \, dr \, dp \, dP, \tag{3.3.28a}$$

$$M^m(R, t) = \frac{1}{c} \int [P^m(R, P, t) V] \, dP. \tag{3.3.28b}$$

Let us examine the meaning of each term in the above formulae. The first term in the right-hand side of (3.3.26) is determined by the dipole moment, and the second term by the quadrupole moment. In (3.3.27) the first term describes the contribution of the movement of separate atoms to the magnetization vector, and the second term is determined by the magnetic moments of separate atoms

$$m = \frac{1}{2c} \sum_{l=a,b} e_l [r_l \times v_l] \,. \tag{3.3.29}$$

By virtue of (3.3.8) and the equation $p = \mu v$, the magnetic moment of an atom can be expressed as

$$m = \frac{1}{2c} \left(\frac{e_a}{m_a^2} + \frac{e_b}{m_b^2} \right) \mu [r \times p] \equiv \frac{1}{2c} \left(\frac{e_a}{m_a^2} + \frac{e_b}{m_b^2} \right) \mu L \tag{3.3.30}$$

where L is the mechanical moment of relative motion of particles within an atom, and μ is the equivalent mass. In the approximation $m_a = m_+ = \infty$, $\mu = m_-$, the magnetic moment of an atom is determined by electron motion

$$m = \frac{e_-}{2c} [r \times v] \,. \tag{3.3.31}$$

Finally, we shall present the polarization vector in a simpler form in dipole approximation. From (3.3.8, 28a) we find

$$P^m(R, t) = e \int r N_{ab}(X, t) \, dr \, dp \, dP \,. \tag{3.3.32}$$

The above formulae for magnetization and polarization vectors are based on the assumption that atoms consist of two charged particles. The results, however, also hold for more complex atoms. In (3.3.32), for example, r_+ and r_- should in such a case be taken for radius vectors of the "centers" of positive and negative charges of the atom.

3.4 Lagrange Function

In discussing the system of free charged particles and a field we have employed two different expressions for the Lagrange function (2.6.1, 6). By virtue of the definition of phase density these two expressions are equivalent. To describe the interaction of pairs of charged particles with the field we shall employ the second form from the start, expressed through the function N_{ab}:

$$L = \int \left[\frac{m_a v'^2}{2} + \frac{m_b v''^2}{2} + \frac{e_a}{c} v' A^m(r', t) + \frac{e_b}{c} v'' A^m(r'', t) \right.$$

$$\left. - e_a \varphi^m(r', t) - e_b \varphi^m(r'', t) \right] N_{ab}(x', x'', t) \, dx' \, dx''$$

$$+ \frac{1}{8\pi} \int \left[\left(-\frac{1}{c} \frac{\partial A^m}{\partial t} \right)^2 - (\text{rot } A^m)^2 \right] dr, \quad a \neq b. \tag{3.4.1}$$

The scalar potential here is defined by

$$\varphi^m(R, t) = \int \frac{q^m(R', t)}{|R - R'|} dR'. \tag{3.4.2}$$

Charge density is linked to phase density N_{ab} (3.3.2). In (3.4.1) A is a rotational vector.

In order to describe bound states of pairs of particles (i. e., atoms) we shall replace the variables x' and x'' by r, p, R, and P using formulae (3.1.4, 6). If $\mu_{at} \ll 1$, the Lagrange function in the integrand can be expanded with respect to the atomic parameter. Retaining only first-order terms, we get

$$L = \int \left[\frac{MV^2}{2} + \frac{\mu v^2}{2} - \Phi(r) + \frac{e_a}{c} v A^m(R, t) \right.$$

$$\left. - \frac{e_a}{c} V (r \, \text{grad}_R) A^m(R, t) - e_a (r \, \text{grad}_R) \varphi^m(R, t) \right] N_{ab}(X, t) dX$$

$$+ \frac{1}{8\pi} \int \left[\left(-\frac{1}{c} \frac{\partial A^m}{\partial t} - \text{grad}_R \varphi^m \right)^2 - (\text{rot } A^m)^2 \right] dR \tag{3.4.3}$$

where Φ is the potential energy of interaction of particles within an atom.

In order to obtain the equations of motion in the space of variables r, p, R, and P, we use the Lagrange equations

$$\frac{d}{dt} \frac{\partial}{\partial V} \frac{\delta L}{\delta N_{ab}} = \frac{\partial}{\partial R} \frac{\delta L}{\delta N_{ab}}, \quad \frac{d}{dt} \frac{\partial}{\partial v} \frac{\delta L}{\delta N_{ab}} = \frac{\partial}{\partial r} \frac{\delta L}{\delta N_{ab}}, \tag{3.4.4}$$

and after conventional transformations get

$$M \frac{dV}{dt} = e_a (r \, \text{grad}_R) \left\{ E^m(R, t) + \frac{1}{c} [V \times B^m(R, t)] \right\} + \frac{e_a}{c} [v \times B^m(R, t)], \tag{3.4.5}$$

$$\mu \frac{dv}{dt} = -\frac{\partial \Phi}{\partial r} + e_a \left\{ E^m(R, t) + \frac{1}{c} [V \times B^m(R, t)] \right\}, \tag{3.4.6}$$

where

$$E^m = -\text{grad } \varphi^m - \frac{1}{c} \frac{\partial A^m}{\partial t}.$$

These are the characteristic equations for equations for phase density N_{ab} (3.2.7).

Consider that the right-hand side of (3.4.6) contains only zero-order terms, and the right-hand side of (3.4.5) first-order terms with respect to the atomic parameter μ_{at}. This is the consequence of the fact that in the Lagrange set (3.4.4) the right-hand side of the second equation contains a derivative $\partial/\partial r$, while the right-hand side of the first equation contains $\partial/\partial R$ — and $\partial/\partial R \sim \mu_{at}\, \partial/\partial r$.

In order to obtain field equations, one has to utilize Lagrange field equations (2.6.5) once more, with charge and current densities determined now by (3.3.6, 16), and P^m and M^m by (3.3.28), which corresponds to zero approximation with respect to the atomic parameter μ_{at}. We shall not go into detail here because the equations of motion and field equations will be derived in the next section with help of the Hamiltonian function.

Let us rewrite the expression of the Lagrange function (3.4.3) in a form which will be more convenient for future use. It is convenient because the interaction of particles with field is expressed through field strengths. We take advantage of the fact that the equations of motion, derived with the help of the Lagrangian, remain intact if we add a time derivative of any arbitrary function to the Lagrangian.

From the integrand in (3.4.3) we subtract the following expression

$$\frac{e_a}{c}\frac{d}{dt}(r\cdot A^m(R,\,t)) = \frac{e_a}{c}\,r\,\frac{\partial A^m}{\partial t} + \frac{e_a}{c}\,r\,(V\,\mathrm{grad}_R)\,A^m(R,\,t) + \frac{e_a}{c}\,(v\cdot A^m(R,\,t))$$

and use the vector identity

$$r\,(V\,\mathrm{grad}_R)\,A\,(R,\,t) - V\,(r\,\mathrm{grad}_R A) = V\,[r\,\mathrm{rot}_R A]\,. \tag{3.4.7}$$

As a result, we get an expression which can be written in the form

$$L = \int\left\{\frac{MV^2}{2} + \frac{\mu v^2}{2} - \Phi(|r|) - \frac{e_a}{c}\,V\,[r\times B^m] + e_a r E^m(R,\,t)\right\}N_{ab}\,dX$$

$$+ \frac{1}{8\pi}\int\{[E^m(R,\,t)]^2 - [B^m(R,\,t)]^2\}dR\,. \tag{3.4.8}$$

Using Lagrange's equation (3.4.4), we again obtain equations of motion (3.4.5, 6).

In (3.4.8) the interaction of particles with the field is described by two terms, which can be united in one,

$$-\frac{e_a}{c}\,V\,[r\times B^m] + e_a r E^m = e_a r\left\{E^m(R,\,t) + \frac{1}{c}\,[V\times B^m(R,\,t)]\right\}. \tag{3.4.9}$$

Thus, the interaction is determined by the potential energy of a dipole in the field

$$E^m(R,\,t) + \frac{1}{c}\,[V\times B^m]\,.$$

We can suggest one more possible form of the Lagrangian, utilizing the expressions for polarization and magnetization vectors in the zero approximation with respect to atomic parameter (3.3.28). The Lagrangian can then be written in the form

$$L = \int \left[\frac{MV^2}{2} + \frac{\mu v^2}{2} - \Phi(|r|) \right] N_{ab}(X, t) dX$$

$$+ \int (P^m \cdot E^m + M^m \cdot B^m) dR + \frac{1}{8\pi} \int [(E^m)^2 - (B^m)^2] dR . \qquad (3.4.10)$$

It can be seen that the terms which account for the interaction correspond to dipole electric and dipole magnetic interactions. In the chosen approximation (zero-order approximation with respect to the atomic parameter μ_{at}), however, the magnetization vector M^m is defined by (3.3.28 b). It follows that the magnetization is not determined by the magnetic moment of an atom, but by the movement of the electric dipole moment.

From (3.4.1) one can also obtain better approximations for the Lagrangian with respect to the atomic parameter. If, for example, we take into account second-order terms in expansions of A^m and φ^m, the Lagrangian can again be expressed in the form of (3.4.10). In this case, however, the polarization and magnetization vectors will be defined by more precise formulae (3.3.26, 27), which account for quadrupole electric and dipole magnetic interactions.

3.5 Hamiltonian Function

Let us find the form of the Hamiltonian function which corresponds to the Lagrange function (3.4.8). First we write down the expressions for generalized momenta,

$$P = \frac{\partial}{\partial V} \frac{\delta L}{\delta N_{ab}} = MV - \frac{e_a}{c} [r \times B^m(R, t)] , \qquad (3.5.1)$$

$$p = \frac{\partial}{\partial v} \frac{\delta L}{\delta N_{ab}} = \mu v . \qquad (3.5.2)$$

The generalized momentum of the field will be designated by Π^m. From (3.4.8) we get

$$\Pi^m = \frac{\delta L}{\delta \dot{A}^m} = \frac{1}{4\pi c^2} \frac{\partial A^m}{\partial t} - \frac{e_a}{c} (\int r N_{ab} dr \, dp \, dP)^{\perp} = - \frac{1}{4\pi c} D^m . \qquad (3.5.3)$$

We can see that the generalized momentum of the field is proportional to the electric induction vector D^m. Since $\operatorname{div} D^m = 0$, both vectors are rotational.

For a system of free charged particles the generalized momentum of the field is determined by the rotational component of the vector of electric field strength, and not by the vector of electric induction (2.7.2). This dissimilarity arises in the transition from the Lagrangian (3.4.3) to (3.4.8).

The Hamiltonian function is defined by

$$H = \int (P \cdot V + p \cdot v) N_{ab}(X, t) dX + \int \Pi^m \dot{A}^m dR - L . \tag{3.5.4}$$

Substituting (3.4.8, 5.1 – 3) and utilizing (3.3.28 a), we get the sought expression,

$$H = \int \left(\frac{1}{2M} \left\{ P + \frac{e_a}{c} [r \times B^m(R, t)] \right\}^2 + \frac{p^2}{2\mu} + \Phi \right) N_{ab}(X, t) dX$$

$$- \frac{1}{4\pi} \int 4\pi P^m D^m dR + \frac{1}{8\pi} \int (4\pi P^m)^2 dR$$

$$+ \frac{1}{8\pi} \int [(D^m)^2 + (B^m)^2] dR , \quad B^m = \mathrm{rot}\, A^m . \tag{3.5.5}$$

When using the expression, one should keep in mind that

$$P^m(R, t) = e \int r N_{ab}(X, t) dr\, dp\, dP , \quad -4\pi (P^m)^\| = - \mathrm{grad}\, \varphi^m .$$

If we replace the induction vector in (3.5.5) by the noncanonical momentum E^m, and P by the noncanonical momentum $MV \equiv \mathscr{P}$, the Hamiltonian function becomes

$$H = \int \left(\frac{\mathscr{P}^2}{2M} + \frac{p^2}{2\mu} + \Phi \right) N_{ab}(X, t) dX + \frac{1}{8\pi} \int [(E^m)^2 + (B^m)^2] dR . \tag{3.5.6}$$

In this presentation the Hamiltonian appears as a sum of the energy of atoms and the energy of the electromagnetic field.

The Hamiltonian (3.5.5) can be used to obtain the equations of motion of particles and the field equations. First we shall write down Hamilton's equations for functions A^m and Π^m. Making use of (3.5.5) we get

$$\frac{\partial A^m}{\partial t} = \frac{\delta H}{\delta \Pi^m} = - c [D^m - (4\pi P^m)^\perp] , \tag{3.5.7}$$

$$\frac{\partial \Pi^m}{\partial t} = - \frac{\delta H}{\delta A} = - \frac{e_a}{c} \int [V (r\, \mathrm{grad}_R) - r (V\, \mathrm{grad}_R)] N_{ab}\, dr\, dp\, dP$$

$$+ \frac{1}{4\pi} \Delta A^m \tag{3.5.8}$$

with the designation

$$V = \frac{1}{M} \left\{ p + \frac{e_a}{c} [r \times B^m] \right\}. \tag{3.5.9}$$

Equation (3.5.7) coincides with the definition of the canonical field momentum (3.5.3). The equation for vector potential follows from (3.5.7, 8),

$$\frac{\partial^2 A^m}{\partial t^2} - c^2 \Delta A^m = - c \frac{\partial}{\partial t} \mathrm{grad}\, \varphi^m + 4\pi c j^m, \tag{3.5.10}$$

where

$$j^m = \frac{\partial P^m}{\partial t} + c \, \mathrm{rot}\, M^m.$$

In deriving this equation we have employed the vector identity (3.3.14). Polarization and magnetization vectors in (3.5.10) are defined by (3.3.28). Using the equations that link potentials with field strengths, one can go from (3.5.10) to the equations of the microscopic field.

In order to obtain the equation of motion of particles we shall use the second pair of Hamilton's equations. Taking into account that D^m is a rotational vector and that $-4\pi P^{\parallel} = - \mathrm{grad}\, \varphi$, we get from (3.5.5) the following equations:

$$\frac{\partial}{\partial t} M V = \frac{e_a}{c} [v \times B^m] + \frac{e_a}{c} \left[r \frac{dB^m}{dt} \right] - \frac{e_a}{c} (V \, \mathrm{grad}_R) [r \times B^m]$$

$$- \frac{e_a}{c} [V \, \mathrm{rot}_R [r \times B^m]] + e_a \, \mathrm{grad}_R (r \cdot E^m), \tag{3.5.11}$$

$$\frac{dp}{dt} = - \frac{\partial \Phi}{\partial r} + e_a E^m(R, t) - (V \, \mathrm{grad}_r)[r \times B^m] - [V \, \mathrm{rot}_r [r \times B^m]]. \tag{3.5.12}$$

Here we have taken advantage of the dependence between velocity V and canonical momentum.

To transform these equations further we apply the vector equations

$$\mathrm{rot}_R [r \times B(R, t)] = - (r \, \mathrm{grad}_R) B,$$

$$- \frac{e_a}{c} (V \, \mathrm{grad}_R) [r \times B] + e_a \, \mathrm{grad}_R (r \cdot E(R, t))$$

$$= - \frac{e_a}{c} \left[r, \frac{\partial B}{\partial t} + (V \, \mathrm{grad}_R) B \right] + e_a (r \, \mathrm{grad}_R) E$$

to (3.5.11), and

$$\text{rot}_r [r \times B] = -2B, \quad [V \text{rot}_R [r \times B]] = -2[V \times B],$$

$$(V \text{grad}_r) [r \times B] = [V \times B]$$

to (3.5.12). As a result we get a set of equations which is identical to (3.4.5, 6).

In conclusion we shall present an expression of the Hamiltonian which corresponds to the Lagrangian (3.4.1),

$$H = \int \left\{ \frac{1}{2m_a} \left[P' - \frac{e_a}{c} A^m(r', t) \right]^2 + \frac{1}{2m_b} \left[P'' - \frac{e_b}{c} A^m(r'', t) \right]^2 \right.$$

$$+ e_a \varphi^m(r', t) + e_b \varphi^m(r'', t) \bigg\} N_{ab}(x', x'', t) \, dx' dx''$$

$$+ \frac{1}{8\pi} \int \{[E^m(r, t)]^2 + [B^m(r, t)]^2\} dr . \tag{3.5.13}$$

In order to get (3.5.5) from (3.5.13), one must go over to variables r, p, R, and P, carry out the expansion with respect to the atomic parameter, retain zero- and first-order terms, and carry out the canonical transform, which for a Lagrange function would correspond to the transition from (3.4.3) to (3.4.8).

3.6 The Closed Equation for Phase Density of Atoms

There may be cases when the electromagnetic field is entirely determined by the pattern of particles' distribution. This means that it will be quite sufficient to use just the stationary solution of field equations at zero initial conditions. In such a case the field strengths can be excluded from equations, and one can get a closed equation for phase density $N_{ab}(X, t)$. We shall illustrate it for immobile atoms.

This approximation allows us to use, instead of (3.2.8), an equation for a simpler phase density $N_{ab}(r, p, R, t)$ which is independent of momenta. Let us consider a model with atoms oscillating with eigenfrequency ω_0; phase density then will be described by

$$\left[\frac{\partial}{\partial t} + v \frac{\partial}{\partial r} - \mu \omega_0^2 r \frac{\partial}{\partial p} + e_a E^m(R, t) \frac{\partial}{\partial p} \right] N_{ab}(r, p, R, t) = 0 . \tag{3.6.1}$$

Now we use the connection between strength and induction vectors

$$D^m = E^m + 4\pi P^m = (E^m + 4\pi P^m)^{\perp} . \tag{3.6.2}$$

Here we take into account that

$$(E^m)^{\|} = -\text{grad}\, \varphi^m(R, t) = -(4\pi P^m)^{\|}. \tag{3.6.3}$$

Assuming immobile atoms, $M^m(R, t) = 0$ in the field equations, and instead of using the set of equations (3.3.21, 22) it is sufficient to use only one equation for the electric induction vector. Substituting the stationary solution of this equation into (3.6.2) we get the expression for microscopic electric field strength

$$E^m(R, t) = \text{rot}_R \, \text{rot}_R \int \frac{[er' N_{ab}(X', t)]_{t - |R - R'|/c}}{|R - R'|} dX' - 4\pi P^m(R, t),$$

$$X = (r, p, R).$$
(3.6.4)

Substituting this expression into the equation for phase density (3.6.1), we get a closed equation for the function N_{ab}. In turn, this equation can be reduced to the equation for microscopic polarization vector

$$P^m(R, t) = e \int r N_{ab}(X, t) \, dr \, dp,$$
(3.6.5)

$$\frac{\partial^2 p^m}{\partial t^2} + \omega_0^2 P^2 = \frac{e^2}{\mu} n^m E_0 + \frac{e^2 n^m}{\mu} \left[\text{rot}_R \, \text{rot}_R \int \frac{P^m(R', t - |R - R'|/c)}{|R - R'|} dR' \right.$$

$$\left. - 4\pi P^m(R, t) \right].$$
(3.6.6)

This equation, however, includes together with function P^m the microscopic density of atoms $n^m = \sum_{1 \leqslant i \leqslant N} \delta(R - R_i(t))$. In (3.6.6) E_0 is the external field.

In Sect. 6.3 we shall average (3.6.6) obtaining the equations for the averaged polarization vector accounting for the effective Lorentz field and the influence of the correlations of atoms' positions on the radiative broadening of the emission spectra of oscillating atoms. The corresponding problem within the framework of quantum theory will be discussed in Chap. 11.

3.7 The Interaction of Atoms

By neglecting the time delay the interaction of atoms in our model is essentially dipole – dipole interaction. In order to take into consideration the short-range forces not belonging to dipole – dipole interaction, we shall designate additional potential energy of interaction between atoms,

$$U = U(r_i, R_i, r_j, R_j) \equiv U_{ij}.$$
(3.7.1)

In the equation for phase density $N_{ab}(r, p, R, P, t)$ in zero-order approximation with respect to the atomic parameter (3.2.8), additional terms will appear, and the equation will become

$$\left(\frac{\partial}{\partial t}+V\frac{\partial}{\partial R}+v\frac{\partial}{\partial r}-\frac{\partial \Phi}{\partial r}\frac{\partial}{\partial p}+e_a\left\{E^m(R,t)+\frac{1}{c}[V\times B^m]\right\}\frac{\partial}{\partial p}\right)N_{ab}$$

$$+\left[-\frac{\partial}{\partial R}\int U(r,R,r',R')N_{ab}(X',t)dX'\frac{\partial}{\partial P}\right.$$

$$\left.-\frac{\partial}{\partial r}\int U(r,R,r',R')N_{ab}(X',t)dX'\frac{\partial}{\partial p}\right]N_{ab}(X,t)=0. \qquad (3.7.2)$$

This equation accounts for both elastic and nonelastic interactions of atoms because it describes noncentral interactions. For a system of particles with only central interactions, when the potential energy only depends on $|R-R'|$, (3.7.2) is reduced to an equation for a simpler phase density

$$N(R,p,t)=\sum_{1\leqslant i\leqslant N}\delta(R-R_i(t))\,\delta(P-P_i(t)), \qquad (3.7.3)$$

which satisfies the equation

$$\frac{\partial N}{\partial t}+V\frac{\partial N}{\partial R}-\frac{\partial}{\partial R}\int U(|R-R'|)N(R',P',t)dR'\,dP'\frac{\partial N}{\partial P}=0. \qquad (3.7.4)$$

This equation can be used as a foundation for a kinetic theory of a gas of structureless particles [3.4].

3.8 Atoms and Field Oscillators

Dealing with a system made up of atoms and a field, as in the case of free charged particles and a field (Sects. 2.5, 11), we shall expand the vector potential in a Fourier series (2.5.1, 6). Since for a system composed of atoms and a field the canonical momentum is proportional to the electric induction vector, we also use expansions

$$D(R,t)=-\sqrt{\frac{4\pi}{V}}\sum_{k,\alpha}e_k P_k^\alpha\left\{\begin{matrix}\sin kR\\\cos kR\end{matrix}\right\},$$

$$P_k^\alpha=-\frac{1}{\sqrt{4\pi V}}\int(e_k D)\left\{\begin{matrix}\sin kR\\\cos kR\end{matrix}\right\}dR. \qquad (3.8.1)$$

Equation (2.5.10) is valid only for free field.

Using (2.5.6, 3.8.1) one can obtain the Hamiltonian function under the assumption of immobile atoms,

$$H = \int \left(\frac{p^2}{2\mu} + \Phi \right) N_{ab}(X, t) dX$$

$$+ \sqrt{\frac{4\pi}{V}} \sum_{k,\alpha} \int e_k P_k^\alpha P^m(R, t) \begin{Bmatrix} \sin kR \\ \cos kR \end{Bmatrix} dR + 2\pi \int [P^m(R, t)]^2 dR$$

$$+ \frac{1}{2} \sum_{k,\alpha} (P_{k,\alpha}^2 + c^2 k^2 Q_{k,\alpha}^2) . \tag{3.8.2}$$

From Hamilton's equations for a field

$$\dot{Q}_k^\alpha = \frac{\partial H}{\partial P_k^\alpha} = P_k^\alpha + \sqrt{\frac{4\pi}{V}} \int (e_k \cdot P^m(R, t)) \begin{Bmatrix} \sin kR \\ \cos kR \end{Bmatrix} dR ,$$

$$\dot{P}_k^\alpha = -\frac{\partial H}{\partial Q_k^\alpha} = -c^2 k^2 Q_k^\alpha, \quad e_k K = 0 , \tag{3.8.3}$$

we find the equation for field oscillators

$$\ddot{P}_k^\alpha + c^2 k^2 P_k^2 = -c^2 k^2 \sqrt{\frac{4\pi}{V}} \int (e_k \cdot P^m(R, t)) \begin{Bmatrix} \sin kR \\ \cos kR \end{Bmatrix} dR . \tag{3.8.4}$$

This equation, naturally, could be obtained directly from (3.3.21) with $M^m(R, t) = 0$ (i.e., for immobile atoms). Doing this, one should utilize the equations

$$\int \text{rot}_R \text{rot}_R P(R, t) \begin{Bmatrix} \sin kR \\ \cos kR \end{Bmatrix} dR = -\int [k [k \times P(R, t)]] \begin{Bmatrix} \sin kR \\ \cos kR \end{Bmatrix} dR$$

$$= k^2 \int e_k (e_k \cdot P(R, t)) \begin{Bmatrix} \sin kR \\ \cos kR \end{Bmatrix} dR . \tag{3.8.5}$$

Let us find the electric induction vector by solving (3.8.4). We can write the solution in the form

$$P_k^\alpha(t) = P_{k,\alpha(t)}^{(0)} - \sqrt{\frac{4\pi}{V}} \int_0^t \frac{\sin ck\tau}{ck} (e_k \cdot p^m(R, t-\tau)) \begin{Bmatrix} \sin kR \\ \cos kR \end{Bmatrix} dR \, d\tau . \tag{3.8.6}$$

The first term is determined by the initial (at $t = 0$) values of $P_{k,\alpha}(0)$, and the second is the solution of the nonuniform equation at zero initial conditions.

Let us substitute this solution into the Fourier expansion (3.8.1) and switch from summation over k to integration,

$$\frac{1}{V}\sum_k \rightarrow \frac{1}{(2\pi)^3}\int dk \, , \tag{3.8.7}$$

and utilize (3.8.5). As a result we get

$$D(R, t) = D_0(R, t)$$

$$+ \frac{4\pi c^2}{(2\pi)^3}\int_0^t d\tau \int dR' \int dk \, \frac{\sin ck\tau}{ck}\cos k\,(R-R')\,\mathrm{rot}_{R'}\,\mathrm{rot}_{R'}P\,(R', t-\tau)\,. \tag{3.8.8}$$

Now we calculate the integral

$$\frac{1}{(2\pi)^3}\int \frac{\sin ck\tau}{ck}\cos[k \cdot (R-R')]\,dk$$

$$= \frac{1}{4\pi c^2}\frac{\delta\left(\tau - \dfrac{|R-R'|}{c}\right) - \delta\left(\tau + \dfrac{|R-R'|}{c}\right)}{|R-R'|}\,, \tag{3.8.9}$$

substitute it in (3.8.8), and carry out integration over τ. The contribution of the second δ function in (3.8.9) to (3.8.8) is zero. The contribution of the first δ function is nonzero only provided that

$$|R-R'| \leqslant ct \, ,$$

since the argument of the δ function cannot otherwise assume a zero value. The final result can be written in the form

$$D(R, t) = D_0(R, t) + \int_{|R-R'|\leqslant ct} \frac{[\mathrm{rot}_{R'}\,\mathrm{rot}_{R'}P'(R', t)]_{t-|R-R'|/c}}{|R-R'|}\,dR'\,. \tag{3.8.10}$$

At $t = 0$ the second term vanishes because the range of integration is reduced to zero, and the first term describes the induction distribution at the initial time moment.

The solution (3.8.10) can also be obtained directly from (3.3.21) with $M^m = 0$ (immobile atoms).

The equations for oscillators that we have employed are valid for zero approximation with respect to two parameters μ_{at} and μ_T. If we wish to account for thermal motion still clinging to zero approximation with respect to the atomic parameter, instead of (3.8.4) we get

$$\ddot{P}_k^\alpha + c^2 k^2 P_k^\alpha = -\sqrt{\frac{4\pi}{V}}c^2 k^2 \int (e_k \cdot P^m(R, t)) \begin{Bmatrix} \sin kR \\ \cos kR \end{Bmatrix} dR$$

$$-\sqrt{\frac{4\pi}{V}}c\frac{\partial}{\partial t}\int e_k[k \times M^m(R, t)] \begin{Bmatrix} -\cos kR \\ \sin kR \end{Bmatrix} dR\,. \tag{3.8.11}$$

In this approximation the polarization and magnetization vectors are defined by formulae which correspond to (3.3.28),

$$P^m(R, t) = \sum_i er_i \delta(R - R_i(t)) ,$$

$$M^m(R, t) = \sum_i \frac{e}{c}[r_i V_i] \, \delta(R - R_i(t)) . \tag{3.8.12}$$

As in the case of completely ionized plasma (Sect. 2.8), it is convenient to present the state of the system of atoms and a field by giving the phase densities of atoms and field oscillators

$$N_{ab}(r, p, R, P, t) , \quad N_{k,\alpha}(Q_k^\alpha, P_k^\alpha, t) . \tag{3.8.13}$$

In place of the canonical variables Q_k^α and P_k^α [the Fourier components of vector potential and of electric induction − see expansions (2.5.6, 3.8.1)], one could use of the noncanonical variables Q_k^α, $\dot{Q}_k^\alpha \equiv \tilde{P}_k^\alpha$, where P_k^α is a Fourier component of rotational electric field strength

$$(E^m)^\perp = - \sqrt{\frac{4\pi}{V}} \sum_{k,\alpha} e_k \tilde{P}_k^\alpha \begin{Bmatrix} \sin kR \\ \cos kR \end{Bmatrix}. \tag{3.8.14}$$

Let us return to the zero approximation with respect to parameters μ_{at} and μ_{T}. The equations of motion of atom oscillators in accordance with (3.1.16) become

$$\frac{\partial p_i}{\partial t} = \frac{p_i}{\mu} = v_i ,$$

$$\frac{dp_i}{dt} + \mu \omega_0^2 r_i = - e \sqrt{\frac{4\pi}{V}} \sum_{k,\alpha} e_k P_k^\alpha \begin{Bmatrix} \sin kR \\ \cos kR \end{Bmatrix}$$

$$+ \int \frac{\text{grad}_{R'} \, \text{div}_{R'} \sum_j er_j \delta(R' - R_j)}{|R_i - R'|} \, dR' . \tag{3.8.15}$$

In the latter equation the first term in the right-hand side determines the contribution of the rotational electric field, and the second term that of a potential electric field. We also omit the sign $\sim$ over P_k^α.

In noncanonical variables the equations of oscillators motion (3.8.3) will become

$$\dot{Q}_k^\alpha = P_k^\alpha ,$$

$$\dot{P}_k^\alpha + c^2 k^2 Q_k^\alpha = \sqrt{\frac{4\pi}{V}} \int (e_k \cdot J^m(R, t)) \begin{Bmatrix} \sin kR \\ \cos kR \end{Bmatrix} dR , \tag{3.8.16}$$

where

$$J^m = \frac{\partial P^m}{\partial t} = e \sum_i v_i \delta(R - R_i(t)) \tag{3.8.17}$$

is the current due to inner motion of charged particles within atoms.

The equations for phase densities of atoms and field oscillators, which would correspond to the equations of motion (3.8.15, 16), are

$$\left[\frac{\partial}{\partial t} + v \frac{\partial}{\partial r} - \mu \omega_0^2 r \frac{\partial}{\partial p} - e \sqrt{\frac{4\pi}{V}} \sum_{k,\alpha} \int e_k P_k^\alpha \mathcal{N}_{k,\alpha}(X, t) dX \begin{Bmatrix} \sin kR \\ \cos kR \end{Bmatrix} \frac{\partial}{\partial p} \right.$$

$$\left. + \int \frac{\mathrm{grad}_R \cdot \mathrm{div}_{R'} \, er' N_{ab}(x', t)}{|R - R'|} dx' \frac{\partial}{\partial p} \right] N_{ab}(r, p, R, t) = 0, \tag{3.8.18}$$

$$\left[\frac{\partial}{\partial t} + P_k^\alpha \frac{\partial}{\partial Q_k^\alpha} - c^2 k^2 Q_k^\alpha \frac{\partial}{\partial p_k^\alpha} + \sqrt{\frac{4\pi}{V}} \int (e_k \cdot v) \begin{Bmatrix} \sin kR \\ \cos kR \end{Bmatrix} N_{ab}(x, t) dx \frac{\partial}{\partial p_k^\alpha} \right]$$

$$\times \mathcal{N}_{k,\alpha}(X, t) = 0. \tag{3.8.19}$$

In (3.8.18) there are two nonlinear terms. One of them describes the interaction of atoms with the rotational electromagnetic field, and the second, which contains the product of two functions $N_{ab}(x, t)$ and $N_{ab}(x', t)$, describes the interaction of atoms via a potential field.

On the basis of (3.8.18, 19), as in Sect. 2.11, one can obtain a chain of equations for the moments of random functions N_{ab} and $\mathcal{N}_{k,\alpha}$. It will be put to use later for obtaining the kinetic equations.

3.9 The Method of Distribution Functions for a System of Atoms and Field Oscillators

As in the kinetic theory of plasmas and gases (Chap. 2, [3.4]), there are two ways of constructing statistical theory. The equations for microscopic phase densities of atoms and field oscillators (Sect. 2.8) can serve as initial equations, along with the Liouville equation for the distribution function for microstates of the system of atoms and a field

$$f(x_i, \ldots, x_N \ldots, X_k^\alpha, \ldots, t), \quad \int f \prod_i dx_i \prod_{k,\alpha} dX_k^\alpha = 1. \tag{3.9.1}$$

As above, here we use the terms

$$x_i = (r_i, p_i, R_i, P_i), \quad X_k^\alpha = (Q_k^\alpha, P_k^\alpha).$$

The Liouville equation for a system of stationary atoms and a field is

$$\frac{\partial f}{\partial t} + \sum_{1 \leqslant i \leqslant N} \left(v_i \frac{\partial f}{\partial r_i} - \mu \omega_0^2 r_i \frac{\partial f}{\partial p_i} - e \sqrt{\frac{4\pi}{V}} \sum_{k,\alpha} e_k P_k^\alpha \begin{Bmatrix} \sin kR_i \\ \cos kR_i \end{Bmatrix} \frac{\partial f}{\partial p_i} \right.$$

$$+ \int \frac{\mathrm{grad}_{R'} \, \mathrm{div}_{R'} \, \sum_j er_j \delta(R' - R_j)}{|R_i - R'|} \, dR' \, \frac{\partial f}{\partial p_i} \Bigg)$$

$$+ \sum_{k,\alpha} \left[P_k^\alpha \frac{\partial f}{\partial Q_k^\alpha} - c^2 k^2 Q_k^\alpha \frac{\partial f}{\partial P_k^\alpha} + \sqrt{\frac{4\pi}{V}} \sum_i e \, (e_k v_i) \begin{Bmatrix} \sin kR_i \\ \cos kR_i \end{Bmatrix} \frac{\partial f}{\partial P_k^\alpha} \right] = 0 .$$

$$(3.9.2)$$

Based on this equation one can construct two independent chains of equations for a succession of distribution functions of atoms and a field, as was done for a system of free charged particles and a field (Sects. 2.9, 10). One of these chains begins with two equations for one-particle functions of the distribution of atoms and a field, $f_{ab}(r, p, R, P, t)$ and $\mathcal{F}_{k,\alpha}(X, t)$.

The other chain of equations starts from the equation for the two-particle distribution function for field oscillators $\mathcal{F}_{k_1, \alpha_1, k_2, \alpha_2}(X', X'', t)$. As in the case of a system of free charged particles and a field, the equations for the distribution functions of atoms and a field are nonsymmetric since two-particle distribution functions for field oscillators do not enter the equations for one-particle distribution functions for particles and field oscillators. It is due to this that two independent chains of equations arise. We shall write the equations for distribution functions when required. They are derived from the Liouville equation in a conventional way, and obtaining them presents no fundamental problem.

4. The Kinetic Equations for a System of Free Charged Particles and a Field

This chapter deals with two major problems: 1) The construction of kinetic equations for one-particle distribution functions on the basis of the equations for microscopic phase densities of the free charged particles and electromagnetic field strengths, stated in Sect. 2. 2) The construction of the theory of large-scale (kinetic) fluctuations of the distribution functions and the electromagnetic field strengths.

4.1 The Principal Parameters (Free Charged Particles)

Now we have reached the main subject of this book: the kinetic theory of electromagnetic processes. We shall begin by discussing a relatively "simple" case, i. e., when all particles are free and constitute a completely ionized plasma. Those acquainted with monograph [4.1] may skip Sects. 4.4, 5, which contain the necessary background on the kinetic theory of plasma.

First we shall list the principal parameters of a Coulomb plasma, characteristic of the processes in a system of charged particles (free in Sect. 4.1, and bound in Sect. 4.2) and a field. This will quickly convince the reader of the complexity and diversity of the processes in these systems. We shall also see that the given parameters, numerous as they are, are still insufficient to analyze many important problems, i. e., those concerning phase transitions and self-oscillatory processes in systems of atoms and a field (Chaps. 5, 12, 13).

The principal parameters of length of a Coulomb plasma are the average distance between charged particles $r_{av} \sim n^{-1/3}$, the Debye radius r_D, and the shortest of the mean free paths, which is the mean free path for electron – electron collisions l_{ee} (in the future $l_{ee} \equiv l$).

A plasma is rarefied if the plasma parameter [4.1 – 3]

$$\mu = \frac{1}{n r_D^3} \sim \left(\frac{r_{cp}}{r_D}\right)^3 \sim \frac{e^2}{r_D k T} \ll 1 \ . \tag{4.1.1}$$

In a rarefied plasma

$$l \sim r_D/\mu \gg r_D. \tag{4.1.2}$$

Besides the plasma parameter μ, a significant role is also played by the interaction parameter. It is defined as a ratio of the potential energy at $r = r_{av}$ to the mean kinetic energy,

$$\xi = \frac{e^2}{r_{av}kT} \sim \mu \frac{r_D}{r_{av}} \sim \mu^{2/3}. \tag{4.1.3}$$

If $\xi \ll 1$, one can use the perturbation theory for interactions. It should be kept in mind that the actual interaction depends on the distance between particles, and perturbation theory is only valid for distances

$$r > \frac{e^2}{kT} \equiv l_L \sim r_{av}\xi \sim r_D\mu, \tag{4.1.4}$$

where l_L is the Landau length.

These definitions (4.1.1 – 4) at $\mu \ll 1$ lead to the following inequalities:

$$l_L \ll r_{av} \ll r_D \ll l. \tag{4.1.5}$$

Let us introduce the corresponding time parameters,

$$\tau_L \sim l_L/v_T, \quad t_{av} \sim r_{av}/v_T, \quad 1/\omega_L \sim r_D/v_T, \quad \tau \sim l/v_T. \tag{4.1.6}$$

Here ω_L is the Langmuir frequency (plasma frequency), defined as

$$\omega_L = \sqrt{\sum_a \frac{4\pi e_a^2 n_a}{m_a}} = \sqrt{\frac{4\pi e_a^2 n_a}{\mu_{ab}}}. \tag{4.1.7}$$

In constructing the kinetic equations for a plasma the concepts of physically infinitesimal time intervals τ_{ph} and lengths l_{ph} are employed. For a rarefied plasma they can be defined as follows [Ref. 4.1, Sect. 33]:

$$\frac{\tau}{n l_{ph}^3} = \tau_{ph} = \frac{l_{ph}}{v_T};$$

hence

$$l_{ph} = \sqrt[4]{\frac{l}{n}}. \tag{4.1.8}$$

Taking (4.1.1, 2) into account, we find

$$l_{ph} \sim r_D, \quad r_{ph} \sim \frac{1}{\omega_L}, \quad n l_{ph}^3 \sim \frac{1}{\mu} \gg 1 . \tag{4.1.9}$$

Thus a physically infinitesimal time interval is much less than mean-free-path time (relaxation time in a kinetic equation). From definition (4.1.8) it also follows that the magnitudes of physical infinitesimals τ_{ph} and l_{ph} depend on the relaxation process in question. For example, these intervals are greater for processes determined by ion – ion collisions.

4.2 Principal Parameters for a System of Atoms

We shall discuss both the quantum and the corresponding classical parameters of a system of atoms and a field. For a classical model we shall choose an atom oscillator with eigenfrequency ω_0 and radius r_0. The matching quantum characteristics are

$$r_0 \leftrightarrow \frac{\hbar^2}{me^2}, \quad \omega_0 \leftrightarrow \frac{me^4}{\hbar^3} . \tag{4.2.1}$$

The dipole moment is designated $p = er$. In quantum theory r corresponds to the magnitude of the electron displacement matrix element for an atom.

4.2.1 Interaction Parameters

In Chap. 3 we introduced the atomic parameter μ_{at} (3.1.9). In radiation theory a small parameter characteristic of the interaction between atoms and radiation is used,

$$\mu_{int} \sim e^2/\hbar c \tag{4.2.2}$$

where $e^2/\hbar c$ is the fine structure coefficient.
Replacing c in (4.2.2) with $\omega_0 \lambda$ (λ is the radiation wavelength) and using (4.2.1) we find

$$\mu_{int} \sim \mu_{at} \sim r_0/\lambda \sim v/c . \tag{4.2.3}$$

Now let us introduce the perturbation theory parameter for the interaction of atoms. It can be derived from the conditions governing the applicability of Born's approximation of dipole – dipole interaction [4.4],

$$\frac{e^2 r^2}{R^3} < \frac{\hbar v_T}{R} ;$$

hence

$$R > \sqrt{\frac{e^2 r^2}{\hbar v_T}} \equiv \rho_W \tag{4.2.4}$$

where ρ_W is the Weisskopf radius. Perturbation theory is thus valid for distances greater than the Weisskopf radius, which here replaces the Landau length l_L in (4.1.4). The parameter of the interaction of charged particles is represented here by the parameter of dipole – dipole interaction (subscript d)

$$\xi_d \sim \frac{e^2 r^2}{\hbar v_T r_{av}^2} \sim \frac{\rho_W^2}{r_{av}^2}. \tag{4.2.5}$$

The classical analog to the Weisskopf radius is given by the quantity

$$\rho_W \sim r_0 \sqrt{\frac{v}{v_T}} \gg r_0 \qquad v \gg v_T. \tag{4.2.6}$$

It is thus possible to introduce three characteristic parameters of length, r_0, ρ_W, and r_{av}. When perturbation theory is applicable ($\xi_d \ll 1$), the inequalities

$$r_0 \ll \rho_W \ll r_{av} \tag{4.2.7}$$

follow from (4.2.5, 6).

We can now introduce the "density parameter" for dipole – dipole interaction. Formula (4.2.4) implies that the Weisskopf radius can serve as a boundary between strong and weak interactions. Therefore we can introduce a dimensionless "density parameter",

$$\varepsilon_d = n \rho_W^3. \tag{4.2.8}$$

From (4.2.5, 8) it follows that $\varepsilon_d \sim \xi_d^{2/3}$, and so the condition governing the applicability of perturbation theory also implies that $\varepsilon_d \ll 1$.

4.2.2 The Relaxation Processes Parameters for Atoms

In the kinetic theory of gases composed of structureless atoms the latter can be imagined as spheres of diameter r_0. The mean free path is $l \sim 1/n r_0^2$, and for a rarefied gas

$$r_0 \ll r_{av} \ll l. \tag{4.2.9}$$

The principal dimensionless parameter for such a gas is the density parameter,

$$\varepsilon = n r_0^3. \tag{4.2.10}$$

The physically infinitesimal element of length in this case is defined as follows [4.1]:

$$l_{ph} = \sqrt[4]{\frac{l}{n}} \sim \sqrt{\bar{\varepsilon}}\, l \sim \frac{r_0}{\sqrt{\varepsilon}} \,, \quad n l_{ph}^3 \sim \frac{1}{\sqrt{\varepsilon}} \gg 1 \,. \tag{4.2.11}$$

In mean-free-path time $\tau = l/v$, relaxation occurs with respect to the translational degrees of freedom of atoms. In other words, in time τ the velocities of the atoms' centers of mass assume a Maxwell distribution.

The dipole–dipole interaction determines the relaxation processes with respect to the inner degrees of freedom in atoms. For small distances, when perturbation theory is no longer valid ($R < \rho_W$), estimates can be based on atoms considered as spheres of diameter ρ_W, and the density parameter (4.2.8) can be used. Then the existing relaxation length is

$$l_d \sim 1/n\rho_W^2 \,, \quad \tau_d = l_d/v_T \,, \tag{4.2.12}$$

and the physically infinitesimal length, defined according to (4.1.8), is

$$l_{ph} \sim \sqrt{\bar{\varepsilon}_d}\, l_d \sim \frac{\rho_W}{\sqrt{\varepsilon_d}} \,, \quad n l_{ph}^3 \sim \frac{1}{\sqrt{\varepsilon_d}} \gg 1 \,. \tag{4.2.13a}$$

For great distances ($R > \rho_W$) perturbation theory can be used. In this approximation the interaction due to the resonance transmission of radiation from excited atoms to nonexcited ones can also be accounted for (Sects. 10.6, 8).

Let us introduce the designations for the time and length of the relaxation of this process. In Sect. 10.6 we shall show that the inverse relaxation time

$$\gamma_{res} \equiv 1/\tau_{res} \sim e^2 r^2 n/\hbar \,. \tag{4.2.14}$$

From definition (4.2.4) of the Weisskopf radius we find the expression for the corresponding relaxation length,

$$l_{res} \sim \frac{v_T}{\gamma_{res}} \sim \frac{\hbar v_T}{n e^2 r^2} \sim \frac{1}{n\rho_W^2} \,. \tag{4.2.15}$$

From a comparison of (4.2.12, 15) it follows that the relaxation lengths l_d and l_{res} are of the same order of magnitude, and so, according to (4.2.13a),

$$l_{ph} \sim \sqrt{\bar{\varepsilon}_d}\, l_{res} \,. \tag{4.2.13b}$$

The relaxation processes in atoms due to spontaneous emission of radiation are determined by the parameters

$$\tau_{em} \sim \frac{1}{\gamma} \,, \quad l_{em} \sim v_T \tau_{em} \,, \quad \gamma \sim \frac{e^2 \omega_0^2}{m c^3} \sim \frac{e^2 r^2 \omega_0^2}{\hbar c^3} \,,$$

$$\tau_{\mathrm{ph}}^{\mathrm{sp}} \sim \frac{1}{\gamma} \sqrt[4]{\frac{1}{n l_{\mathrm{em}}^{3}}} \, . \tag{4.2.16}$$

Comparing the expressions for γ_{res} and γ (4.2.14, 16) we find

$$\gamma_{\mathrm{res}} \sim \gamma n \lambda^{3} \equiv \gamma \varepsilon_{\mathrm{em}} \, . \tag{4.2.17}$$

Here we have introduced one more parameter, the "optical density parameter"

$$\varepsilon_{\mathrm{em}} = n \lambda^{3} \, , \tag{4.2.18}$$

where λ is the wavelength of resonant radiation.
Comparing (4.2.12, 15, 17), we find

$$l_{\mathrm{d}} \sim l_{\mathrm{res}} \sim l_{\mathrm{em}} / \varepsilon_{\mathrm{em}} \, ,$$

and hence at small optical densities ($n \lambda^{3} \ll 1$) the shortest relaxation length will be l_{em}.

4.2.3 Comparison of the Density Parameters ε_{d} and $\varepsilon_{\mathrm{em}}$

We designate the ratio of the Doppler breadth of the spectral line $k v_{\mathrm{T}}$ to its natural breadth γ with η,

$$\eta = k v_{\mathrm{T}} / \gamma \, . \tag{4.2.19}$$

If $\eta > 1$, the spectral line is said to be nonuniformly broadened (Sect. 12.1).
From the definitions for ρ_{W} and η (4.2.4, 19) we find the connection between ρ_{W} and λ,

$$\rho_{\mathrm{W}} \sim \lambda / \sqrt{\eta} \, ;$$

hence

$$\varepsilon_{\mathrm{d}} \sim \varepsilon_{\mathrm{em}} / \eta^{3/2} \ll \varepsilon_{\mathrm{em}} \qquad \eta \gg 1 \, . \tag{4.2.20}$$

Finally, from the inequalities (4.2.7, 20) it follows that

$$\varepsilon \ll \varepsilon_{\mathrm{d}} \ll \varepsilon_{\mathrm{em}} \qquad \eta \ll 1 \, . \tag{4.2.21}$$

Thus in systems with $\eta \gg 1$ a gas might be rarefied with respect to parameters ε and ε_{d}, while at the same time being dense with respect to the optical density parameter. It is also possible that a gas is only rarefied with respect to parameter ε.

In estimating the influence of an effective Lorentz field upon an atom's eigenfrequency ω_0 and in studying the dielectric permittivity of a system of atoms we encounter the parameter

$$\chi = \frac{4\pi e^2 n}{m\omega_0^2} = \frac{\omega_L^2}{\omega_0^2}. \tag{4.2.22}$$

Here we use (4.1.7) for ω_L.

By virtue of (4.2.16) for γ and (4.2.18) we can get the estimate

$$\chi \sim \frac{\gamma}{\omega_0} n\lambda^3 = \frac{\gamma}{\omega_0}\varepsilon_{em}, \tag{4.2.23}$$

whence it follows that the value of χ can be substantially greater than the ratio γ/ω_0 when the values of the optical density parameter ε_{em} are large.

These parameters give the reader some notion of the diversity of processes in the systems of free charged particles and atoms in question. In Sects. 12.1 – 3, after we have gathered all the relevant information, we shall introduce the principal parameters which determine, in particular, the coherent and non-coherent states.

Now we come to the essential part of this chapter – the description of the statistical theory of electromagnetic processes in a system comprising free charged particles and a field.

4.3 First Moments Approximation

We shall start by discussing various ways to approximate first moments for a system of free charged particles and a field, disregarding correlations on all scales, i.e., both small- and large-scale correlations.

For the initial microscopic equations we first choose the set for the random functions E^m, B^m, and N_a (2.2.1, 3, 4.1). In the first moments approximation in (2.11.2) the correlations of fluctuations can be neglected, and (2.11.2) replaced by

$$\overline{E^m N_a} = E\bar{N}_a, \quad \overline{B^m N_a} = B\bar{N}_a. \tag{4.3.1}$$

Equation (2.11.4) now becomes

$$\hat{L}_a f_a = 0. \tag{4.3.2}$$

Together with the equations for mean field strengths E and B (2.11.6), it forms a closed set of equations, the so-called Vlasov equations.

Vlasov equations form the base of the theory of a "collisionless" plasma. This term means that the characteristic scale values T and L are smaller than the minimal relaxation time τ (for electron – electron collisions) and the corresponding relaxation length l. We use the term "collisions" in analogy to the kinetic theory of gases, though this term is rather inadequate for a plasma since a charged particle in plasma interacts with a great number of particles simultaneously ($N_D \sim nr_D^3 \gg 1$).

Let us now take the equations for the microscopic phase densities of particles and field oscillators for the initial set of equations (2.8.3, 4). We average these equations and use the first moments approximations for N_a and $\mathcal{N}_{k,a}$. In this approximation we have, instead of (2.11.11),

$$\overline{N_a(x, t) N_b(x', t)} = n_a n_b f_a f_b,$$

$$\overline{N_a(x, t) \mathcal{N}_{k,a}(x, t)} = n_a f_a \mathcal{F}_{k,a}.$$

Consequently, the collision integrals (2.11.13, 15) become equal to zero, and the equations for the distribution functions of charged particles and field oscillators take the form

$$\hat{L}_a f_a = 0, \quad \hat{L}_{k,a} \mathcal{F}_{k,a} = 0 \tag{4.3.3}$$

where the operators are defined by (2.10.2, 5). Equations (4.3.3) comprise a closed set of equations for one-particle distribution functions f_a and $\mathcal{F}_{k,a}$.

Let us compare the two sets of equations (2.11.6, 4.3.2) and (4.3.3). In the latter the Coulomb interaction of charged particles is accounted for directly, i.e., without the aid of the field. Of course, the same can also be done in (2.11.6, 4.3.2). A more significant feature of (4.3.3) is that they give an explicit description of the rotational electromagnetic field.

Indeed, by averaging (2.8.6 − 8) we find the relation between the mean values of the field strengths and the distribution function of coordinates and momenta of field oscillators. Using this relation one can easily find all the equations for mean strengths of the rotational field (2.11.6) from (4.3.3). Vlasov equations (2.11.6, 4.3.2) therefore follow from the set of equations (4.3.3). The reverse, of course, is not true because (4.3.3) provide us with more information on the evolution of field oscillators. In particular, the second equation of (4.3.3) enables us to obtain equations for any desired moment of Fourier components Q_k^α and P_k^α of vector potential and rotational field $(E)^\perp$.

Naturally, the equations for the one-particle distribution functions for particles and field oscillators (4.3.3) can also be obtained from the chain of equations for the sequence of distribution functions for particles and a field.

The Vlasov set of equations (2.11.6, 4.3.2) is commonly used in the theory of collisionless plasma [4.1 − 3, 5]. A more general set of equations for the first moments of random functions N_a and $\mathcal{N}_{k,a}$ [the set (4.3.3)] has been poorly investigated, though it is potentially more informative regarding the distribution of electromagnetic field variables in collisionless plasma.

There is a possibility of obtaining yet another closed set of equations for first moments, this time for the mean phase density of pairs of charged particles and mean field strengths. In order to do this one should use the microscopic equations for the phase density of pairs of particles and microscopic field strengths

(2.2.1, 3.2.6, 3.6, 16). For a Coulomb plasma this set can be simplified, taking the form of (3.3.17), and (3.3.18) in dipole approximation. The mean value of phase density is linked to the appropriate distribution function,

$$\langle N_{ab}(r, p, R, P, t)\rangle = n f_{ab} . \tag{4.3.4}$$

Let us, for example, discuss the first moments approximation for the set of equations (3.3.18). Taking (4.3.4) into account, after averaging and dropping the second moments we get a closed set of equations for the distribution function $f_{ab}(r, p, R, t)$ and the mean potential field,

$$\frac{\partial f_{ab}}{\partial t} + v \frac{\partial f_{ab}}{\partial r} - \frac{\partial \Phi_{ab}}{\partial r} \frac{\partial f_{ab}}{\partial p} + e_a E(R, t) \frac{\partial f_{ab}}{\partial p} = 0 ,$$

$$\operatorname{rot} E = 0 , \quad \operatorname{div} E = -4\pi \operatorname{div} P , \quad P = en \int r f_{ab}\, dr\, dp . \tag{4.3.5}$$

The corresponding equation for the polarization vector $P(R, t)$ is

$$\frac{d^2 P}{dt^2} = -\frac{e^2 n}{\mu_{ab}} \int \frac{\partial \Phi_{ab}}{\partial r} f_{ab}\, dr\, dp + \frac{e^2 n}{\mu_{ab}} E . \tag{4.3.6}$$

Consider the first term in the right-hand side of this equation. If we take an atom to be a harmonic oscillator, the equation for the polarization vector becomes

$$\frac{d^2 P}{dt} + \omega_0^2 P = \frac{e^2 n}{\mu} E . \tag{4.3.7}$$

The potential field approximation is insufficient to calculate the polarization (Sects. 9.1, 2).

To estimate the first term in the right-hand side of (4.3.6) for a system composed of free charged particles and a field, we note that the eigenfrequency of an atom is connected to the radius of the orbit (the interaction radius) by the expression $\omega_0^2 = e^2/m r_0^3$. In a plasma of free charged particles r_0 is the Debye radius, and so

$$\omega_0^2 \rightarrow \frac{e^2}{m r_D^3} \sim \mu \omega_L^2 , \tag{4.3.8}$$

where μ is the plasma parameter (4.1.1).

Since the first moments approximation only holds for the zero-order plasma parameter approximation (Sect. 4.4), we have to concede that according to (4.3.8) $\omega_0^2 = 0$. Consequently, the equation for the polarization vector becomes

$$\frac{d^2 P}{dt^2} = \frac{e^2 n}{\mu_{ab}} E . \tag{4.3.9}$$

It is then clear that the response to the mean field is determined by dielectric permittivity

$$\varepsilon = 1 - \omega_L^2 / \omega^2 . \tag{4.3.10}$$

The permittivity in our approximation [dipole approximation, neglect of thermal motion of pairs of charged particles $(m_i = \infty)$] does not depend on the wave vector; in other words, there is no spatial dispersion.

4.4 The Kinetic Equations for a Coulomb Plasma

The kinetic equations for a Coulomb plasma have been thoroughly investigated in [4.1 – 3, 6, 7]. Here we shall only review the results relevant to our presentation.

In the case of a Coulomb plasma one can start from the equations for phase density N_a and potential electric field strength (2.3.1, 5). After averaging, one gets the equations for the distribution function f_a and mean field E,

$$\frac{\partial f_a}{\partial t} + v \frac{\partial f_a}{\partial r} + e_a E(r, t) \frac{\partial f_a}{\partial p} = I_a(r, p, t) , \tag{4.4.1}$$

$$\operatorname{rot} E = 0 , \quad \operatorname{div} E = 4\pi \sum_a e_a n_a \int f_a(x, t) dp . \tag{4.4.2}$$

The collision integral is expressed by correlating the fluctuations δN_a and δE [cf. (2.11.5)],

$$I_a(x, t) = - \frac{e_a}{n_a} \frac{\partial \overline{(\delta N_a \delta E)}_{x,r,t}}{\partial p} . \tag{4.4.3}$$

The equations for f_a and E are not closed. In order to obtain the equations for second moments we employ the equations for deviations δN_a and δE. They have the form

$$\left(\frac{\partial}{\partial t} + v \frac{\partial}{\partial r} + e_a E \frac{\partial}{\partial p} \right) \delta N_a + e_a \delta E \frac{\partial n_a f_a}{\partial p}$$

$$= - e_a \frac{\partial}{\partial p} (\delta N_a \delta E - \overline{\delta N_a \delta E}) ,$$

$$\operatorname{rot} \delta E = 0 , \quad \operatorname{div} \delta E = 4\pi \sum_a e_a \int \delta N_a(x, t) dp . \tag{4.4.5}$$

From this it follows that the equation for the second moment $\overline{(\delta N_a \delta N_b)}_{x, x', t}$ will include third moments, and therefore the three-particle correlation function g_{abc}.

If the plasma parameter μ is small, it is possible to break the chain of equations for moments. For this a so-called polarization approximation can be instrumental, with

$$g_{abc} = 0 , \quad g_{ab} \ll f_a f_b . \tag{4.4.6}$$

The name "polarization approximation" alludes to the fact that this approximation, as compared to the perturbation theory, accounts for the effects due to the polarization of a plasma.

In the polarization approximation the chain of equations for the moments is finite. The equation for the second moment does not include higher moments (to be more precise, higher correlation functions), and can be written in the form [Ref. 4.1, Sect. 27]

$$\left(\frac{\partial}{\partial t} + v \frac{\partial}{\partial r} + v' \frac{\partial}{\partial r'} + e_a E \frac{\partial}{\partial p} + e_b E \frac{\partial}{\partial p'} \right) [(\overline{\delta N_a \delta N_b})_{x,x',t}$$

$$- (\overline{\delta N_a \delta N_b})_{x,x',t}^{\text{source}}] + e_a (\overline{\delta E \, \delta N_b})_{r,x',t} \frac{\partial n_a f_a}{\partial p}$$

$$+ e_b (\overline{\delta N_a \delta E})_{x,r',t} \frac{\partial n_b f_b}{\partial p} = 0 . \tag{4.4.7}$$

Here we have used the notation

$$(\overline{\delta N_a \delta N_b})_{x,x',t}^{\text{source}} = n_a \delta_{ab} (\delta(x - x') f_a(x, t) - \frac{1}{V} f_a(x, t) f_b(x', t)) . \tag{4.4.8}$$

The superscript "source" emphasizes the fact that this expression is wholly determined by the one-particle distribution function f_a.

If we drop the "source" term in (4.4.7) this equation will correspond to the second moments approximation. Indeed, we can then omit all terms nonlinear with respect to fluctuations from (4.4.5) so that it becomes

$$\left(\frac{\partial}{\partial t} + v \frac{\partial}{\partial r} + e_a E \frac{\partial}{\partial p} \right) \delta N_a + e_a \delta E \frac{\partial n_a f_a}{\partial p} = 0 . \tag{4.4.9}$$

From (4.4.9) we get (4.4.7) without the source term. Thus the presence of the source term in (4.4.7) corresponds to a partial consideration of the third-moments contribution. This ought to be expected because in the polarization approximation we assume the three-particle correlation function g_{abc} to be equal to zero, but not the third moments.

Let us pay attention to the fact that the source term (4.4.8) coincides with the second term in the right-hand side of (2.12.5). It appears that the source (4.4.8) is determined by the difference between the second moment and two-particle correlation function $n_a n_b g_{ab}$.

From the preceding discussion it follows that the equation for δN_a in the polarization approximation can be presented in the form [4.1]

$$\left(\frac{\partial}{\partial t} + v\frac{\partial}{\partial r} + e_a E\frac{\partial}{\partial p}\right)(\delta N_a(x, t) - \delta N_a^{\text{source}}(x, t)) + e_a \delta E\frac{\partial n_a f_a}{\partial p} = 0 .$$

(4.4.10)

It differs from the second moments approximation (4.4.9) by the source term $\delta N_a^{\text{source}}$. The two-time correlation of fluctuations of this source is determined by

$$\left[\frac{\partial}{\partial t} + v\frac{\partial}{\partial r} + e_a E(r, t)\frac{\partial}{\partial p}\right](\overline{\delta N_a \delta N_b})_{x, t, x', t'}^{\text{source}} = 0 .$$

(4.4.11)

The expression (4.4.8) serves as an initial condition for this equation: the value of the correlation at $t = t'$.

Taken together with equations for field fluctuations

$$\text{rot }\delta E = 0, \quad \text{div }\delta E = 4\pi \sum_a e_a \int \delta N_a dp ,$$

(4.4.12)

the set of equations (4.4.10 – 12) can serve as a basis for calculating the fluctuations (or the corresponding spectral densities) and constructing the kinetic equations for a Coulomb plasma in the polarization approximation [Ref. 4.1, Part II].

The collision integral (4.4.3) can be presented in the form

$$I_a(x, t) = -\frac{e_a}{n_a}\frac{\partial}{\partial p}\frac{1}{(2\pi)^4}\int \text{Re}\{\delta N_a \delta E\}_{\omega, k, r, p, t} d\omega\, dk ,$$

(4.4.13)

and in this manner the calculation of the collision integral leads essentially to the determination of the nonstationary and nonuniform spectral density of fluctuations δN_a and δE. Using the definitions of physically infinitesimal intervals l_{ph} and τ_{ph} (4.1.9), we can sort out the fluctuations into small- and large-scale ones: for the former $r_{\text{cor}} < r_{\text{D}}$, and for the latter $r_{\text{cor}} > r_{\text{D}}$.

When studying the processes where large-scale fluctuations are of little significance, one can express the spectral densities of fluctuations through one-particle distribution functions and obtain closed equations for f_a (the kinetic equations) by solving (4.4.10 – 12) [4.1]. The simplest of all is the expression for the spectral density of field fluctuations, which runs as follows:

$$(\delta E\, \delta E)_{\omega, k} = \frac{(\delta E\, \delta E)_{\omega, k}^{\text{source}}}{|\varepsilon(\omega, k)|^2}$$

(4.4.14)

where

$$\varepsilon(\omega, k) = 1 + \sum_a \frac{4\pi e_a^2 n_a}{k^2 \omega}\int \frac{(kv)\left(k\dfrac{\partial f_a}{\partial p}\right)}{\omega - kv + i\Delta} dp \quad (\Delta \to 0)$$

(4.4.15)

is the dielectric permittivity of plasma, which determines the response (the Fourier component $\delta D(\omega, k)$ to the field fluctuations $\delta E(\omega, k)$ with correlation radius $r_{cor} < l_{ph} \sim r_D$, that is, to the small-scale fluctuations.

The numerator in the fraction (4.4.14) is the spectral density of the source of the field fluctuations,

$$\delta E^{source}(\omega, k) \equiv \delta D^{source}(\omega, k) = -\frac{ik}{k^2} \sum_a 4\pi e_a \int \delta N_a^{source}(\omega, k, p) \, dp \, . \tag{4.4.16}$$

The expression for the spectral density of source fluctuations can be found by solving (4.4.11) [Ref. 4.1, Sect. 34],

$$(\delta N_a \delta N_b)^{source}_{\omega, k} = \delta_{ab} n_a \delta(p - p') 2\pi \delta(\omega - kv) f_a \, . \tag{4.4.17}$$

The distribution function is a slow function of coordinates and time ($l \sim r_D/\mu$, $\tau \sim 1/\omega_L \mu$). In finding the solution of (4.4.11) we have assumed a zero-order approximation with respect to μ. In this approximation the distribution function can be taken to be independent of coordinates and time. From (4.4.16, 17) we find the spectral density

$$(\delta E \, \delta E)^{source}_{\omega, k} = \sum_a \frac{(4\pi)^2 e_a^2 n_a}{k^2} \int 2\pi \delta(\omega - kv) f_a dp \, . \tag{4.4.18}$$

For the equilibrium state this expression becomes

$$(\delta E \, \delta E)^{source}_{\omega, k} \equiv (\delta D \, \delta D)^{source}_{\omega, k} = \frac{8\pi}{\omega} \mathrm{Im}\{\varepsilon(\omega, k)\} \mathscr{H} T \, . \tag{4.4.19}$$

The imaginary part of the permittivity can be found from (4.4.15). Formula (4.4.19) coincides in form with the general result of the theory of equilibrium electromagnetic fluctuations [4, 9] (see also [4.1 − 3]).

The same scheme can be used for finding the spectral density which defines the collision integral. Skipping the conventional calculations [Ref. 4.1, Sect. 35], the final result is

$$I_a = \frac{e_a^2}{16\pi^3} \frac{\partial}{\partial p_i} \int \frac{k_i}{k^2} \delta(\omega - kv) \left[(\delta E \, \delta E)_{\omega, k} k \frac{\partial f_a}{\partial p} + \frac{8\pi \mathrm{Im}\{\varepsilon\}}{|\varepsilon(\omega, k)|^2} f_a \right] d\omega \, dk \, . \tag{4.4.20}$$

Thus we have expressed the collision integral through the distribution function f_a, spectral density of the field [which is determined by (4.4.14, 18), and dielectric permittivity (4.4.15). This presentation of the collision integral is very advantageous, and later we shall discuss the physical meaning of the separate terms in (4.4.20). There is still another useful form of the collision integral for a plasma, which appears when we substitute the expressions for the spectral density of the field fluctuations and for the imaginary part of permittivity in (4.4.20),

$$I_a(x, t) = \sum_b 2e_a^2 e_b^2 n_b \frac{\partial}{\partial p_i} \int \frac{k_i k_j}{k^4} \frac{\delta(kv - kv')}{|\varepsilon(kv, k)|^2} \left(\frac{\partial f_a}{\partial p_j} f_b - \frac{\partial f_b}{\partial p_j'} f_a \right) dk\, dp' .$$

(4.4.21)

This expression was first worked out by *Balescu* and *Lenard* [4.1 – 3, 6]; the kinetic equation with collision integral (4.4.21) is therefore called the Balescu – Lenard kinetic equation.

When we search for the collision integral, not assuming the polarization approximation, but using the approximation of perturbation theory with respect to parameter ξ (4.1.3), we get an expression which differs from (4.4.21) in that $\varepsilon(\omega, k) \to 1$, i.e., in which the polarization of plasma is neglected. Integrating this expression over k, however, we encounter a logarithmic divergence at small and large k values. Small values of k correspond to great distances between particles, while big values of k correspond to small distances.

The divergence for great distances (small values of k) can be avoided by considering polarization. For large values of k, however, (4.4.21) coincides with the result from perturbation theory since polarization has no effect at small distances. Due to this, the divergence of (4.4.21) remains for small distances.

This divergence is due to violation of the disturbance theory for interactions at small distances. Indeed, when charged particles come close enough to one another, their potential energy becomes comparable to their kinetic energy. This induces us to restrict the range of integration over k in (4.4.21) by the condition $k < 1/l_L$. Such a rough method of regularizing a diverging integral is justified by the fact that the principal contribution to the final result is given by the logarithm of large arguments, and the final result depends but little on the actual position of the limit of integration range.

In the Balescu – Lenard integral the main contribution is from the range of values $k > 1/r_D$. Smaller values of k contribute comparatively little because of polarization effects. This enables us to simplify considerably the collision integral (4.4.21). Namely, we can assume $\varepsilon = 1$ and the same time restrict the integration range from the side of small k values, i.e., $k > 1/r_D$. As a result we get the collision integral in the form first obtained by *Landau* [4.10], long before the works of *Balescu* and *Lenard*. This expression is called the Landau collision integral and appears in the following form,

$$I_a(x, t) = \sum_b 2e_a^2 e_b^2 n_b \frac{\partial}{\partial p_i} \int \frac{k_i k_j}{k^4} \delta(kv - kv') \left(\frac{\partial f_a}{\partial p_j} f_b - \frac{\partial f_b}{\partial p_j'} f_a \right) dk\, dp' .$$

(4.4.22)

The range of integration over k is defined as

$$1/l_L > k > 1/r_D .$$

(4.4.23)

We can see that in the Balescu – Lenard and Landau integrals the contribution from the interaction of particles at small ($r < l_L$) distances, where perturbation theory no longer holds, is neglected. This is only justified in calculations of dissipative characteristics (viscosity, conductivity, etc.); however for nondissipative characteristics the interactions at small distances can play a significant role. One possible approach to the calculation of kinetic processes in a nonideal plasma is the application of a kinetic Boltzmann equation, which describes the collective interactions of charged particles by means of an effective potential [Ref. 4.1, Sect. 56].

The estimate of relaxation time and length for the Balescu – Lenard and Landau integrals yields

$$l \sim r_D/\mu, \quad \tau \sim 1/\omega_L \mu.$$

Consequently, in the zero approximation in respect to plasma parameter μ these values are infinite, which corresponds to first moments approximation (Sect. 4.3).

4.5 Electromagnetic Interaction in the Kinetic Equations of a Plasma

Along the lines drawn in the previous section, one can calculate the spectral densities of fluctuations in a nonequilibrium plasma involving electromagnetic interactions ([Ref. 4.1, Chap. 7] and [4.2, 11]). For an isotropic plasma, where the distribution function only depends on the absolute value of momentum, one can again utilize (4.4.10) to correlate the fluctuations δE, δB, and δN_a, this time, however, together with the complete set of equations for field fluctuations (2.11.8).

In the polarization approximation and assuming low field strengths E and B, field fluctuations can be described by means of two spectral densities of fluctuations, $\delta E^{\parallel}$ for potential and $\delta E^{\perp}$ for rotational fields. The expression for the spectral density of fluctuations of a potential field coincides with (4.4.14, 18). Expression (4.4.15), which can now be denoted as $\varepsilon^{\parallel}$ determines the response to fluctuations $\delta E^{\parallel}(\omega, k)$. The spectral density of the rotational field fluctuations is given by

$$(\delta E^{\perp} \delta E^{\perp})_{\omega, k} = \frac{\omega^4 (\delta E^{\perp} \delta E^{\perp})^{\text{source}}_{\omega, k}}{|\omega^2 \varepsilon^{\perp}(\omega, k) - c^2 k^2|^2}, \tag{4.5.1}$$

where

$$(\delta E^{\perp} \delta E^{\perp})^{\text{source}}_{\omega, k} = \sum_a \frac{(4\pi)^2 e_a^2 n_a}{k^2} \int 2\pi \delta(\omega - kv) \frac{[k \times v]^2}{\omega^2} f_a dp \tag{4.5.2}$$

is the spectral density of the source fluctuations, and

$$\varepsilon^{\perp}(\omega, k) = 1 + \sum_a \frac{2\pi e_a^2 n_a}{k^2 \omega} \int \frac{[[k \times v] \times k] \frac{\partial f_a}{\partial p}}{\omega - kv + i\Delta} dp \ (\Delta \to 0) \tag{4.5.3}$$

is the dielectric permittivity, which determines the response to fluctuations $\delta E^{\perp}(\omega, k)$.

The expression for the collision integral accounting for the fluctuations of both potential and rotational fields is

$$I_a(x, t)$$

$$= \frac{e_a^2}{16\pi^3} \frac{\partial}{\partial p_i} \int \delta(\omega - kv) \left\{ [(\delta E_i^{\parallel} \delta E_j^{\parallel})_{\omega, k} + (\delta E_i^{\perp} \delta E_j^{\perp})_{\omega, k}] \frac{\partial f_a}{\partial p_j} \right.$$

$$\left. \times \left[\frac{8\pi k_i \operatorname{Im}\{\varepsilon^{\parallel}\}}{k^2 |\varepsilon^{\parallel}(\omega, k)|^2} + \frac{8\pi [[k \times v] \times k]_i \omega^3 \operatorname{Im}\{\varepsilon^{\perp}\}}{k^2 |\omega^2 \varepsilon^{\perp}(\omega, k) - c^2 k^2|^2} \right] f_a \right\} d\omega \, dk . \qquad (4.5.4)$$

The field fluctuation tensors entering these equations are expressed through the spectral densities (4.4.14, 5.2)

$$(\delta E_i^{\parallel} \delta E_j^{\parallel})_{\omega, k} = \frac{k_i k_j}{k^2} (\delta E^{\parallel} \delta E^{\parallel})_{\omega, k} ,$$

$$(\delta E_i^{\perp} \delta E_j^{\perp})_{\omega, k} = \frac{1}{2} \left(\delta_{ij} - \frac{k_i k_j}{k^2} \right) (\delta E^{\perp} \delta E^{\perp})_{\omega, k} . \qquad (4.5.5)$$

Naturally, (4.5.4) can be presented in a form similar to (4.4.21).

The kinetic equation with the collision integral (4.5.4) is derived under the same assumptions as the kinetic equations for a Coulomb plasma. Both it and (4.4.20) neglect the influence of the mean field upon collision processes. This naturally imposes restrictions on the magnitude of the field strengths E and B. With sufficiently strong fields the collision integrals and the kinetic coefficients begin to depend directly on the magnitude of E and B. Nondissipative characteristics are also additionally contributed to; for instance, the mean field acting on charged particles is replaced by the effective field [Ref. 4.1, Sects. 48, 76].

Let us compare the relaxation times for the Coulomb interaction of charged particles and for interaction via the rotational electromagnetic field. The collision integral (4.5.4) can be presented as a sum of two parts,

$$I_a = I_a^{\parallel} + I_a^{\perp} , \qquad (4.5.6)$$

determined respectively by fluctuations of the potential and rotational fields. The right-hand side coincides, of course, with the collision integral for a Coulomb plasma (4.4.20).

In Sect. 4.4 we have already mentioned that in obtaining the collision integrals (4.4.20, 5.4) we only take into account the contribution of small-scale fluctuations. For a Coulomb plasma the border between "small scale" and "large scale" is set by the Debye radius (4.1.9). The range of integration over k in the collision integrals (4.4.20, 5.4) should therefore be restricted by the condition $k > 1/r_D$. The collision integral (4.4.20) then essentially coincides with the

Landau collision integral (4.4.22), since polarization plays no significant role in the region $k > 1/r_D$ and one can adopt $\varepsilon^{\parallel} = 1$.

The characteristic relaxation times for the Landau collision integral are well known. There are three relaxation times, τ_{ee}, τ_{ii}, and τ_{ei}. The respective collision rates are defined by the expressions

$$\nu_{ee} \sim \omega_L L_{\mu}, \quad \nu_{ii} \sim \sqrt{\frac{m_e}{m_i}}\, \nu_{ee}, \quad \nu_{ei} \sim \frac{m_e}{m_i}\, \nu_{ee} \tag{4.5.7}$$

where $L = \ln(r_D/l_L) \sim \ln(1/\mu)$ is the Coulomb logarithm.

From the aforesaid it follows that the region of wave numbers with $k < 1/r_D$ ought to be excluded from the collision integral for a Coulomb plasma. For the Landau collision integral this condition is satisfied (4.4.23). In the Balescu – Lenard integral, however, the restriction $k > 1/r_D$ is not imposed explicitly. The integration range is restricted naturally by taking the polarization of plasma into consideration; nevertheless it includes the region $k < 1/r_D$, where the assumptions under which the collision integral is obtained are no longer plausible. Let us estimate the contribution to the collision integral coming from this "outlaw" region of wave numbers. Later we shall compare this estimate with the corresponding result obtained taking large-scale fluctuations into account.

In a plasma for the wave numbers $k < 1/r_D$ there exist waves whose Landau damping decrement is much smaller than the corresponding eigenfrequencies ω_k. For this region the expression (4.4.14) can be presented in the form

$$(\delta E\, \delta E)_{\omega,k} = \pi[\delta(\omega - \omega_k) + \delta(\omega + \omega_k)](\delta E\, \delta E)_k, \tag{4.5.8}$$

where $(\delta E\, \delta E)_k$ is the spatial spectral density of the field fluctuations.

Let us present the Balescu – Lenard collision integral in Fokker – Planck form (Chap. 5)

$$I_a = \frac{\partial}{\partial p_i} D_{ij}^a \frac{\partial f_a}{\partial p_j} + \frac{\partial}{\partial p_i} A_i^a f_a. \tag{4.5.9}$$

The coefficients of diffusion and friction for the region $k < 1/r_D$ are defined by the expressions [Ref. 4.1, Sect. 57]

$$D_{ij}^a = \frac{e_a^2}{16\pi^2} \int_{k < 1/r_D} \frac{k_i k_j}{k^2} [\delta(kv - \omega_k) + \delta(kv + \omega_k)](\delta E\, \delta E)_k\, dk,$$

$$A_i^a = \frac{e_a^2}{4\pi} \int \frac{k_i}{k^2} \omega_k [\delta(kv - \omega_k) - \delta(kv + \omega_k)]\, dk. \tag{4.5.10}$$

Here we have used (4.5.8) for the spectral density of field fluctuations and the relevant approximation for the second term in the collision integral (4.4.20). In the equilibrium state

$$I_a = 0, \quad (\delta E\, \delta E)_k = 4\pi \mathscr{K}T. \tag{4.5.11}$$

We can present the coefficient of friction (4.5.10) in the form

$$A_a = m_a v_a^{\text{em}} v .$$ (4.5.12)

The collision rate introduced here for the radiation emission region is defined by

$$v_a^{\text{em}} = \frac{e_a^2}{2\pi m_a v^2} \int \frac{\omega_k^2}{k^2} \delta(k v - \omega_k) dk$$

$$= \frac{e_a^2 \omega_L^2}{m_a |v|^3} \ln \frac{|v|}{\omega_L r_D} \qquad \omega_k = \omega_L , \quad |v| \gg \omega_L r_D .$$ (4.5.13)

Comparing (4.5.13) and (4.5.7) we see that the radiation emission region contributes much less than the collision region, and therefore the kinetic coefficients calculated on the basis of the Landau integral differ little from those obtained using the Balescu – Lenard integral.

Considering now the second term in (4.5.6), in nonrelativistic plasma for wave numbers $k > 1/r_D$ the relative contribution of transverse field fluctuations is of the order $v_T^2/c^2 \ll 1$ [Ref. 4.1, Sect. 40]. Let us show that the contribution of wave fluctuations ($k < 1/r_D$) in our approximation is zero.

The spectral density of transverse field fluctuations is, in form similar to (4.5.8),

$$(\delta E^\perp \delta E^\perp)_{\omega,k} = \pi[\delta(\omega - \omega_k) + \delta(\omega + \omega_k)](\delta E^\perp \delta E^\perp)_k .$$ (4.5.14)

The eigenfrequencies are now defined by

$$\omega_k^2 = \omega_L^2 + c^2 k^2 .$$ (4.5.15)

It follows that the phase velocity of the waves is greater than the speed of light. Therefore the argument of the δ function in the expressions for the diffusion and friction coefficients (4.5.10) cannot become equal to zero. We conclude that the contribution from the radiation emission region to the second term in (4.5.6) is zero.

From the above discussion if follows that for nonrelativistic plasma the collision integral (4.5.4) actually coincides with the Landau collision integral. Of course, this does not mean that the role of large-scale fluctuations is insignificant. In order to understand the role of these fluctuations, however, one has to exceed the limits of convenient kinetic theory.

4.6 The Polarization Approximation for the System of Charged Particles and Field Oscillators

The kinetic equations accounting for dissipative processes can also be constructed on the basis of the equations for the microscopic phase densities $N_a(x, t)$ and $\mathcal{N}_{k,a}(X, t)$ (2.8.3, 4). In order to obtain a closed set of kinetic equations for one-particle distribution functions f_a and $\mathscr{F}_{k,\alpha}$ which also account for dissipative processes [in contrast to the set (4.3.3)], we shall again turn to the polarization approximation.

The example of a Coulomb plasma (Sect. 4.4) reveals that the spectral densities of nonequilibrium fluctuations can be calculated in either of two ways. The first method is based on getting the solution for the equations (4.4.7) for one-time moments of fluctuations δN_a and the relevant equations for two-time moments [4.11]. The second method is based on solving (4.4.10, 11) for the fluctuations proper, δN_a and δE. Let us first consider the former.

In the polarization approximation one must utilize the set of equations for three one-time second moments of fluctuations δN_a and $\delta \mathscr{N}_{k,\alpha}$ instead of (4.4.7). The first of these is similar to (4.4.7) and can be written in the form

$$\hat{L}_{ab}[\overline{(\delta N_a \delta N_b)}_{x,x',t} - (\delta N_a \delta N_b)_{x,x',t}^{\text{source}}]$$

$$+ \overline{(\delta F_a \delta N_b)}_{x,x',t} \frac{\partial n_a f_a}{\partial p} + \overline{(\delta N_a \delta F_b)}_{x,x',t} \frac{\partial n_b f_b}{\partial p'} = 0 . \quad (4.6.1)$$

The operator $\hat{L}_{ab}$ is defined by (2.10.1). The fluctuations δE and δB in the expression for the Lorentz force fluctuation are <u>linked to</u> δN_a and $\delta \mathscr{N}_{k,\alpha}$ by (2.8.7, 8). Hence the combined second moment $\overline{\delta N_a \delta \mathscr{N}_{k,\alpha}}$ also enters (4.6.1), in which the source is given by (4.4.8) and is therefore determined by the one-particle distribution function f_a.

The equation for the combined moment in the polarization approximation is

$$\hat{L}_{a,k,\alpha} \overline{(\delta N_a \delta \mathscr{N}_{k,\alpha})}_{x,X,t} + \overline{\delta F_a \delta \mathscr{N}_{k,\alpha}} \frac{\partial n_a f_a}{\partial p} + 4\pi e_k \overline{\delta j_k^\alpha \delta N_a} \frac{\partial \mathscr{F}_{k,\alpha}}{\partial P_k^\alpha} = 0 . \quad (4.6.2)$$

The source term, of course, does not enter this equation since one-particle distribution functions do not contribute to the combined moment.

Finally we can write the equation for the second moment of fluctuations $\delta \mathscr{N}_{k,\alpha}$ as

$$\hat{L}_{k,\alpha;k',\alpha'}[\overline{(\delta \mathscr{N}_{k,\alpha} \delta \mathscr{N}_{k',\alpha'})}_{X,X',t} - \overline{(\delta \mathscr{N}_{k,\alpha} \delta \mathscr{N}_{k',\alpha'})}_{X,X',t}^{\text{source}}]$$

$$+ 4\pi \overline{\delta j_k^\alpha \delta \mathscr{N}_{k',\alpha'}} \frac{\partial \mathscr{F}_{k,\alpha}}{\partial P_k^\alpha}$$

$$+ 4\pi \overline{\delta \mathscr{N}_{k,\alpha} \delta j_{k'}^{\alpha'}} \frac{\partial \mathscr{F}_{k',\alpha'}}{\partial P_{k'}^{\alpha'}} = 0 , \quad (4.6.3)$$

where the operator is defined by (2.10.5), and the source is defined by

$$\overline{(\delta \mathscr{N}_{k,\alpha} \delta \mathscr{N}_{k',\alpha'})}_{X,X',t}^{\text{source}} = \delta_{k,k'} \delta_{\alpha,\alpha'} [\delta(X - X') \mathscr{F}_{k,\alpha}(X, t) - \mathscr{F}_{k,\alpha}(X, t) \mathscr{F}_{k',\alpha'}(X', t)] . \quad (4.6.4)$$

Comparing this with (2.12.11) we notice that the source (4.6.4) appears because of the difference between the two-particle correlation of field oscillators and the second central moment of fluctuations $\delta \mathcal{N}_{k,\alpha}$. The source is now determined [cf. (4.4.8)] by the one-particle distribution function of field oscillators.

Equations (2.11.12, 14) together with (4.6.1 – 3) form a closed set. In order to obtain the kinetic equations, i. e., the closed set of equations for one-particle distribution functions f_a and $\mathcal{F}_{k,\alpha}$, one must utilize the partial solutions of nonuniform equations (4.6.1 – 3) which express the second moments through one-particle functions f_a and $\mathcal{F}_{k,\alpha}$. Examples of such solutions for a less complicated case, i. e., starting from the set of equations for functions N_a, E^m, and B^m (2.2.1, 4.1) are found in [4.11] and in [Ref. 4.1, Chaps. 7, 8]. Finding the solutions of (4.6.1 – 3) is a far more arduous task.

Another approach, which is similar to the one discussed in Sect. 4.4 for the case of a Coulomb plasma, seems more effective. Following the scheme outlined there, we write the equations for fluctuations δN_a and $\delta \mathcal{N}_{k,\alpha}$ in the polarization approximation. The first has the form

$$(\hat{L}_a + \Delta)[\delta N_a(x, t) - \delta N_a^{\text{source}}(x, t)] = \sum_b \int \frac{\partial \Phi_{ab}}{\partial r} \delta N_b(x', t) dx' \frac{\partial n_a f_a}{\partial p}$$

$$- \frac{e_a}{c} \sum_{k,\alpha} \int \left\{ -\frac{\partial A_{k,\alpha}}{\partial t} + [v \operatorname{rot} A_{k,\alpha}] \right\} \delta \mathcal{N}_{k,\alpha}(X, t) dX \frac{\partial n_a f_a}{\partial p}$$

$$\equiv -\delta F_a \frac{\partial n_a f_a}{\partial p} \tag{4.6.5}$$

where the operator $\hat{L}_a$ is defined by (2.10.2).

In order to single out the contribution of small-scale fluctuations, which determines the collision integral, a dissipative term with $\Delta \sim 1/\tau_{\text{ph}}$ is introduced into (4.6.5) (for more detail see [Ref. 4.1, Chap. 7]. In the final results $\Delta \to 0$.

Equation (4.6.5) is similar to (4.4.10) for a Coulomb plasma. The two-time correlation of source fluctuations $\delta N_a^{\text{source}}$ satisfies the equation

$$(\hat{L}_a + \Delta)\overline{(\delta N_a \delta N_b)}_{x,t,x',t'}^{\text{source}} = 0, \tag{4.6.6}$$

which ought to be solved at the initial condition (4.4.8). The source in (4.6.5) corresponds to the source in (4.6.1) for the one-time moment of fluctuations of phase density of particles.

The equation for fluctuation $\delta \mathcal{N}_{k,\alpha}$ can be written in the form

$$\hat{L}_{k,\alpha}[\delta \mathcal{N}_{k,\alpha}(X, t) - \delta \mathcal{N}_{k,\alpha}^{\text{source}}(X, t)]$$

$$= -\sum_a e_a \sqrt{\frac{4\pi}{V}} \mathfrak{f}(e_k v) \begin{Bmatrix} \sin kr \\ \cos kr \end{Bmatrix} \delta N_a(x, t) dx \frac{\partial \mathcal{F}_{k,\alpha}}{\partial P_k^\alpha}$$

$$\equiv -4\pi \delta j_k^\alpha \frac{\partial \mathcal{F}_{k,\alpha}}{\partial P_k^\alpha}. \tag{4.6.7}$$

The operator entering this equation is defined by (2.10.5). The two-time correlation of source fluctuations is determined by the solution of the equation

$$\hat{L}_{k,\alpha}(\overline{\delta \mathcal{N}_{k,\alpha}\delta \mathcal{N}_{k';\alpha'}})^{\text{source}}_{X,t,X';t'} = 0 .$$

(4.6.8)

The solution of this equation is sought at a so-to-say "initial" condition $(t = t')$ (4.6.4).

Hence the source in (4.6.7) appears due to the presence of a source in (4.6.3) for the one-time moment of fluctuations $\delta \mathcal{N}_{k,\alpha}$, and its appearance reflects the structure of the field as a system of oscillators. This source is similar to that in (4.6.1, 5), which is brought about by the discrete structure of the system of charged particles.

Let us point out that in the description of fluctuant processes in the polarization approximation for random deviations δN_a, δE, and δB on the basis of (2.11.8, 4.4.10, 11), we only take the source into consideration which reflects the discrete structure of a system of charged particles. As a consequence, the calculations of the correlations of transverse field fluctuations, and in particular (4.5.2), are not complete.

The boundaries of all real systems are more or less absorptive. The absorption, which results in the loss of energy of an electromagnetic field within the volume under consideration, can be due to the escape of radiation. In order to account for this, we introduce a dissipative term

$$\hat{I}_{k,\alpha} = \gamma_k P_k^\alpha \frac{\partial}{\partial P_k^\alpha}$$

(4.6.9a)

into the operator $\hat{L}_{k,\alpha}$. With the help of the Fokker – Planck operator the same can be given in another form,

$$\hat{I}_{k,\alpha} = D_k \frac{\partial}{\partial P_{k,\alpha}^2} + \frac{\partial}{\partial P_k^\alpha}(\gamma_k P_k^\alpha \dots) ,$$

(4.6.9b)

where $D_k = \gamma_k k T$ is the relevant diffusion coefficient. The dissipative coefficient γ_k depends on the oscillator frequency ck. By using (4.6.9b) the normalization condition for the function $\mathcal{N}_{k,\alpha}$ is retained.

To calculate the fluctuation sources of the electromagnetic field with (4.6.7, 8) we must find the equations for coordinates and momenta of field oscillators, taking into account the relations (2.8.10). In place of equations for the real functions $Q_k^\alpha(t)$ and $P_k^\alpha(t)$ it is more practical to utilize the corresponding equations for complex functions. From (2.5.5, 9) we introduce the designations

$$\delta E_k = -\sqrt{4\pi}e_k(\delta P_k^2 - i\delta P_k^1) ,$$

$$\delta Q_k \equiv -\frac{1}{c}\delta A_k = -\sqrt{4\pi}e_k(\delta Q_k^2 - i\delta Q_k^1)$$

(4.6.10)

for the Fourier components of the field fluctuations and the corresponding coordinates. The equations for these functions follow from (4.6.7, 9b),

$$\frac{d}{dt}(\delta Q_k - \delta Q_k^{\text{source}}) = \delta E_k - \delta E_k^{\text{source}},$$

$$\left(\frac{d}{dt} + \gamma_k\right)(\delta E_k - \delta E_k^{\text{source}}) + c^2 k^2 (\delta Q_k - \delta Q_k^{\text{source}}) = -4\pi \delta j_k^\perp \qquad (4.6.11)$$

where

$$\delta j_k^\perp = \sum_a e_a \int \frac{[k \times [k \times v]]}{k^2} \delta N_a(x, t) e^{-ikr} dx \qquad (4.6.12)$$

is the Fourier component of the current fluctuations.

The corresponding equations for two-time correlations of source fluctuations are, from (4.6.8) and taking (4.6.9b) into account,

$$\frac{d}{dt}(\delta Q \, \delta E)_{k,t,t'}^{\text{source}} = (\delta E \, \delta E)_{k,t,t'}^{\text{source}},$$

$$\left(\frac{d}{dt} + \gamma_k\right)(\delta E \, \delta E)_{k,t,t'}^{\text{source}} + c^2 k^2 (\delta Q \, \delta E)_{k,t,t'}^{\text{source}} = 0. \qquad (4.6.13)$$

These equations should be considered together with initial conditions ($t = t'$).

From (4.6.4) it is clear that one-time correlators of source fluctuations are determined by the one-particle distribution function $\mathcal{F}_{k,\alpha}$, which is an unknown and can generally only be found by solving the set of kinetic equations for the functions f_a and $\mathcal{F}_{k,\alpha}$. Here we shall discuss the relaxation process at a stage when a "local" Gauss distribution is already established, i.e.,

$$\mathcal{F}_{k,\alpha}(X, t) = A \exp\left[-\frac{[P_k^\alpha - \bar{P}_k^\alpha(t)]^2 + c^2 k^2 (Q_k^\alpha - \bar{Q}_k^\alpha)^2}{2\mathcal{K} T_k(t)}\right]. \qquad (4.6.14)$$

Factor A is found from the normalization condition. In general, the mean values of coordinates and momenta of oscillators not only depend on time, but also on the frequency.

Provided condition (4.6.14) is fulfilled, the initial conditions for the set of equations (4.6.13) are

$$(\delta Q \, \delta E)_{k,t}^{\text{source}} = 0,$$

$$(\delta E \, \delta E)_{k,t}^{\text{source}} = 4\pi \sum_\alpha (\delta P_k^\alpha \delta P_k^\alpha)^{\text{source}} = 8\pi \mathcal{K} T_k(t). \qquad (4.6.15)$$

Let us now employ the equations obtained in this section to find the spectral density of field fluctuations taking the two types of field fluctuation sources introduced above into consideration.

4.7 The Equilibrium Fluctuations of an Electromagnetic Field

In the state of equilibrium the mean values $\bar{Q}_k^\alpha$ and $\bar{P}_k^\alpha$ are zero in (4.6.14), and the temperature is constant. Furthermore, in the equilibrium state the normal distribution of coordinates and momenta also holds in quantum theory [4.12, 13]. Then

$$\mathscr{K} T_k = \hbar c k \left(\frac{1}{2} + \bar{n}_k \right), \quad \bar{n}_k = \frac{1}{\exp(\hbar c k / \mathscr{K} T) - 1}. \tag{4.7.1}$$

If $\hbar = 0$, then naturally $\mathscr{K} T_k = \mathscr{K} T$.

In the state of equilibrium the two-time correlations depend only on $\tau = |t - t'|$, and therefore the spectral density can be defined by the expression

$$(xy)_\omega = 2 \operatorname{Re} \{ (xy)_\omega^+ \} = 2 \operatorname{Re} \left\{ \int_0^\infty (xy)_\tau e^{i\omega\tau} d\tau \right\}. \tag{4.7.2}$$

A unilateral Fourier transform then has to be carried out in the set of equations (4.6.13) to determine the spectral density. As a result we get a set of algebraic equations

$$-i\omega(\delta Q \, \delta E)_{\omega,k}^+ = (\delta E \, \delta E)_{\omega,k}^+,$$
$$(-i\omega + \gamma_k)(\delta E \, \delta E)_{\omega,k}^+ + c^2 k^2 (\delta Q \, \delta E)_{\omega,k}^+ = 8\pi \mathscr{K} T_k \tag{4.7.3}$$

from which we find the expression for the spectral density of field source fluctuations

$$(\delta E \, \delta E)_{\omega,k}^{\text{source}} = \frac{16\pi \gamma_k \omega^2 \mathscr{K} T_k}{(\omega^2 - c^2 k^2)^2 + \omega^2 \gamma_k^2}. \tag{4.7.4}$$

To obtain (4.6.15) it is necessary use the definition

$$(\delta E \, \delta E)_k = \frac{1}{2\pi} \int_{-\infty}^{\infty} (\delta E \, \delta E)_{\omega,k} d\omega \tag{4.7.5}$$

and carry out integration over ω.

In the following we shall often present the equations for fluctuations in the Langevin form. For instance, (4.6.11) can be written in the form

$$\frac{d\delta Q_k}{dt} = \delta E_k, \tag{4.7.6}$$

$$\left(\frac{d}{dt} + \gamma_k\right)\delta E_k + c^2 k^2 \delta Q_k = -4\pi \delta j_k^\perp + icky_{f,k} .$$ (4.7.7)

The subscript f in $y_{f,k}$ emphasizes the fact that this source is brought forth by the structure of the field.

From the set of equations (4.7.6, 7) we find

$$(\delta E\, \delta E)_{\omega,k}^{\text{source}} = \frac{\omega^2 c^2 k^2 (y_f y_f)_{\omega,k}}{(\omega^2 - c^2 k^2)^2 + \omega^2 \gamma_k^2} .$$ (4.7.8)

Comparing this expression with (4.7.4) we obtain the equations for the spectral density of the source,

$$(y_f y_f)_{\omega,k} = \frac{16\pi\gamma_k}{c^2 k^2} \mathcal{K} T_k , \quad (y_f y_f)_{\tau,k} = \frac{16\pi\gamma_k}{c^2 k^2} \mathcal{K} T_k \delta(\tau) .$$ (4.7.9)

Thus the Langevin source is δ correlated.

Let us now single out in (4.7.6, 7) another source, determined by the distribution of charged particles. We shall use (4.6.5) to find the Fourier component of current density (4.6.12). The first term in the right-hand side of (4.6.5) gives a zero contribution, as it is proportional to the longitudinal component of field fluctuation. The contribution of the last term is also zero if the function f_a is assumed to depend only on the absolute value of p. This allows us to employ a simpler equation to calculate the current,

$$\left(\frac{\partial}{\partial t} + \Delta\right)\delta N_a + v\frac{\partial \delta N_a}{\partial r} + e_a \delta E^\perp \frac{\partial n_a f_a}{\partial p} = 0 .$$ (4.7.10)

By virtue of this equation we find

$$\delta j^\perp(\omega, k) = (\delta j^\perp(\omega, k))^{\text{source}}$$

$$-i\sum_a \frac{e_a^2 n_a}{k^2} \int \frac{[k \times [v \times k]]}{\omega - kv + i\Delta} \frac{\partial f_a}{\partial p} dp\, \delta E^\perp(\omega, k) .$$ (4.7.11)

Here we employ the designation

$$[\delta j^\perp(\omega, k)]^{\text{source}} \equiv -i\omega(\delta P^{\text{source}})^\perp = \sum_a e_a \int \frac{[k \times [v \times k]]}{k^2} \delta N_a^{\text{source}}(\omega, k, p) dp ,$$ (4.7.12)

where δP^{source} is the source due to the polarization of a system of charged particles.

Using (4.5.2) we now find the spectral density in the form

$$(\delta P^\perp \delta P^\perp)_{\omega,k}^{\text{source}} = \sum_a e_a^2 n_a 2\pi \int \frac{[k \times v]^2}{\omega^2 k^2} \delta(\omega - kv) f_a dp .$$ (4.7.13)

In the equilibrium state the right-hand side can be expressed through the imaginary part of dielectric permittivity $\varepsilon^\perp(\omega, k)$ [see (4.5.3)],

$$(\delta P^\perp \delta P^\perp)^{\text{source}}_{\omega,k} = \frac{1}{\pi\omega} \, \text{Im}\{\varepsilon^\perp(\omega, k)\} \, \mathscr{K}T . \tag{4.7.14}$$

So from (4.7.6, 7, 11, 12) it follows that the equation for field fluctuations can be presented in the form

$$\{[\omega^2\varepsilon^\perp(\omega, k) - c^2 k^2] + i\omega\gamma_k\}\delta E(\omega, k)$$
$$= -\omega^2 4\pi[\delta P^\perp(\omega, k)]^{\text{source}} + \omega c k y_f(\omega, k) \equiv \omega^2 y(\omega, k) . \tag{4.7.15}$$

The spectral density of field fluctuations is now determined by the formula

$$(\delta E \, \delta E)_{\omega,k} = \frac{\omega^4 (yy)_{\omega,k}}{|\omega^2\varepsilon(\omega, k) + i\omega\gamma_k - c^2 k^2|^2} . \tag{4.7.16}$$

Here we have introduced the designation for the spectral density of the total source of fluctuations

$$(yy)_{\omega,k} = (4\pi)^2(\delta P^\perp \delta P^\perp)^{\text{source}}_{\omega,k} + \left(\frac{ck}{\omega}\right)^2 (y_f y_f)_{\omega,k}$$

$$= \left[\frac{16\pi}{\omega} \, \text{Im}\{\varepsilon^\perp(\omega, k)\} + 16\pi\frac{\gamma_k}{\omega^2}\right] \mathscr{K}T . \tag{4.7.17}$$

In order to obtain our former results (4.5.1, 2) from this, one should assume $\gamma_k = 0$ in (4.7.17) and thus only take into account the fluctuations due to polarization of a system of charged particles. Expressions (4.5.1, 2) are also approximately true with small values of γ_k. In some cases, however, the situation may be different, e.g., the inverse case presents thermal radiation from a cavity when the principal loss is connected with γ_k. In active media, as in lasers (Chap. 12), the imaginary part of the permittivity at the frequency of generation is negative, and the loss due to escape of radiation from the cavity determines the generation threshold.

Thus for the equilibrium state we have obtained the expression for the spectral density of field fluctuations which accounts for two types of fluctuation sources. This accounts for the fact that the studied system is composed of two subsystems: charged particles and field oscillators.

In (4.7.17) the contribution of each type of source differs. The first term, determined by the structure of the system of charged particles, is proportional to the imaginary part of permittivity and, therefore, to the square of the electric charge. It accounts, consequently, for the interaction of charged particles. The second term, being proportional to γ_k, has nothing to do with the losses within the system, but is wholly determined by the escape of radiation from the system.

These losses are similar to those in a collisionless plasma, where the interactions (collisions) of particles are negligible and losses depend solely upon the border situation.

It should be recalled that the contribution determined by the structure of the charged particles is calculated for the case that the fluctuations δN_a satisfy (4.6.5), i.e., are small-scale ones. Large-scale fluctuations will be dealt with in the next section. Taking them into account in the state of equilibrium, however, does not change the structure of (4.7.17); only the expression for the dielectric permittivity is altered. We shall discuss this matter once more in Chap. 11, devoted to the fluctuation dissipation theorem (FDT).

Whether the fluctuations described by the second term in (4.7.17) are small or large scale depends upon the actual relative values of the dissipative parameters. In the first case one can assume $\gamma_k = 0$ in (4.7.17), and still obtain the results (4.5.1, 2). The fluctuations for which the source y_f is responsible should then be taken into account in the calculation of large-scale fluctuations; we shall now turn to this for a system of free charged particles and a field for nonequilibrium states.

4.8 The Kinetic Theory of Fluctuations

In Sects. 4.4, 5 we have discussed certain results of the theory of nonequilibrium small-scale fluctuations in a system made up of free charged particles and a field. We attributed those correlations to small-scale fluctuations, the characteristic times of which are much smaller than the relaxation times of one-particle distribution functions. As we have seen, these correlations determine the collision integrals.

In deriving the kinetic equations we used the principle concerning the damping of initial correlations (Bogolyubov's principle). This principle implies that large-scale fluctuations (with $\tau_{cor} > \tau_{ph}$) which do not have time to fade out, play no role in the kinetic theory. This neglect of large-scale fluctuations shows up in the deterministic (not random) nature of the distribution functions in the kinetic equations. In order to account for the large-scale fluctuations in the kinetic equations the principle of complete attenuation of initial correlations must be replaced by the principle of partial attenuation [4.1]. This implies that only small-scale fluctuations are damped, thus allowing for the fluctuations of distribution functions.

There is possibly a dual approach to the construction of the kinetic theory of large-scale fluctuations [Ref. 4.1, Chaps. 3, 4, 11].

1) Under the condition of partial attenuation of the initial correlations, the transition is carried out from a Liouville equation for the distribution function of the studied system (2.9.1) to the approximate equation for a smoothed − with regard to fine structure − distribution function [Ref. 4.1, Sect. 18]. On the basis of this equation a chain of equations for smoothed distribution functions is built. As opposed to the $B-B-G-K-Y$ chain of equations this chain is dissipative, the small-scale fluctuations being responsible for dissipation.

2) Instead of the exact set of equations for random functions $N_a(x, t)$ and $\mathscr{N}_{k,a}(X, t)$ (2.8.3, 4), that for the random functions $\tilde{N}_a$ and $\tilde{\mathscr{N}}_{k,a}$, smoothed with regard to fine structure, serve as the initial set of equations. This set is similar in appearance to the kinetic equations, whose collision integrals are determined by small-scale fluctuations.

Both approaches are equivalent [Ref. 4.1, Sect. 22, 61]; the latter, however, seems simpler and more comprehensive.

Let us write the equations for the smoothed functions $\tilde{N}_a$ and $\tilde{\mathscr{N}}_{k,a}$. The estimate made in Sect. 4.5 shows that when only small-scale fluctuations are taken into consideration, the main contribution to the collision integral (4.5.4) comes from the Coulomb interactions. This is sufficient reason to replace the collision integral (4.5.4) for nonrelativistic plasma by a far simpler Landau collision integral, (4.4.22). The equation for the function $\tilde{N}_a$ can then be written in the form

$$\left(\frac{\partial}{\partial t} + v \frac{\partial}{\partial r} + \tilde{F}_a \frac{\partial}{\partial p}\right) \tilde{N}_a(x, t) = \sum_b 2 e_a^2 e_b^2 \frac{\partial}{\partial p_i} \int \frac{k_i k_j}{k^4} \delta(k v - k v')$$

$$\times \left[\frac{\partial N_a(x, t)}{\partial p_j} \tilde{N}_b(x', t) - \frac{\partial \tilde{N}_b}{\partial p_j'} \tilde{N}_a\right] dk \, dp' .$$

(4.8.1)

The normalization condition for $\tilde{N}_a$ has the form

$$\int \tilde{N}_a(x, t) dx = N_a ,$$

(4.8.2)

where N_a is the total number of particles.

In (4.8.1) $\tilde{F}_a$ is the Lorentz force, which is determined by smoothed electric and magnetic field strengths. As before, two ways of presenting the field are possible here − either via Lorentz equations for smoothed field strengths, or using the equations for $\tilde{\mathscr{N}}_{k,a}$ with all permissible values of k and α. Let us discuss them in the same order.

We use the expression for the Lorentz force

$$\tilde{F}_a = e_a \left(\tilde{E} + \frac{1}{c} [v \times \tilde{B}]\right)$$

(4.8.3)

and the Lorentz equations for smoothed field strengths

$$\text{rot } \tilde{B} = \frac{1}{c} \frac{\partial \tilde{E}}{\partial t} + \frac{4\pi}{c} \sum_a e_a \int v \tilde{N}_a(x, t) dp ,$$

$$\text{rot } \tilde{E} = -\frac{1}{c} \frac{\partial \tilde{B}}{\partial t} ,$$

$$\text{div } \tilde{B} = 0$$

$$\text{div } \tilde{E} = 4\pi \sum_a e_a \int \tilde{N}_a(x, t) dp .$$

(4.8.4)

Equations (4.8.1 – 4) make up a closed set of equations for random functions $\tilde{N}_a$, $\tilde{E}$, and $\tilde{B}$. In contrast to (2.2.1, 3, 4.1) this set of equations is approximate; it accounts for dissipation caused by small-scale fluctuations.

In the second method the Lorentz force is presented in the form

$$\tilde{F}_a = e_a \left(\tilde{E}^{\parallel} + \tilde{E}^{\perp} + \frac{1}{c} [v \operatorname{rot} \tilde{B}] \right). \tag{4.8.5}$$

Field strengths are connected with functions $\tilde{N}_a$ and $\tilde{\mathcal{N}}_{k,\alpha}$ through (2.8.7 – 9). The equation for the smoothed phase density of field oscillators follows from (2.8.4) when the dissipative contribution (4.6.9 b) is taken into account,

$$\left(\frac{\partial}{\partial t} + P_k^{\alpha} \frac{\partial}{\partial Q_k^{\alpha}} - c^2 k^2 Q_k^{\alpha} \frac{\partial}{\partial P_k^{\alpha}} + 4 \pi e_k \tilde{j}_k^{\alpha} \frac{\partial}{\partial P_k^{\alpha}} \right) \tilde{\mathcal{N}}_{k,\alpha} = \hat{I}_{k,\alpha} \tilde{\mathcal{N}}_{k,\alpha}. \tag{4.8.6}$$

The Fourier component of the current $\tilde{j}_k^{\alpha}$ is linked to the function $\tilde{N}_a$ by an expression similar to (2.10.6). The operator $\hat{I}_{k,\alpha}$ is defined by (4.6.9b).

Based on the above equations it is again possible to build a chain of equations for the moments of smoothed functions $\tilde{N}_a$ and $\tilde{\mathcal{N}}_{k,\alpha}$. Using (2.11.10) (this time though for smoothed functions $\tilde{N}_a$ and $\tilde{\mathcal{N}}_{k,\alpha}$) we write the equations for one-particle distribution functions

$$\hat{L}_a f_a = I_a + \tilde{I}_a, \quad \hat{L}_{k,\alpha} \mathcal{F}_{k,\alpha} = I_{k,\alpha} + \tilde{I}_{k,\alpha}. \tag{4.8.7}$$

In these equations the operators $\hat{L}_a$ and $\hat{L}_{k,\alpha}$ are defined (2.10.2, 5). The integral I_a in the first equation is the Landau collision integral (4.4.22) where $\overline{\tilde{N}_a \tilde{N}_b}$ is replaced by $n_a n_b f_a f_b$; $\tilde{I}_a$ is an additional collision integral determined by large-scale fluctuations

$$\tilde{I}_a = - \frac{1}{n_a} \frac{\partial \overline{\delta \tilde{F}_a \delta \tilde{N}_a}}{\partial p}. \tag{4.8.8}$$

In the second equation (4.8.7) the operator $\hat{I}_{k,\alpha}$ is defined by (4.6.9b), while $\tilde{I}_{k,\alpha}$ is also an additional collision integral determined by large-scale fluctuations

$$\tilde{I}_{k,\alpha} = - 4 \pi \frac{\partial}{\partial P_k^{\alpha}} (e_k \overline{\delta \tilde{j}_k^{\alpha} \delta \tilde{\mathcal{N}}_{k,\alpha}}). \tag{4.8.9}$$

Equations (4.8.7) are naturally nonclosed since they involve second moments of large-scale fluctuations. They are the beginning of the new chain of equations for the moments of smoothed phase densities $\tilde{N}_a$ and $\tilde{\mathcal{N}}_{k,\alpha}$.

As before, analysis of this chain of equations can be started with the first moments approximation. In this approximation the additional collision integrals are assumed equal to zero. As a result we get a set of kinetic equations, one of which is the Landau kinetic equation for the distribution function of particles f_a (4.4.22), and the other is the equation for the distribution function of field oscillators

$$\hat{L}_{k,\alpha}\,\mathscr{F}_{k,\alpha}=D_k\,\frac{\partial^2\,\mathscr{F}_{k,\alpha}}{\partial P_{k,\alpha}^2}+\frac{\partial}{\partial P_k^\alpha}\,(\gamma_k P_k^\alpha\,\mathscr{F}_{k,\alpha})\,. \tag{4.8.10}$$

Let us compare this set of equations with (4.3.3). These latter equations have been obtained with considering correlations on any scales, and, therefore, they do not consider the dissipative processes in the system. In particular, this can be seen from the fact that the total entropy of a closed system, calculated from (4.3.3), does not change with time.

Equations (4.4.22, 8.10) account for the contribution of small-scale correlations determined by the interaction of particles, as well as the losses due to the escape of radiation from the system. These equations are therefore dissipative, and the total entropy of a closed system, calculated on this basis, tends to increase with time, see, e.g., [Ref. 4.1, Sects. 14, 37].

Later we shall see that it is possible to construct an actual hierarchy of first moments approximations taking into consideration correlations on ever increasing scale. The examples discussed here represent the first two steps in such a succession.

In a similar way a succession of polarization approximations can be built if correlations are incorporated on a larger and larger scale. The first step in this succession of approximations has already been discussed in Sect. 4.4.5. The collision integrals (4.4.20) [or (4.4.21)] and (4.5.4) account for polarization effects due to small-scale fluctuations. We have shown that for a nonrelativistic plasma these collision integrals can be replaced by a simpler Landau collision integral. One may therefore consider the kinetic equations (4.4.22, 8.10) to be the polarization approximation equations (to a good approximation) for a chain of equations constructed on the basis of the exact equations for phase densities N_a and $\mathscr{N}_{k,\alpha}$ (2.8.3, 4). On the other hand, these are the equations of the first moments approximation for the chain constructed on the basis of (4.8.1, 6).

Everything that has been said here applies equally well if the equation for smoothed phase densities $\tilde{N}_a$ (4.8.1), together with those for smoothed field strengths $\tilde{E}$ and $\tilde{B}$ (4.8.4), is taken as the initial equation instead of (4.8.1, 2). In this case we also get successions of sets of equations which correspond to first moments approximations on different levels, as well as a succession of polarization approximations.

The first stage in the succession of first moments approximations is in this case represented by Vlasov equations [the set of equations (2.11.6, 4.3.2)], in which fluctuations on all scales are neglected. The second stage is represented by the set of equations (4.4.1, 22) together with (2.11.6) for mean field strengths. This set accounts for small-scale correlations in a system of charged particles.

We have already noted in Sect. 4.3 that the expressions for collision integrals obtained here do not account for the influence of mean fields upon the spectral densities of small-scale fluctuations. This, naturally, imposes restrictions on the magnitude of mean field strengths. However, this simplification is not fundamental. In certain cases it is possible to obtain more general expressions for collision integrals which remain true for plasmas with strong fields as well (see [4.2] and [Ref. 4.1, Sect. 8]).

Consider now the polarization approximation for the chain of equations obtained on the basis of the equations for smoothed functions $\tilde{N}_a$ and $\tilde{\mathcal{N}}_{k,\alpha}$. The equations for the fluctuations δN_a and $\delta \mathcal{N}_{k,\alpha}$ in the polarization approximation are similar to (4.6.5, 7),

$$(\hat{L}_a + \delta \hat{I}_a)[\delta \tilde{N}_a(x, t) - \delta \tilde{N}_a^{\text{source}}(x, t)] = - \delta \tilde{F}_a \frac{\partial n_a f_a}{\partial p},$$

$$(\hat{L}_{k,\alpha} + \delta \hat{I}_{k,\alpha})[\delta \tilde{\mathcal{N}}_{k,\alpha}(X, t) - \delta \tilde{\mathcal{N}}_{k,\alpha}^{\text{source}}] = - 4\pi e_k \delta j_k^\alpha \frac{\partial \mathscr{F}_{k,\alpha}}{\partial P_k^\alpha}. \tag{4.8.11}$$

They formally differ from (4.6.5, 7) in that the dissipative constant Δ is replaced by the operator of linearized collision integrals for charged particles, i.e.,

$$\Delta \to \delta \hat{I}_a, \tag{4.8.12}$$

and by the inclusion of an additional dissipative term with operator (4.6.9b) in the equation for fluctuations $\delta \tilde{\mathcal{N}}_{k,\alpha}$. Of course, there is also a difference since the Lorentz force fluctuations and current fluctuations in (4.8.11) are determined this time by large-scale fluctuations of the field and of the phase density of charged particles.

For the Landau collision integral the action of the operator $\delta \hat{I}_a$ upon an arbitrary function $F_a(p)$ is defined by

$$\delta \hat{I}_a F_a(p) = - \sum_b 2 e_a^2 e_b^2 n_b \frac{\partial}{\partial p_i} \int \frac{k_i k_j}{k^1} \delta(kv - kv')$$

$$\times \left(\frac{\partial}{\partial p_j} - \frac{\partial}{\partial p_j'} \right) [F_a(p) f_b(r, p', t) + f_a(r, p, t) F_b(p')] dk \, dp'. \tag{4.8.13}$$

The operator $\delta \hat{I}_{k,\alpha}$ is linked to operator (4.6.9b) by the equation

$$\delta \hat{I}_{k,\alpha} = - \hat{I}_{k,\alpha}. \tag{4.8.14}$$

The minus sign in this definition emphasizes the dissipative nature of the corresponding terms in (4.8.11).

Equations (4.8.11) can be presented in the Langevin form,

$$(L_a + \delta \hat{I}_a) \delta \tilde{N}_a(x, t) + \delta \tilde{F}_a \frac{\partial n_a f_a}{\partial p} = y_a(x, t), \tag{4.8.15}$$

$$(\hat{L}_{k,\alpha} + \delta \hat{I}_{k,\alpha}) \delta \tilde{\mathcal{N}}_{k,\alpha} + 4\pi e_k \delta j_k^\alpha \frac{\partial \mathscr{F}_{k,\alpha}}{\partial P_k^\alpha} = y_{k,\alpha}(X, t). \tag{4.8.16}$$

The Langevin sources are connected with the sources $\delta N_a^{\text{source}}$ and $\delta \mathcal{N}_{k,\alpha}^{\text{source}}$ by the equations

$$(\hat{L}_a + \delta \hat{I}_a) \delta N_a^{\text{source}} = y_a,$$

$$(\hat{L}_{k,\alpha} + \delta \hat{I}_{k,\alpha}) \mathcal{N}_{k,\alpha}^{\text{source}} = y_{k,\alpha}. \tag{4.8.17}$$

Let us now find the connection between the intensities of Langevin sources and the distribution functions f_a and $\mathcal{F}_{k,\alpha}$. For this purpose we present the correlations of random Langevin sources in the form

$$\overline{(y_a y_b)}_{x,t;x',t'} = A_{ab}(x, x', t) \delta(t - t'),$$

$$\overline{(y_{k,\alpha} y_{k',\alpha'})}_{X,t;X',t'} = A_{k,\alpha,k',\alpha'}(X, X', t) \delta(t - t'). \tag{4.8.18}$$

This implies the assumption that the random sources are δ correlated, while their intensities are functions of all coordinates and time. The final result will confirm our assumption.

From the first equation in (4.8.17) we find

$$\delta N_a^{\text{source}}(x, t) = \int_0^\infty e^{-\delta \hat{I}_a \tau} y_a(t - \tau, r - v\tau, p) d\tau.$$

Hence the expression for the one-time moment follows,

$$\overline{(\delta N_a \delta N_b)}_{x,x',t}^{\text{source}} = \int_0^\infty d\tau \int_0^\infty d\tau' e^{-\delta \hat{I}_a \tau - \delta \hat{I}_b \tau'} \overline{(y_a y_b)}_{t-\tau, r-v\tau, p; t-\tau', r'-v'\tau', p'}.$$

Now we substitute here the first expression in (4.8.18) and turn to a differential form. As a result we get

$$(\hat{L}_{ab} + \delta \hat{I}_a + \delta \hat{I}_b) \overline{(\delta \tilde{N}_a \delta \tilde{N}_b)}_{x,x',t}^{\text{source}} = A_{ab}(x, x', t), \tag{4.8.19}$$

which connects the intensity of the Langevin source y_a with the second one-time moment of fluctuations $\delta \tilde{N}_a^{\text{source}}$. This equation coincides with [Ref. 4.1, Eq. (62.12)]. We can therefore take advantage of the formulae obtained there which establish the connection between the function $A_{ab}(x, x', t)$ (the intensity of the Langevin source) and one-particle distribution functions f_a.

The function A_{ab} can be presented as a sum of two parts [Ref. 4.1, Eq. (62.17)],

$$A_{ab}(x, x', t) = \tilde{A}_{ab} + A_{ab}^{\text{L}}. \tag{4.8.20}$$

The first term in the right-hand side is determined by the collision integral $\tilde{I}_a$, that is, by that part of the total collision integral in the first equation of (4.8.7) which is expressed through the large-scale fluctuations

$$\tilde{A}_{ab}(x, x', t) = n_a \delta_{ab} \{\delta(x-x')\tilde{I}_a(x, t) - \frac{1}{V}[\tilde{I}_a(x, t) f_b(x', t) + \tilde{I}_b f_a]\}. \quad (4.8.21)$$

The second term reflects the discrete nature of collision processes: the shot noise. Its superscript is L instead of B in [4.1], the Landau collision integral here taking the place of the Boltzmann collision integral. In [4.1] here are a number of equivalent expressions for A_{ab}^{B}. Here we adopt the one which most clearly demonstrates the meaning of this contribution to the intensity of the Langevin source y_a,

$$A_{ab}^{L}(x, x', t) \equiv A_{ab}^{L}(r, t, p, p') \delta(r-r')$$

$$= [(\delta\hat{I}_a + \delta\hat{I}_b) - (\delta\hat{I}_a + \delta\hat{I}_b)_o]] n_a \delta_{ab} \delta(r-r') \delta(p-p') f_a(r, p, t). \quad (4.8.22)$$

The subscript o on the right-hand side indicates that the operators only act upon the distribution function f_a and not upon δ functions.

The "width" of the one-dimensional spatial δ function in kinetic theory is of the order of the corresponding physically infinitesimal length element l_{ph}. Let us recall that in the kinetic gas theory $l_{ph} \sim \sqrt{\varepsilon}l$ [Ref. 4.1, Eq. (16.3)], where $\varepsilon = nr_0^3$ is the density parameter, and l is the mean free path. For a plasma (4.1.9) $l_{ph} \sim r_D \sim \mu l$, and the number of particles in the physically infinitesimal volume is $nl_{ph}^3 \sim 1/\mu$; μ is the plasma parameter (4.1.1). Thus when $\mu \ll 1$ the number of particles in the physically infinitesimal volume is great but not infinitely great, as is assumed in conventional kinetic theory. Hence the result (4.8.22) is reached by taking into account the fact that the number of particles in a physically infinitesimal volume (also the width of δ function) is finite.

Using the definition of the operator $\delta\hat{I}_a$ (4.8.13) and the corresponding expression for the operator $\delta\hat{I}_b$, this result can be expressed in the form

$$A_{ab}^{L} = - \delta(r-r') \left\{ 2n_a n_b \frac{\partial^2}{\partial p_i \partial p_j'} D_{ij}^{ab}(p-p') f_a(r, p, t) f_b(r, p', t) \right.$$

$$+ \delta_{ab} n_a \left[\frac{\partial}{\partial p_i} D_{ij}^{a}(p) \frac{\partial}{\partial p_j} + \frac{\partial}{\partial p_i'} D_{ij}^{b}(p') \frac{\partial}{\partial p_j'} \right]$$

$$\times [\delta(p-p') - \delta_0(p-p')] f_a(r, p, t) \bigg\}, \quad (4.8.23)$$

where

$$D_{ij}^{a}(p) \equiv \sum_c n_c \int dp'' D_{ij}^{ac}(p-p'')$$

$$= 2 \sum_c e_a^2 e_c^2 n_c \int \frac{k_i k_j}{k^4} \delta(kv - kv'') f_c(r, p'', t) dp'' dk \quad (4.8.24)$$

is the diffusion coefficient in the Landau kinetic equation, and $D_{ij}^{ab}(p, p')$ is the local diffusion coefficient. The limits of integration over k are determined by the conditions (4.4.23). The subscript o to the δ function in (4.8.23) warns the operator not to act upon it.

In practical calculations model collision integrals are often used, e. g., the Batnagar – Gross – Krook collision integral [4.14 – 17].

Concerning the intensity of the random source in the equation for fluctuations of phase density of field oscillators (4.8.16), the intensity of the Langevin source $y_{k, \alpha}$ can again be presented as the sum of two parts

$$A_{k, \alpha, k', \alpha'}(X, X', t) = \tilde{A}_{k, \alpha, k', \alpha'} + A_{k, \alpha, k', \alpha'}^{F-P}. \qquad (4.8.25)$$

The first term is determined by the collision integral $\tilde{I}_{k, \alpha}$ [cf. (4.8.21)],

$$\tilde{A}_{k, \alpha, k', \alpha'} = \delta_{k, k'} \delta_{\alpha, \alpha'} \{ \delta(X - X') \tilde{I}_{k, \alpha}$$
$$- [\tilde{I}_{k, \alpha}(X, t) \, \mathscr{F}_{k', \alpha'}(X', t) + \tilde{I}_{k', \alpha'}(X', t) \, \mathscr{F}_{k, \alpha}(X, t)] \}. \qquad (4.8.26)$$

The second term has the superscript $F - P$, from the collision integral (4.6.9b) with the Fokker – Planck operator, instead of L in (4.8.20). Instead of (4.8.22) the result is

$$A_{k, \alpha, k', \alpha'}^{F-P} = 2 \delta_{kk'} \delta_{\alpha\alpha'} D_k \frac{\partial^2}{\partial P_{k, \alpha} \partial P_{k, \alpha}'} \delta(X - X') \, \mathscr{F}_{k, \alpha}(X, t). \qquad (4.8.27)$$

In the state of equilibrium $\tilde{I}_\alpha = 0$, and (4.8.21) becomes zero. The second term in (4.8.22) also vanishes, leading to

$$A_{ab}(x, x') = (\delta \hat{I}_a + \delta \hat{I}_b) n_a \delta_{ab} \delta(r - r') \delta(p - p') f_a(p) \qquad (4.8.28)$$

where $f_a(p)$ is the Maxwell distribution. This expression was first obtained by *Kadomtsev* [4.18].

In the state of equilibrium the collision integral $\tilde{I}_{k, \alpha}$ also equals zero, as does (4.8.26). Therefore we get

$$A_{k, \alpha, k', \alpha'}(X, X') = 2 \delta_{kk'} \delta_{\alpha\alpha'} D_k \frac{\partial^2}{\partial P_{k, \alpha} \partial P_{k, \alpha}'} \delta(X - X) \, \mathscr{F}_{k, \alpha}(X), \qquad (4.8.29)$$

where $\mathscr{F}_{k, \alpha}(X)$ is the Gauss distribution (4.6.14) with zero mean values and constant temperature. In particular, this leads to the expression (4.7.9) for the spectral density of the Langevin source in the field equations (4.7.7), when (4.6.10) is observed and the factor ck in the right-hand side of (4.7.7) is taken into account.

In general, in the nonequilibrium state the intensities of Langevin sources y_a and $y_{k, \alpha}$ cannot be linked to the one-particle distribution functions f_a and $\mathscr{F}_{k, \alpha}$. Indeed, the characteristic evolution times of functions f_a and $\mathscr{F}_{k, \alpha}$ [see (4.8.7)] and fluctuations [see (4.8.10, 11)] are usually of the same order. Therefore the collisions integrals $\tilde{I}_a$ and $\tilde{I}_{k,a}$ cannot be expressed through the one-particle distribution functions f_a and $\mathscr{F}_{k,a}$ without special allowances. The situation becomes significantly simpler if the kinetic stage of the relaxation process for one-particle distribution functions f_a and $\mathscr{F}_{k,a}$ is over and the distributions f_a and $\mathscr{F}_{k,a}$ approach those of local equilibrium, which for $\mathscr{F}_{k,a}$ is the Gauss distribution (4.6.14), and for the charged particles is the local Maxwell distribution

$$f_a(r, p, t) = A_a \exp\left(-\frac{[p - m_a u_a(r, t)]^2}{2m_a \mathscr{X} T_a}\right).$$ (4.8.30)

We shall call the stage of the relaxation process following the kinetic stage hydrodynamic. The corresponding relaxation times at the hydrodynamic stage will be designated by τ_a^H and τ_k^H. They are much greater than those at the kinetic stage, i.e.,

$$\tau_a^H \gg \tau_a, \quad \tau_k^H \gg \tau_k \equiv 1/\gamma_k,$$ (4.8.31)

enabling us to introduce two small parameters τ_a/τ_a^H and τ_k/τ_k^H. We shall next turn to the spectral densities of large-scale fluctuations in zero approximation in respect to these parameters.

4.9 Nonequilibrium Fluctuations of the Field

In order to calculate the correlations of field fluctuations we employ (4.8.15, 16). In a zero-order approximation with respect to the parameters τ_a/τ_a^H and τ_k/τ_k^H, the distribution functions f_a and $\mathscr{F}_{k,a}$ can be considered invariable.

From (4.8.15) for nonrelativistic plasma we find the equation for the Fourier component of current fluctuation,

$$\delta j_i(\omega, k) = \sigma_{ij}(\omega, k) \delta E_j(\omega, k) + \delta j_i^{\text{source}}(\omega, k).$$ (4.9.1)

The electrical conductivcity tensor introduced here is defined as

$$\delta_{ij}(\omega, k) = -i \sum_a e_a^2 n_a \int \frac{v_i \dfrac{\partial f_a}{\partial p_i}}{\omega - kv + i\delta \hat{I}_a} \, dp.$$ (4.9.2)

It is linked to the dielectric permittivity tensor by the equation

$$\varepsilon_{ij}(\omega, k) = \delta_{ij} + \frac{4\pi i}{\omega} \sigma_{ij}.$$ (4.9.3)

The operator $\delta \hat{I}_a$ in (4.9.2) is defined by (4.8.13).

The source of fluctuation in (4.9.1) is connected with the Langevin source

$$\delta j^{\text{source}}(\omega, k) = i \sum_a e_a \int \frac{v y_a(\omega, k, p)}{\omega - kv + i\delta \hat{I}_a} \, dp \, . \tag{4.9.4}$$

In an isotropic plasma the permittivity tensor can be presented in the form

$$\varepsilon_{ij} = \left(\delta_{ij} - \frac{k_i k_j}{k^a} \right) \varepsilon^{\perp}(\omega, k) + \frac{k_i k_j}{k^4} \varepsilon^{\parallel}(\omega, k) \tag{4.9.5}$$

and is, therefore, determined by the two functions $\varepsilon^{\parallel}$ and $\varepsilon^{\perp}$ for the longitudinal and transverse permittivities. In the "small-scale" region these functions are determined by (4.4.15, 5.3). For the domain of large-scale fluctuations we find, from (4.9.2, 3, 5),

$$\varepsilon^{\parallel}(\omega, k) = 1 + \sum_a \frac{4\pi e_a^2 n_a}{k^2 \omega} \int kv \frac{k \, \partial f_a / \partial p}{\omega - kv + i\delta \hat{I}_a} \, dp \tag{4.9.6}$$

$$\varepsilon^{\perp}(\omega, k) = 1 + \sum_a \frac{2\pi e_a^2 n_a}{k^2 \omega} \int [[k \times v] \times k] \frac{\partial f_a / \partial p}{\omega - kv + i\delta \hat{I}_a} \, dp \, . \tag{4.9.7}$$

When $\delta \hat{I}_a$ is replaced by Δ, these equations coincide with (4.4.15, 5.3).

Following from the Poisson equation, another useful expression for $\varepsilon^{\parallel}$ is

$$\varepsilon^{\parallel}(\omega, k) = 1 + \sum_a \frac{4\pi e_a^2 n_a}{k^2} \int \frac{k \, \partial f_a / \partial p}{\omega - kv + i\delta I_a} \, dp \, . \tag{4.9.8}$$

Comparing (4.9.6, 9.8), we find the helpful equation

$$\sum_a \frac{e_a^2 n_a}{\omega} \int kv \frac{k \, \partial f_a / \partial p}{\omega - kv + i\delta \hat{I}_a} \, dp = \sum_a e_a^2 n_a \int \frac{k \, \partial f_a / \partial p}{\omega - kv + i\delta \hat{I}_a} \, dp \, . \tag{4.9.9}$$

The validity of this equation for the region of small-scale fluctuations, when $\delta \hat{I}_a \rightarrow \Delta$ and $\Delta \rightarrow 0$, is evident. Generally this equation follows from the charge conservation law. For the Fourier components of fluctuations this law can be written in the form

$$\delta q(\omega, k) = \frac{1}{\omega} [k \, \delta j(\omega, k)] \, . \tag{4.9.10}$$

Another important equation following from this equation as applied to the source fluctuations is

$$\sum_a \frac{e_a}{\omega} \int kv \frac{y_a(\omega, k, p)}{\omega - kv + i\delta \hat{I}_a} \, dp = \sum_a e_a \int \frac{y_a(\omega, k, p)}{\omega - kv + i\delta \hat{I}_a} \, dp \, . \tag{4.9.11}$$

These equations play a significant role in the investigation of the properties of the collision integral $\tilde{I}_a$ determined by large-scale fluctuations (see [Ref. 4.1, Sect. 64]).

The spectral densities of field fluctuations can also be presented in the form (4.4.14, 5.1). This time, however, the expressions for the sources of the fluctuations will be different. In order to find these expressions we turn to the equations for smoothed field strengths (4.8.4), from which we obtain the equations for the Fourier components of fluctuations

$$i[k\,\delta\tilde{B}(\omega,k)] = -\frac{i\omega}{c}\,\delta\tilde{E}(\omega,k) + \frac{4\pi}{c}\,\delta j(\omega,k)\,,$$

$$i[k\,\delta\tilde{E}(\omega,k)] = \frac{i\omega}{c}\,\delta\tilde{B}(\omega,k)\,,$$

$$i(k\,\delta\tilde{B}(\omega,k)) = 0\,,$$

$$i(k\,\delta\tilde{E}(\omega,k)) = 4\pi\,\delta q(\omega,k) = \frac{4\pi k\,\delta j(\omega,k)}{\omega}\,. \tag{4.9.12}$$

In the last equation the law of charge conservation (4.9.10) is utilized.

From (4.9.1, 12) it follows that the sources in the field equations are determined by the source δj^{source}. It is convenient to use the relevant expression for δE^{source}, determined by

$$\delta E^{\text{source}}(\omega,k) = -\frac{4\pi i}{\omega}\,\delta j^{\text{source}}(\omega,k)\,. \tag{4.9.13}$$

From (4.9.4, 13) we find the connection between the source δE^{source} and the Langevin source y_a,

$$(\delta E^{\parallel})^{\text{source}} = \frac{4\pi k}{k^2}\sum_a e_a \int \frac{kv}{\omega}\,\frac{y_a(\omega,k,p)}{\omega - kv + i\delta\hat{I}_a}\,dp\,. \tag{4.9.14}$$

The right-hand side can also be written in another form by virtue of (4.9.11).

If we accept the local Maxwell distribution as f_a, then the second term with the zero subscript in (4.8.22) can be omitted. In this case the spectral density of the Langevin source is determined by the formula

$$(y_a y_b)_{\omega,k,p,p'} = \delta_{ab}n_a(\delta\hat{I}_a + \delta\hat{I}_b)\,\delta(p-p')\,f_a(p) \tag{4.9.15}$$

using the first expression in (4.8.18) for the correlation of the source $y_a(x,t)$.

From (4.9.14, 15) we find the spectral density of field fluctuations $(\delta E^{\parallel})^{\text{source}}$,

$$(\delta E^{\parallel}\delta E^{\parallel})^{\text{source}}_{\omega,k} = \sum_a \frac{(4\pi)^2 e_a^2 n_a}{k^2}\int\frac{(k\cdot v)^2}{\omega^2}\,2\,\text{Re}\left\{\frac{i f_a}{\omega - kv + i\delta\hat{I}_a}\right\}dp\,. \tag{4.9.16}$$

The corresponding expression for the spectral density of fluctuations $\delta E^\perp$ has the form

$$(\delta E^\perp \delta E^\perp)_{\omega,k}^{\text{source}} = \sum_a \frac{(4\pi)^2 e_a^2 n_a}{k^2} \int \frac{[k \times v]^2}{\omega^2} 2\,\text{Re} \left\{ \frac{i f_a}{\omega - kv + i \delta \hat{I}_a} \right\} dp \,. \tag{4.9.17}$$

By substituting $\delta \hat{I}_a \to \Delta$ and $\Delta \to 0$, these equations coincide with (4.4.18, 5.2), respectively. By virtue of (4.9.11), (4.9.16) can also be presented in another form by substituting $(k \cdot v)^2/\omega^2 \to 1$.

The results obtained for the spectral densities are presented in operator form. The explicit presentation of these functions is possible either with the aid of model collision integrals, or in the hydrodynamic approximation.

For the state of equilibrium, (4.9.16) can be written in the form of (4.4.19), and (4.9.17) in the form of (4.7.17), where $\gamma_k = 0$. The expressions for the imaginary parts of permittivities will, of course, then be different. They follow from (4.9.6, 7). This demonstrates once more the versatility of the form accepted here to present the spectral densities of the sources of field fluctuations in the equilibrium state. We shall refer to this matter again in Chap. 11.

5. Brownian Motion

This chapter starts with the review of the classical theory of Brownian movement, developed in the works of Langevin, Einstein, Smoluchowsky, Fokker, and Planck. These results are employed for a description of Brownian motion in nonlinear systems, which are divided into two classes. The first includes systems with nondissipative nonlinearity. The discussion is based on the Ginzburg – Landau equation with a Langevin source. The Brownian particles are represented by the parameters which characterize phase transitions of the second kind. The second class includes systems with dissipative nonlinearity. The discussion is based on the example of a self-oscillatory system.

The theory of large-scale (kinetic) fluctuations, developed in Chap. 4, is employed to describe fluctuations of distribution functions which satisfy the relevant Fokker – Planck and Einstein – Smoluchowsky equations.

5.1 The Langevin Equations

We have seen that the equations for fluctuations of phase density may be considered as Langevin equations. The intensity of random (Langevin) sources is determined by the distribution functions of particles and field oscillators, and the set of equations for fluctuations and distribution functions is closed.

In kinetic theory the Langevin method plays an auxiliary role. Nevertheless, the utilization of this method permits simplifying the calculation of nonequilibrium fluctuations and the construction of kinetic equations.

There is still another class of problems for which the Langevin source enters the equations of motion of the system to depict the fluctuation processes in the "environment." These problems arose from the study of the motion of small though macroscopic particles in a liquid. Later it was discovered that this kind of motion can take place in very different systems. It is in this general sense that we use the term "Brownian motion."

Considering the Langevin equation, we imagine a Brownian particle to be a small sphere of radius R. The Stokes force exerted on such a sphere whose motion corresponds to small Reynolds numbers is [5.1]

$$f = -m\gamma v, \quad \gamma = 6\pi\eta R/m, \tag{5.1.1}$$

where m is the mass of the sphere, and η is the dynamic viscosity coefficient. This is not the only force since the conventional equations of fluid dynamics are not even complete for the nonturbulent case. Let us illustrate this of the equations of gas dynamics.

Gas dynamic equations can be obtained from the kinetic Boltzmann equation. If the latter contains random sources, they will also appear in the equations of gas dynamics (i.e., if large-scale fluctuations are taken into account) ([5.2, 3] and [Ref. 5.4, Sect. 23]). They enter in the form of the additional terms $\delta\pi_{ij}$ and δS_i in the expressions for the viscous stress tensor and the thermal flux vector. If the deviations from local equilibrium are small, the spectral densities of these sources are determined by the expressions [5.2 – 4]

$$(\delta\pi_{ij}\delta\pi_{kl})_{\omega,k} = 2\eta\,\varkappa\,T\left(\delta_{il}\delta_{jk} + \delta_{ik}\delta_{jl} - \frac{2}{3}\delta_{ij}\delta_{kl}\right),$$

$$(\delta S_i\delta S_j)_{\omega,k} = 2\,\varkappa\,T^2\lambda\delta_{ij}\,, \tag{5.1.2}$$

where λ is the thermal conductivity coefficient.

The presence of the random source $\delta\pi_{ij}$ leads to the appearance of additional force in the equation of motion of the Brownian particle ([Ref. 5.3, Sect. 11.4] and [5.5]). As a result the following equation of motion is obtained:

$$\frac{dr}{dt} = v\,, \quad \frac{dp}{dt} + \gamma p = y(t)\,,$$

$$\langle y_i\rangle = 0\,, \quad \langle y_i(t)y_j(t')\rangle = 2D\delta_{ij}\delta(t-t')\,, \quad D = \gamma m\,\varkappa\,T\,. \tag{5.1.3}$$

In the next section we shall see that D is the diffusion coefficient in the space of momenta.

In this example the Langevin equation is linear. When applied to the study of fluctuation processes in self-oscillating systems and to phase transitions, the Langevin equations are nonlinear.

Let us carry out the transition from the Langevin equation to the corresponding kinetic equation, the Fokker – Planck equation. Of the various ways to do this [5.1, 6, 7], we shall employ that which is a natural consequence of the method of obtaining the kinetic equations presented in the foregoing chapters. This method is easily generalized for the case of a random source with a finite correlation time.

5.2 The Fokker – Planck Equation

We introduce the phase density of Brownian particles in six-dimensional space $x = (r,p)$ [cf. (2.1.5)],

$$N(x, t) = \sum_{1\leqslant i\leqslant N} \delta(x-x_i(t)) = \sum_{1\leqslant i\leqslant N} \delta(r-r_i(t))\,\delta(p-p_i(t))\,. \tag{5.2.1}$$

If the total number of Brownian particles is conserved, i.e.,

$$\int N(x, t)\,dx = N = \text{const}\,, \tag{5.2.2}$$

then the phase density satisfies the continuity equation

$$\frac{\partial N}{\partial t} + v\frac{\partial N}{\partial r} + \frac{\partial}{\partial p}[(-\gamma p + y)N] = 0. \tag{5.2.3}$$

The distribution function is connected with the mean value of phase density [cf. (2.11.1)],

$$nf(x, t) = \langle N(x, t)\rangle, \quad \frac{1}{V}\int f(x, t)dx = 1. \tag{5.2.4}$$

From (5.2.3) we find after averaging

$$\frac{\partial f}{\partial t} + v\frac{\partial f}{\partial r} = \frac{\partial}{\partial p}(\gamma p f) - \frac{1}{n}\frac{\partial}{\partial p_i}\langle y_i \delta N\rangle; \tag{5.2.5}$$

in the second term on the right-hand side it has been taken into account that

$$\langle yN\rangle = \langle y\delta N\rangle, \tag{5.2.6}$$

because $\langle y\rangle = 0$. In order to find the unknown function $\langle y\delta N\rangle$ we utilize the equation for δN,

$$\left(\frac{\partial}{\partial t} + v\frac{\partial}{\partial r}\right)\delta N - \frac{\partial}{\partial p}(\gamma p\,\delta N) = -\frac{\partial}{\partial p_j}(y_j nf) - \frac{\partial}{\partial p_j}(y_j\delta N - \langle y_j\delta N\rangle). \tag{5.2.7}$$

Since y is a δ-correlated random process ($\tau_{cor} = 0$), knowing the solution of (5.2.7) for time intervals smaller than all the other characteristic times is sufficient to find the correlation $\langle y_i\delta N\rangle$. Thanks to this, all but the first terms on both sides of (5.2.7) can be neglected. Then

$$\frac{\partial \delta N}{\partial t} = -\frac{\partial}{\partial p_j}(y_j nf),$$

whence

$$\delta N(x, t) = -\frac{\partial}{\partial p_j}\int_{-\infty}^{t} y_j(t')nf(x, t')dt'. \tag{5.2.8}$$

Using this solution and (5.1.3) to correlate the Langevin source, we find

$$-\frac{1}{n}\frac{\partial}{\partial p_i}\langle y_i\delta N\rangle = D\frac{\partial^2 f}{\partial p^2}. \tag{5.2.9}$$

As a result we get a closed equation fot the distribution function f, the Fokker – Planck equation

$$\frac{\partial f}{\partial t} + v \frac{\partial f}{\partial r} = D \frac{\partial^2 f}{\partial p^2} + \frac{\partial}{\partial p} (\gamma p f) \equiv I. \tag{5.2.10}$$

Drawing an analogy to gas theory we shall call I the collision integral.

The Maxwell distribution

$$f(p) = \frac{1}{(2\pi m \,\mathscr{K} T)^{3/2}} \exp \left(- \frac{p^2}{2m \,\mathscr{K} T} \right) \tag{5.2.11}$$

should serve as the equilibrium solution of the Fokker–Planck equation (5.2.10). The latter is satisfied if the diffusion coefficient is determined by the Einstein formula

$$D = \gamma m \,\mathscr{K} T. \tag{5.2.12}$$

Let us pay attention to the fact that the intensity of the source representing the influence of the thermal motion of the environment is determined by two factors: the temperature and the dissipative coefficient. In Sect. 5.9 we shall see that the intensity of noise in a nonlinear dissipative system is a variable since it depends upon the velocity.

The Fokker–Planck equation (5.2.10) corresponds to the approximation of the δ-correlated random source in the Langevin equation. The method used here to obtain the Fokker–Planck equation allows deriving a more general equation which accounts for the finiteness of correlation time τ_{cor} of a random source (cf. [5.3, 8]). For example, let

$$\langle y_i(t) y_j(t') \rangle = B \delta_{ij} \exp \left(- \frac{|\tau|}{\tau_{\text{cor}}} \right). \tag{5.2.13}$$

At $\tau_{\text{cor}} \to 0$ this expression becomes (5.1.3) with

$$D = B \tau_{\text{cor}}. \tag{5.2.14}$$

In the first approximation with respect to $\tau_{\text{cor}} \gamma$ on the right-hand side of (5.2.10), an additional term appears,

$$- D \tau_{\text{cor}} \frac{\partial}{\partial p_i} \left(\frac{\partial}{\partial t} + v \frac{\partial}{\partial r} - \frac{\partial}{\partial p} \gamma p \right) \frac{\partial f}{\partial p_i}. \tag{5.2.15}$$

Thus we notice that taking the finiteness of the parameter $\tau_{\text{cor}} \gamma$ into account results in the emergence of higher derivatives in the Fokker–Planck equation (5.2.10). This equation holds for the zero-order approximation in respect to $\tau_{\text{cor}} \gamma$, which is a necessary, though not sufficient condition.

Indeed, in searching for the solution of (5.2.7) we have dropped nonlinear fluctuation terms proportional to the product of y and δN, which implies assuming that the fluctuations are small. This assumption is justified provided that the concentration of Brownian particles is great enough for the physically in-finitesimal volume l_{ph}^3 to be introduced. The magnitude of this volume can be estimated from a condition similar to (4.1.8).

Finally, there is one more important condition. When solving the equation for δN we neglected the initial correlations on all scales, thus assuming that large-scale correlations (with $\tau_{cor} > 1/\gamma$) are of little import. In Sect. 5.15 we shall see that a theory of large-scale fluctuations in Brownian motion can be developed in a way similar to the kinetic theory of fluctuations in Sect. 4.8 (see also [Ref. 5.4, Chaps. 4, 11]).

5.3 Diffusion of Brownian Particles

Using the Fokker – Planck equation we write the transfer equations for particle density, momentum density and mean kinetic energy density,

$$\frac{\partial \rho}{\partial t} + \frac{\partial \rho u}{\partial r} = 0 , \tag{5.3.1}$$

$$\frac{\partial \rho u_i}{\partial t} + \frac{\partial \rho u_i u_j}{\partial r_j} = - \frac{\partial p}{\partial r_i} - \frac{\partial \pi_{ij}}{\partial r_j} - \gamma \rho u_i , \tag{5.3.2}$$

$$\frac{\partial}{\partial t} \left(\frac{\rho u^2}{2} + \frac{3}{2} \frac{\rho}{m} \mathscr{K} T_B \right) + \frac{\partial}{\partial r_i} \left[u_i \left(\frac{\rho u^2}{2} + \frac{3}{2} \frac{\rho}{m} \mathscr{K} T_B + p \right) + \pi_{ij} + S_i \right]$$
$$= 3 \gamma \rho \left[\frac{\mathscr{K} T}{m} - \left(\frac{u^2}{3} + \frac{\mathscr{K} T_B}{m} \right) \right] \tag{5.3.3}$$

where T_B is the temperature, and $p = (\rho/m) \mathscr{K} T_B$ is the pressure of Brownian particles. The essential terms in the right-hand side of (5.3.2, 3) are moments of the collision integral in the Fokker – Planck equation [see with transfer equations in [Ref. 5.4, Sect. 7]).

From the structure of the transfer equations it follows that for time spans

$$t - t_0 \geq 1/\gamma , \tag{5.3.4}$$

the temperature of Brownian particles

$$T_B = T , \tag{5.3.5}$$

and the momentum density is determined by the density gradient

$$\rho u = - \frac{\mathscr{K} T}{\gamma m} \frac{\partial p}{\partial r} . \tag{5.3.6}$$

Substituting this expression into (5.3.1) we get the equation for the diffusion of Brownian particles,

$$\frac{\partial \rho}{\partial t} = \mathscr{D} \, \Delta_r \rho \tag{5.3.7}$$

or the Einstein – Smoluchowsky equation. Here

$$\mathscr{D} = \frac{\mathscr{K} T}{m \gamma} \tag{5.3.8}$$

is the diffusion coefficient in conventional space.

The solution of the diffusion equation having the initial condition

$$\rho(r, t_0) = mN \delta(r - r_0) , \tag{5.3.9}$$

has the form

$$\rho(r, t) = \frac{mN}{[4 \pi \, \mathscr{D} \, (t - t_0)]^{3/2}} \exp \left[- \frac{(r - r_0)^2}{4 \, \mathscr{D} \, (t - t_0)} \right] . \tag{5.3.10}$$

From this follows, in particular, the expression for the mean square displacement of Brownian particles from the starting point

$$\langle (r - r_0)^2 \rangle = 6 \, \mathscr{D} \, (t - t_0). \tag{5.3.11}$$

Naturally, the diffusion equation has, in turn, a corresponding Langevin equation,

$$\frac{dr}{dt} = \frac{y(t)}{m \gamma} = y_r(t) ,$$

$$\langle y_r \rangle = 0 , \quad \langle y_r(t) y_r(t') \rangle = 2 \, \mathscr{D} \, \delta(t - t') , \tag{5.3.12}$$

which follows from (5.1.3) and the condition

$$\left| \frac{dp}{dt} \right| \ll \gamma |p| . \tag{5.3.13}$$

The Einstein – Smoluchowsky equation can be obtained from the Langevin equation (5.3.12) in the same way as the Fokker – Planck equation (5.2.10) from the Langevin equation (5.1.3).

Brownian motion of particles is alternating. The frequencies of the spectral components of this motion fall within the region

$$\omega \lesssim \gamma . \tag{5.3.14}$$

If the Brownian motion of a particle is harmonic with frequency ω, then the force acting on the particle from the side of the liquid is determined by a far more complicated expression for the Stokes force than (5.1.1) [Ref. 5.9, p. 120]. It only coincides with (5.1.1) for lower frequencies, $\omega \ll \gamma$; this condition is only satisfied in the diffusion approximation (5.3.4). Hence it follows that the Fokker – Planck equation (5.2.10) obtained from the Langevin equation (5.1.3) should be considered just as a model equation, since using (5.1.1) for the force cannot be justified for small times $(t - t_0 \lesssim 1/\gamma)$. In this sense the Einstein – Smoluchowsky equation seems to be more consistent.

5.4 The Brownian Motion of a Harmonic Oscillator. The Nyquist Formula

The Brownian motion of a one-dimensional harmonic oscillator is described by a set of Langevin equations

$$\frac{dx}{dt} = v , \quad \frac{dp}{dt} + \gamma p + m \omega_0^2 x = y(t) \tag{5.4.1}$$

where γ is the coefficient of friction, and ω_0 the eigenfrequency. The correlation of the random source is determined by the expression

$$\langle y \rangle = 0 , \quad \langle y(t) y(t') \rangle = 2 D \delta(t - t') , \tag{5.4.2}$$

similar to (5.1.3).

The Fokker – Planck equation for the distribution function $f(x, p, t)$ can be obtained in the same way as (5.2.10); it has the form

$$\frac{\partial f}{\partial t} + v \frac{\partial f}{\partial r} - m \omega_0^2 x \frac{\partial f}{\partial p} = D \frac{\partial^2 f}{\partial p^2} + \frac{\partial}{\partial p} (\gamma p f) . \tag{5.4.3}$$

The equilibrium solution of this equation is

$$f(x, p) = C \exp \left(- \frac{p^2 + m^2 \omega_0^2 x^2}{2 m \mathcal{K} T} \right) . \tag{5.4.4}$$

Substituting this into (5.4.3) brings us to Einstein's formula again,

$$D = \gamma m \mathcal{K} T . \tag{5.4.5}$$

To apply these results to an electric oscillatory circuit one should avail oneself of an electromechanical analogy. Then y corresponds to the random emf $\mathcal{E}$; m to the inductance L, γ to the ohmic resistance R divided by L. Taking all this into account, we find the correlation of the emf from (5.4.2, 5),

$$\langle \mathscr{E}(t) \mathscr{E}(t') \rangle = 2R \mathscr{K}T\delta(t-t') \equiv 2D_\varepsilon\delta(t-t') , \qquad (5.4.6)$$

or the Nyquist formula. Hence the expression for D_ε follows, similar to Einstein's formula (5.4.5). A solution which describes the evolution of the distribution function at arbitrary initial conditions [5.1, 9] can be found for (5.4.3).

Let us now discuss the examples of Brownian motion described by nonlinear Langevin equations.

5.5 Nondissipative Nonlinearity. Brownian Motion at Phase Transitions [1]

5.5.1 Fokker – Planck and Einstein – Smoluchowsky Equations

Consider the Langevin equation where, different to (5.4.1), the elastic force $F(x)$ is nonlinear,

$$\frac{dx}{dt} = v , \quad \frac{dp}{dt} + \gamma p = F(x) + y(t) . \qquad (5.5.1)$$

The statistical properties of the random source y are also determined by (5.4.2). The function $F(x)$ can be presented in the form

$$F(x) = - \frac{du(x)}{dx} + F , \qquad (5.5.2)$$

where F is the external force. In this example we set the function u in the form

$$u(x) = \alpha x^2/2 + \beta x^4/2 , \quad \beta > 0 . \qquad (5.5.3)$$

For a harmonic oscillator

$$\alpha = m\omega_0^2 , \quad \beta = 0 , \qquad (5.5.4)$$

and (5.5.1) coincides with (5.4.1).

The Fokker – Planck equation corresponding to the Langevin equations (5.5.1) has the form

$$\frac{\partial f}{\partial t} + v \frac{\partial f}{\partial x} + \left(F - \frac{\partial u}{\partial x} \right) \frac{\partial f}{\partial p} = D \frac{\partial^2 f}{\partial p^2} + \frac{\partial}{\partial p} (\gamma p f) . \qquad (5.5.5)$$

---

[1] In Sects. 5.5 – 7 we shall discuss a model of phase transition using the example of a system with one degree of freedom. Certain questions concerning the statistical theory of phase transitions in macroscopic bodies will be treated in Sect. 5.8, as well as in Chaps. 6, 13.

The equilibrium solution of this equation with $F = 0$ is the Maxwell–Boltzmann distribution

$$f(x, v) = c \exp\left(- \frac{p^2/2m + u(x)}{\mathcal{X} T}\right). \tag{5.5.6}$$

Substituting this expression into (5.5.5) brings us back to Einstein's formula (5.4.5).

We see that for a nonlinear system with nondissipative nonlinearity [arbitrary function $u(x)$] the intensity of random movement caused by thermal motion is determined according to (5.4.5) by constants γ and T, and does not depend on the form of the function $u(x)$.

The distribution pattern of the coordinates in (5.5.6) reveals a strong dependence, the sign of α. If $\alpha > 0$ the distribution has a maximum at $x = 0$, while if $\alpha < 0$ there are two maxima at the points

$$x_0 = \pm \sqrt{|\alpha|/\beta} . \tag{5.5.7}$$

The point $\alpha = 0$ can be considered a point of phase transition.

From (5.5.5) we may again reach the Einstein–Smoluchowsky equation for a simpler distribution function $f(x, t)$. We start with the approximation (5.3.13) in the Langevin equations (5.5.1). In place of the latter we then get a single equation,

$$\frac{dx}{dt} + (a + bx^2)x = F_0 + y(t) , \tag{5.5.8}$$

where

$$a = \alpha/m\gamma, \quad b = \beta/m\gamma, \quad F_0 = F/m\gamma,$$

$$\langle y \rangle = 0 , \quad \langle y(t)y(t') \rangle = 2D\delta(t - t') , \quad D = \mathcal{X} T/m\gamma . \tag{5.5.9}$$

The corresponding Einstein–Smoluchowsky equation now has the form

$$\frac{\partial f(x, t)}{\partial t} = D \frac{\partial^2 f}{\partial x^2} + \frac{\partial}{\partial x} \{[(a + bx^2)x - F_0] f\} . \tag{5.5.10}$$

5.5.2 The Equilibrium Distribution

The equilibrium solution for $F_0 = 0$ is given by the Boltzmann distribution, which is conveniently written in the form

$$f = \frac{1}{Z} \exp\left(- \frac{Ax^2 + Bx^4}{\mathcal{X} T}\right), \quad A = \frac{m\gamma a}{2} = \frac{\alpha}{2} ,$$

$$B = \frac{m \gamma b}{4} = \frac{\beta}{4},$$ (5.5.11)

where

$$Z = 2 \int_0^\infty \exp \left(- \frac{A x^2 + B x^4}{\mathcal{X} T} \right) dx = \sqrt{\pi} \left(\frac{\mathcal{X} T}{2B} \right)^{1/4} e^{y^2/4} D_{-1/2}(y),$$

$$y = \frac{A}{\sqrt{2 \mathcal{X} T B}},$$ (5.5.12)

and $D_\nu(y)$ is the parabolic cylinder function.

The formulae for the first moments are

$$\langle x \rangle = 0,$$

$$\langle x^2 \rangle \equiv \langle E \rangle = \frac{1}{2} \sqrt{\frac{\mathcal{X} T}{2B}} \frac{D_{-3/2}(y)}{D_{-1/2}(y)}, \quad y = \frac{A}{\sqrt{2 \mathcal{X} T B}}.$$ (5.5.13)

There are three interesting special cases:

1) The linear harmonic oscillator ($A > 0$, $B = 0$),

$$\langle E \rangle \equiv \langle x^2 \rangle = \frac{\mathcal{X} T}{2A} = \frac{\mathcal{X} T}{\alpha} = \frac{\mathcal{X} T}{m \omega_0^2} = \frac{D}{a}.$$ (5.5.14)

2) The "phase transition point" ($A = 0$),

$$\langle E \rangle = \frac{\Gamma(3/4)}{\Gamma(1/4)} \sqrt{\frac{\mathcal{X} T}{B}} \approx \frac{1}{3} \sqrt{\frac{\mathcal{X} T}{B}} = \frac{2}{3} \sqrt{\frac{D}{b}}.$$ (5.5.15)

3) The system finds itself far below the transition point. From (5.5.13) at $y \gg 1$,

$$\langle E \rangle = \frac{|A|}{2B} = \frac{|\alpha|}{b} \equiv \langle E \rangle_0.$$ (5.5.16)

Now we introduce a dimensionless parameter

$$\varepsilon = \mathcal{X} T / |A|^2 = D b / |a|^2;$$ (5.5.17)

for the third case $\varepsilon \ll 1$. From the above equations it follows that

$$\langle E \rangle_2 / \langle E \rangle_3 \sim \sqrt{\varepsilon} \ll 1 \quad \text{at} \quad \varepsilon \ll 1,$$

and consequently the energy $\langle E \rangle$ is much greater below the transition point than at it.

Consider the equations for moments $\langle x \rangle$ and $\langle x^2 \rangle$. They follow from (5.5.10) and have the form

$$\frac{d \langle x \rangle}{dt} + \langle (a + bx^2)x \rangle = F_0 ,$$

$$\frac{d}{dt} \frac{\langle x^2 \rangle}{2} + \langle (a + bx^2)x^2 \rangle = D + \langle x \rangle F_0 . \qquad (5.5.18)$$

This set of equations is not closed since it includes the moments of higher order along with the two first moments. Let us discuss certain results of the approximate solution of this set of equations. First of all we calculate the fluctuations of "energy" $E = x^2$ at $F_0 = 0$ in the stationary case.

5.5.3 Energy Fluctuations

From the Langevin equation (5.5.8) we find the equation for E. At $F_0 = 0$ it has the form

$$\frac{d}{dt} \frac{E}{2} + (a + bE)E = xy(t) \equiv y_E(t) . \qquad (5.5.19)$$

The mean value of the random source y_E is nonzero, and can be found if we average (5.5.19) and compare the result with the second equation of (5.5.18). We get

$$\langle y_E \rangle = D . \qquad (5.5.20)$$

Now we write the corresponding equation for fluctuation δE,

$$\frac{d\delta E}{dt} + \lambda_E \delta E = 2 \delta y_E \qquad (5.5.21)$$

with designations

$$\lambda_E = 2(a + 2b \langle E \rangle) \equiv \Delta \omega_E \qquad (5.5.22)$$

and

$$\langle \delta y_E \rangle = 0 , \quad \langle \delta y_E(t) \delta y_E(t') \rangle = 2 \langle E \rangle D \delta(t - t') . \qquad (5.5.23)$$

With the aid of (5.5.21) we find the distribution with respect to frequencies

$$\langle (\delta E)^2_\omega \rangle = \frac{8 \langle E \rangle D}{\omega^2 + (\Delta \omega_E)^2} , \qquad (5.5.24)$$

whence

$$\langle(\delta E)^2\rangle = 4\langle E\rangle D/\Delta\,\omega_E\,.\tag{5.5.25}$$

From (5.5.24) it follows that the magnitude

$$\lambda_E \equiv \Delta\,\omega_E\tag{5.5.26}$$

determines the halfwidth of the energy fluctuation spectrum, and that the energy, in turn, is determined by its mean. The value of $\Delta\,\omega_E$ can be found for all the values of the parameters from (5.5.13); e.g., for special cases 1) and 3),

1) $\Delta\,\omega_E = 2a$,

3) $\Delta\,\omega_E = 2b\langle E\rangle_0 = 2|a|.$ (5.5.27)

The formula $\langle E\rangle_0 = |a|/b$ describes the energy in the zero-order approximation with respect to parameter ε. The principal contribution to the dispersion of energy fluctuations with respect to ε is given

$$\langle(\delta E)^2\rangle = 2D/b\,,\tag{5.5.28}$$

which follows from (5.5.25). In order to find the first-order correction to $\langle E\rangle_0$, we utilize the second equation of (5.5.18) for the stationary state $(F = 0)$. Substituting

$$E = \langle E\rangle + \delta E = \langle E\rangle_0 + \langle E\rangle_1 + \delta E$$

and using the (5.5.28), we find

$$\langle E\rangle_1 = -\tfrac{1}{2}\langle(\delta E)^2\rangle/\langle E\rangle_0 = -D/b\langle E\rangle_0\,.\tag{5.5.29}$$

5.5.4 Self-Consistent Approximation with Respect to Energy

Once more we consider the stationary state and set $F_0 = 0$. Then all odd-numbered moments are zero, and the first equation of (5.5.18) becomes an identity.

The second equation of (5.5.18) can be replaced, with accuracy up to second-order terms with respect to ε, by a closed equation for the function

$$\langle E^*\rangle = \langle E\rangle_0 + \langle E\rangle_1 + \langle(\delta E)^2\rangle/\langle E\rangle_0\tag{5.5.30}$$

which represents the mean energy, renormalized for the contribution of fluctuations. This equation has the form

$$(a + b\langle E^*\rangle)\langle E^*\rangle = D\tag{5.5.31}$$

and is closed for the function $\langle E^* \rangle$, the mean energy, renormalized for fluctuations. For our three special cases we find from (5.5.31) [cf. (5.5.15)]

1) $\langle E^* \rangle = D/a = \mathcal{K}T/m\omega_0^2$,

2) $\langle E^* \rangle = \sqrt{D/b}$,

3) $\langle E^* \rangle = |a|/b$.   (5.5.32)

5.5.5 The Order Parameter

In the state of equilibrium and in the absence of an external field the distribution function is described by (5.5.11), and, consequently, the mean value of $\langle x \rangle$ equals zero.

At $a < 0$ the potential energy has two minima at the points $x_0 = \pm\sqrt{|a|/b}$. The probability of each is equal, and they are therefore degenerate. The degeneration, naturally, disappears in the presence of an external field.

After *Bogolyubov* [5.11], in addition to conventional mean values so-called quasi-mean values can be used; these will be designated by $\langle\ \rangle_{F=0}$. This indicates that in order to calculate a quasi-mean value the calculation is first out at $F \neq 0$, and then a limit transition is made. If the force is nonzero, we shall designate the corresponding mean by $\langle\ \rangle_F$.

Let us find the quasi-means for the cases 1) and 3). The function $f(x)$ in 1) is the Gauss distribution, so

$$\langle x \rangle_{F=0} = \langle x \rangle = 0,$$ (5.5.33)

and the quasi-mean coincides with the conventional mean value.

In case 3) (zero approximation with respect to ε) fluctuations are absent, therefore in the first equation of (5.5.18) we use the first moments approximation. As a result we get the equation

$$(a + b\langle x \rangle_F^2)\langle x \rangle_F = F$$ (5.5.34)

(omitting subscript "0" by F_0). This equation has two solutions at $F = 0$,

$$\langle x \rangle = \pm\sqrt{|a|/b}.$$ (5.5.35)

The quasi-mean is the one which is obtained via the limit transition $\lim_{F \to 0} \langle x \rangle_F$ and lies toward F. In (5.5.35) this corresponds to a plus sign. We shall call this value the order parameter and designate it by η. Therefore

$$\eta = \langle x \rangle_{F=0} = \lim_{F \to 0}\langle x \rangle_F \equiv \lim_{F \to 0}\eta_F,$$ (5.5.36)

and in zero approximation in ε

$$\eta = \sqrt{|a|/b}.$$ (5.5.37)

5.5.6 Fluctuations of the Order Parameter

In the Langevin equation (5.5.8) we assume that

$$x = \eta + \delta x_\eta \equiv \eta + \delta \eta, \qquad (5.5.38)$$

and consider a linear approximation with respect to $\delta \eta$. As a result we get

$$\frac{d\delta \eta}{dt} + (a + 3b\eta^2)\delta \eta = y. \qquad (5.5.39)$$

Using (5.5.9) to correlate the Langevin source, we find the spectral density of the fluctuations of the order parameter,

$$(\delta \eta)^2_\omega = 2D[\omega^2 + (\Delta \omega)^2_\eta]^{-1} \qquad (5.5.40)$$

where we use the following designation [cf. (5.5.22) for $\Delta \omega_E$]:

$$\Delta \omega_\eta = a + 3b\eta^2 = 2b\eta^2 = 2|a|. \qquad (5.5.41)$$

With the aid of the spectral distribution we find the expression for the mean-square fluctuations of the order parameter,

$$\langle (\delta \eta)^2 \rangle = \frac{D}{a + 3b\eta^2} = \frac{D}{2|a|} = \frac{D}{2b\eta^2} \sim \varepsilon \eta^2. \qquad (5.5.42)$$

Formula (5.5.37) determines the value of the order parameter in zero-order approximation with respect to ε, see (5.5.17), that is, without taking fluctuations into account. From this formula it follows that the order parameter equals zero at the critical point ($a = 0$). It is then natural to expect that the fluctuations will destroy the ordered state earlier, i.e., at temperatures below the critical one. The limits of this domain can be estimated from the condition $\varepsilon \sim 1$. Since the parameter ε is determined by (5.5.17), order is destroyed at

$$|a| \sim \sqrt{Db}. \qquad (5.5.43)$$

This relation defines that limit above which description is only possible by conventional mean values, i.e.,

$$\eta = 0, \quad \langle (\delta \eta)^2 \rangle = \langle (\delta x)^2 \rangle = \langle x^2 \rangle \equiv \langle E \rangle, \qquad (5.5.44)$$

and where the mean-square value of fluctuations of the order parameter coincides with the mean value $\langle E \rangle$.

5.6 Spectral Distribution of the Mean Energy

Using the Einstein–Smoluchowsky equation (5.5.10) as the equation for the two-time distribution function $f(x, t, x', t')$, we obtain the equation for the two-time moment of function $x(t)$. The following designations are practical:

$$E(\tau) = x(t)x(t-\tau) , \quad \langle E(\tau)\rangle = \langle x(t)x(t-\tau)\rangle . \tag{5.6.1}$$

Hence it follows that

$$\langle E(0)\rangle = \langle E\rangle = \frac{1}{2\pi} \int_{-\infty}^{\infty} E(\omega)d\omega \equiv \frac{1}{2\pi} \int_{-\infty}^{\infty} (x^2)_\omega d\omega \tag{5.6.2}$$

where $E(\omega)$ is the mean energy distribution with respect to frequencies.

The equation for the function $\langle E(\tau)\rangle$ has the form

$$\frac{d}{d\tau}\langle E(\tau)\rangle + \langle (a+bE)E(\tau)\rangle = 0 . \tag{5.6.3}$$

This equation is not closed. The simplest way to unlink the chain is to express the second moment as a product of the first moments. As a result we get

$$\frac{d}{d\tau}\langle E(\tau)\rangle + (a+b\langle E\rangle)\langle E(\tau)\rangle = 0 \tag{5.6.4}$$

to which we must add an equation for the function $\langle E\rangle$. In the same approximation it follows from the second equation of (5.5.18) that

$$(a+b\langle E\rangle)\langle E\rangle = D \tag{5.6.5}$$

in the stationary case at $F = 0$, which coincides with (5.5.31). This reveals that the first moments approximation employed here leads to renormalization of the mean energy, which differs from the mean value $\langle E\rangle_0$ in that the contribution of fluctuational energy is taken into account (5.5.30). Consequently, the functions $\langle E(\tau)\rangle$ and $\langle E\rangle$ in (5.6.4, 5) should be distinguished by an asterisk. We shall omit it so as not to complicate the designations, although it should always be borne in mind.

By dint of (5.6.5), equation (5.6.4) can be written in a simpler form,

$$\frac{d\langle E(\tau)\rangle}{d\tau} + \lambda\langle E(\tau)\rangle = 0 , \quad \lambda = \frac{D}{\langle E\rangle} . \tag{5.6.6}$$

With its help we can find the spectral distribution of the mean energy,

$$E(\omega) \equiv (x^2)_\omega = \frac{2\lambda\langle E\rangle}{\omega^2 + \lambda^2} = \frac{2D}{\omega^2 + (\Delta\omega)^2} . \tag{5.6.7}$$

The halfwidth of this distribution with respect to frequencies is

$$\Delta\omega \equiv \lambda = \frac{D}{\langle E \rangle} = \frac{\mathcal{K}T}{m\gamma\langle x^2 \rangle}. \tag{5.6.8}$$

We can compare the spectral linewidths of the mean energy spectral distribution with the distribution of fluctuations of the order parameter by using (5.5.32, 41, 6.8). Below the critical point

$$\frac{\Delta\omega}{\Delta\omega_\eta} \sim \frac{Db}{a^2} = \varepsilon, \tag{5.6.9}$$

and hence at $a < 0$ and $\varepsilon \ll 1$, the spectrum of mean energy distribution is much narrower than the spectrum of fluctuations of the order parameter.

In this approximation the Langevin equation for the function $x(t)$ (to be more precise, x^*) takes the form

$$\frac{dx}{dt} + \lambda x = y(t), \quad \lambda = \frac{D}{\langle E \rangle}. \tag{5.6.10}$$

The ratio of relaxation times in (5.5.39, 6.10) is inverse to that of linewidths, therefore at $\varepsilon \ll 1$ the fluctuations of the order parameter are fast compared to fluctuations δx.

Let us finally note that the relative fluctuations of energy at $\varepsilon \ll 1$ are described by

$$\frac{\langle (\delta E)^2 \rangle}{\langle E \rangle^2} \sim \frac{Db}{a^2} = \varepsilon. \tag{5.6.11}$$

We have discussed three types of fluctuations and how, at $\varepsilon \ll 1$, to distinguish between fast ($\delta\eta, \delta E$) and slow (δx) ones.

5.7 The Response of the System to External Factors

According to the fluctuation dissipation theorem (FDT), the responses of a system are related to the corresponding spectral densities of fluctuations (Chap. 11). Since there are different types of fluctuations in the system under consideration at $\varepsilon \ll 1$, e.g., fast fluctuations $\delta\eta$ and δE and slow ones δx, one can introduce three respective susceptibilities χ_η, χ_E, χ. Naturally, in the region far above the phase transition, when $a > 0$ and $b = 0$, the susceptibilities χ and χ_η must coincide, and in the region far below the transition point $\chi_E/\chi \sim \varepsilon$ and $\chi_\eta/\chi \sim \varepsilon$.

First we find the static responses

$$\langle x \rangle_F, \quad \eta_F - \eta, \quad \langle E \rangle_F - \langle E \rangle$$

by turning to (5.5.18). In the linear approximation for the stationary state we obtain three equations

$$(a + b\langle E\rangle)\langle x\rangle_F = F, \quad (a + 2b\langle E\rangle)(\langle E\rangle_F - \langle E\rangle) = \langle x\rangle_F F,$$

$$(a + 3b\eta^2)(\eta_F - \eta) = F, \tag{5.7.1}$$

from which we derive three expressions for the static susceptibilities,

$$\chi = \frac{1}{\Delta\omega} \equiv \frac{1}{\lambda}, \quad \Delta\omega = \frac{D}{\langle E\rangle}, \tag{5.7.2}$$

$$\chi_E = \frac{1}{\Delta\omega_E} \equiv \frac{1}{\lambda_E}, \quad \Delta\omega_E = 2(a + 2b\langle E\rangle), \tag{5.7.3}$$

$$\chi_\eta = \frac{1}{\Delta\omega_\eta} \equiv \frac{1}{\lambda_\eta}, \quad \Delta\omega_\eta = a + 3b\eta^2. \tag{5.7.4}$$

Let us consider each of these functions for the three particular cases discussed previously. We start with χ_η and from (5.7.4) find

1) $\chi_\eta = \dfrac{1}{a}$,

2) $\chi_\eta = \infty$,

3) $\chi_\eta = \dfrac{1}{2|a|}$. $\tag{5.7.5}$

We see that the susceptibility increases according to Curie's law approaching the critical point both from above and from below if we assume that

$$a = a_0\tau, \quad \tau = (T - T_C)/T_C. \tag{5.7.6}$$

The three corresponding expressions for χ_E are

1) $\chi_E = (2a)^{-1}$,

2) $\chi_E \sim (\sqrt{Db})^{-1}$,

3) $\chi_E = (2|a|)^{-1}$. $\tag{5.7.7}$

The susceptibility curve here is symmetrical with respect to the critical point, where it has a finite value.

Finally, for χ we get

1) $\chi = \dfrac{1}{a}$,

2) $\chi \sim (\sqrt{Db})^{-1}$,

3) $\chi = \left(\dfrac{Db}{|a|}\right)^{-1}$. $\qquad\qquad\qquad\qquad\qquad\qquad$ (5.7.8)

The susceptibilities introduced here are dimensional since we take the response to the force $F_0 = F/m\gamma$ into consideration. In order to achieve dimensionless functions one has to multiply all the expressions for susceptibilities by $m\gamma$.

5.7.1 Dynamic Response. Spectral of Fluctuations

If the external force is a periodic function of time with frequency ω, we can obtain the following expressions for dynamic susceptibilities from the first equation of (5.5.18):

$$\chi(\omega) = \frac{1}{-i\omega + \lambda}, \quad \chi_E(\omega) = \frac{1}{-i\omega + \lambda_E}, \quad \chi_\eta(\omega) = \frac{1}{-i\omega + \lambda_\eta}. \qquad (5.7.9)$$

In order to find the corresponding spectral densities of fluctuations δx, δE, and $\delta \eta$, we employ the fluctuation dissipation theorem (FDT) (Chap. 11),

$$(x^2)_\omega = \frac{2\mathcal{K}T}{\omega} \operatorname{Im}\{\alpha(\omega)\}. \qquad\qquad\qquad\qquad\qquad (5.7.10)$$

Another form of this expression is more convenient for our purpose. We obtain it by replacing $\mathcal{K}T$ with $m\gamma D$, and $\operatorname{Im}\{\alpha(\omega)\}$ with $\operatorname{Im}\{\chi(\omega)\}m\gamma$. Then,

$$(x^2)_\omega = \frac{2}{\omega} \operatorname{Im}\{\chi(\omega)\}D. \qquad\qquad\qquad\qquad\qquad (5.7.11)$$

Using the expressions for susceptibilities $\chi(\omega)$ and $\chi(0)$, we get

$$(x^2)_\omega = \frac{2D}{\omega^2 + \lambda^2}, \quad \langle x^2 \rangle = \frac{1}{2\pi}\int (x^2)_\omega d\omega = \frac{D}{\lambda} = \chi(0)D. \qquad (5.7.12)$$

Equation (5.7.11) can also be employed to find the spectral densities of fluctuations δE and $\delta \eta$ if the corresponding intensities of Langevin sources are taken for D, which we designate by D_E and D_η. From (5.5.23, 39) it follows that

$$D_E = 4\langle E \rangle D, \quad D_\eta = D. \qquad\qquad\qquad\qquad\qquad (5.7.13)$$

From these expressions, as well as those for the susceptibilities χ_E and χ_η and the FDT, we find the desired expressions,

$$(\delta E)_\omega^2 = \frac{8\langle E \rangle D}{\omega^2 + \lambda_E^2}, \quad \langle(\delta E)^2\rangle = \frac{4\langle E \rangle D}{\lambda_E} = \chi_E(0)D_E, \qquad (5.7.14)$$

$$(\delta\eta)_\omega^2 = \frac{2D}{\omega^2 + \lambda_\eta^2}, \quad \langle(\delta\eta)^2\rangle = \frac{D}{\lambda_\eta} = \chi_\eta(0)D_\eta. \tag{5.7.15}$$

We see that all three expressions for mean squares of fluctuations have the same structure, determined by the product of statistical susceptibility and the intensity of the corresponding Langevin sources.

5.7.2 The Critical Region

At the end of Sect. 5.5 we remarked that the order parameter equals zero for states above the critical point, so that the difference between means and quasi-means vanishes. This means, in particular, that $\chi_\eta \to \chi$ when crossing the critical point from below, while $\chi_\eta \to \infty$ at the critical point itself. Such behavior is observed because the formula for susceptibility χ_η (5.7.4) does not hold at the critical point and in its neighborhood [condition (5.5.43)].

A universal though approximate expression for susceptibility (designated $\tilde\chi$), which coincides with χ_η below the critical point and with χ above it, is

$$\tilde\chi = \frac{1}{\Delta\tilde\omega}, \quad \Delta\tilde\omega = a + 3b\langle E\rangle. \tag{5.7.16}$$

The mean value $\langle E\rangle$ is determined by (5.6.5). It is very clear that $\tilde\chi = \chi$ in state 1), $\tilde\chi = \chi_\eta$ in state 3), and

$$\tilde\chi \sim 1/\sqrt{Db} \tag{5.7.17}$$

at the critical point; therefore (5.7.16) describes the change of susceptibility at all the values of the parameters.

In the next section we shall see that the relative peak value of susceptibility $\tilde\chi$ at the critical point increases significantly in systems with many degrees of freedom.

5.8 Phase Transition in a Distributed System

5.8.1 A Langevin Source in the Ginzburg–Landau Equation

Instead of Langevin equation (5.5.8) we shall now consider the equation for the function $x(R, t)$,

$$\frac{\partial x}{\partial t} + [(a + bx^2)x]_{R,t} - g\frac{\partial^2 x}{\partial R^2} = F(R, t) + y(R, t). \tag{5.8.1}$$

The function $x(R, t)$ can describe, for example, the polarization vector (see Sect. 6.9),

$$P(R, t) = enx(R, t) . \tag{5.8.2}$$

In (5.8.1) g plays the role of the coefficient of diffusion in a conventional space. Nonlinearity is due to the anharmonicity of the internal oscillations of separate atoms. The value of a depends on the concentration of atoms and the temperature (Sect. 6.9), becoming zero at $T = T_c$.

The random source $y(R, t)$ is similar to the random source in the equation for the polarization vector (Sect. 6.8). The correlation of y values for different times and at different points is given by the expression

$$\langle y \rangle = 0, \quad \langle y(R, t)y(R', t') \rangle = 2\frac{D}{n}\delta(R - R')\delta(t - t') , \tag{5.8.3}$$

where $n = N/V$ is the mean concentration of atoms, and $D = \mathscr{K}T/m\gamma$ the intensity of random pushes caused by the atomic structure of the system.

Now we turn from the equation for the function $x(R, t)$ (5.8.1) to that for the spatial Fourier components $x_k(t)$. Since

$$x(R, t) = \sum_k x_k(t)e^{ikR} , \quad x_k(t) = \frac{1}{V}\int x(R, t)e^{-ikR}dR , \tag{5.8.4}$$

we get the following equation,

$$\frac{\partial x_k(t)}{\partial t} + (a + gk^2)x_k(t) + b(x^3)_{k,t} = y_k(t) \tag{5.8.5}$$

with the designation

$$(x^3)_{k,t} = \sum_{k_1 k_2 k_3} \delta_{k, k_1 + k_2 + k_3} x_{k_1}(t) x_{k_2}(t) x_{k_3}(t) . \tag{5.8.6}$$

The expression correlating the random source $y_k(t)$, following from (5.8.3, 4) has the form

$$\langle y_k(t)y_k(t') \rangle = \frac{1}{V}(yy)_{k, t, t'} = 2\frac{D}{N}\delta(t - t') . \tag{5.8.7}$$

The equation for function $x_k(t)$ at $k = 0$, i.e., for the function

$$x_{k=0}(t) = \frac{1}{V}\int x(R, t)dR , \tag{5.8.8}$$

coincides with (5.5.8) (provided $x_k = 0$ when $k \neq 0$). Between these equations, however, there is a significant difference which results from the fact that this one deals with a system consisting of a great number of particles ($N \gg 1$). Therefore the intensity of the random source is now N times smaller than that in (5.5.9), i.e., $D \rightarrow D/N$. Consequently we can take advantage of many results obtained in Sects. 5.5 − 7.

5.8.2 The Landau Theory

The Ginzburg – Landau equation (5.8.1) [or (5.8.5)], providing an example of a nonlinear Langevin equation for a distributed system, can serve as the initial equation to describe the phase transition of the second kind. This transition – at temperatures lower than the critical temperature T_c – results in permanent and uniform spontaneous polarization.

Introducing the dimensionless parameter

$$\tau = (T - T_c)/T_c \tag{5.8.9}$$

enables us to describe the proximity to the critical point. Since it seems impossible to reach the exact solution of the Langevin equation (5.8.1), phase transition theory employs approximate methods. For regions far enough from the critical point (see below) that fluctuations of the parameter of order are negligible, the Landau theory can be used to describe phase transitions of the second kind. Here we shall mention certain results of this theory which are relevant to our further discussion. This theory is presented in greater detail in [Ref. 5.12, Chap. 14]. Generalizing the Landau theory for the critical region is one of the major problems in the contemporary statistical theory of phase transitions, the basis of which is formed by well-known works by *Kadanoff* [5.13], *Patashinsky* and *Pokrovsky* [5.14], *Wilson* [5.15a], *Fisher* [5.15b], and others [5.16 – 18].

From today's viewpoint the dependence of the principal parameters on τ, as in Landau theory, is of the power-type. The power indices (the critical indices), however, differ markedly today from those adopted in the Landau theory. The development of a method to approximately calculate critical indices is a remarkable success of modern theory.

Here we shall discuss some qualitative results from the theory of fluctuations in the critical region including the critical point. Let us start by summarizing some of the results of Landau theory.

5.8.3 Fluctuations of the Order Parameter

We introduce the designation

$$\eta_k(t) = \langle x_k(t) \rangle_{F=0} \tag{5.8.10}$$

and assume the mean value to be nonzero only when

$$\eta_k(t) = \eta \delta_{k,0} \tag{5.8.11}$$

where η is the order parameter. If fluctuations are not taken into account, the order parameter is determined by (5.5.37), which corresponds to the approximation in Landau theory. Next we assume

$$x_k(t) = \eta \delta_{k,0} + \delta \eta_k(t), \quad \eta = \sqrt{|a|/b}. \tag{5.8.12}$$

In the linear approximation with respect to fluctuations, the nonlinear terms is

$$(x^3)_{k,t} = 3\eta^2 \delta\eta_k(t) ,\tag{5.8.13}$$

so the equation for function $\delta\eta_k(t)$ has the form

$$\frac{\partial\delta\eta_k(t)}{\partial t} + (a + 3b\eta^2)\,\delta\eta_k(t) + gk^2\delta\eta_k(t) = y(k, t) .\tag{5.8.14}$$

Hence, the expression for dynamic susceptibility follows,

$$\chi_\eta(\omega, k) = \frac{1}{-i\omega + \lambda_\eta + gk^2} , \quad \lambda_\eta \equiv \Delta\omega_\eta = a + 3b\eta^2 ,\tag{5.8.15}$$

which at $k = 0$ coincides with (5.7.9).

With the aid of (5.8.14) we find the expressions for spatial and temporal spectral densities of fluctuations of the order parameter,

$$(\delta\eta_k)_\omega^2 \equiv \frac{1}{V}(\delta\eta)_{k,\omega}^2 = \frac{2}{\omega}\,\mathrm{Im}\{\chi_\eta(\omega, k)\}\frac{D}{N} ,\tag{5.8.16}$$

$$\langle|\delta\eta_k(t)|^2\rangle \equiv \frac{1}{V}(\delta\eta)_k^2 = \mathrm{Re}\{\chi_\eta(0, k)\}\frac{D}{N} = \frac{D/N}{\lambda_\eta + gk^2} .\tag{5.8.17}$$

Assuming $k = 0$ in the latter equation, we get a formula for the dispersion of fluctuations of the order parameter,

$$\langle|\delta\eta_{k=0}(t)|^2\rangle = \langle(\delta\eta)^2\rangle = \frac{D}{N\lambda_\eta} = \chi_\eta(0, 0)\frac{D}{N} ,\tag{5.8.18}$$

which differs from (5.7.15) by replacing D with D/N.

In a similar way one can calculate the fluctuations of energy taking spatial diffusion into consideration. Here we shall only give the expression for dynamic susceptibility,

$$\chi_E(\omega, k) = \frac{1}{-i\omega + \lambda_E + gk^2} , \quad \lambda_E = 2(a + 2b\langle E\rangle)\tag{5.8.19}$$

where $\langle E\rangle$ is given by (5.6.5) with D replaced by D/N.

Relative fluctuations of the order parameter are given by [cf. (5.5.17, 42)]

$$\langle(\delta\eta)^2\rangle/\eta^2 \sim Db/Na^2 = \varepsilon_N .\tag{5.8.20}$$

5.8.4 Spatial Correlations

In order to find the spatial correlation of fluctuations of the order parameter according to (5.8.17), we employ the integral

$$\frac{1}{(2\pi)^3}\int\frac{1}{1+r_c^2k^2}\,\mathrm{e}^{\mathrm{i}kr}\,\mathrm{d}k=\frac{1}{4\pi r_c^2 r}\,\mathrm{e}^{-r/r_c},\tag{5.8.21}$$

and get

$$\langle(\delta\eta)^2\rangle_r=\frac{V\langle(\delta\eta)^2\rangle}{4\pi r_c^2 r}\,\mathrm{e}^{-r/r_c},\tag{5.8.22}$$

where

$$r_c=\sqrt{g\chi_\eta}=\sqrt{\frac{g}{a+3b\eta^2}}=\sqrt{\frac{g}{2|a|}}\tag{5.8.23}$$

is the correlation radius. Approaching the critical point, the correlation radius increases as $|a|^{-1/2}$.

When comparing these results with those given by Landau theory [Ref. 5.12, Chap. 14] one should be aware of the discrepancy between the accepted designations. Here well shall mark the designations employed in [5.12] by an additional subscript "L−L". The relation between the designations is determined by the dimension of the order parameter as well as that of susceptibility χ. In our present model $[\eta]$ has the dimension of length. In Chaps. 6, 13 we shall discuss examples of phase transitions in a system of atoms and field; the dimension of the order parameter will then coincide with that of the polarization vector P. We denote the corresponding order parameter by η_P. All this taken into account, we get the relations

$$\eta_{L-L}\leftrightarrow en\eta=\eta_P,\quad \chi_{L-L}\leftrightarrow\frac{e^2n}{m\gamma}\chi_\eta,$$

$$\langle(\delta\eta_{L-L})^2\rangle=\frac{\mathscr{K}T}{V}\chi_{L-L}\leftrightarrow e^2n^2\frac{D}{N}\chi_\eta,\quad g_{L-L}\leftrightarrow\frac{m\gamma}{2e^2n}g.\tag{5.8.24}$$

The susceptibility χ_{L-L} is then a dimensionless quantity, and the order parameter η_{L-L}^2 has the dimension of energy density.

The above formulae permit the introduction of two characteristic dimensionless parameters. One of them (ε_N) characterizes the magnitude of the relative dispersion of fluctuations of the order parameter (5.8.20). The other parameter determines the ratio of the correlation function (5.8.22) at the point $r=r_c$ [the maximum point of function $\langle(\delta\eta)^2\rangle_r r^2$] to the square of the order parameter η^2. As the authors of [5.14] we also name this quantity the Ginzburg parameter and denote it G. Thus

$$G\sim\frac{\langle(\delta\eta)^2\rangle_{r=r_c}}{\eta^2}\sim\frac{V}{r_c^3}\frac{\langle(\delta\eta)^2\rangle}{\eta^2}\sim\frac{V}{r_c^3}\varepsilon_N.\tag{5.8.25}$$

From these definitions if follows that the parameter ε_N in the thermodynamic limit is zero, and the Ginzburg parameter at $|a|\neq0$ in the thermodynamic limit has a finite value.

With the aid of these two parameters one can define the limits of the applicability of Landau theory in the following manner. In this theory, fluctuations of the order parameter are considered negligibly small and vanish when the thermodynamic limit transition is made, so that

$$\varepsilon_N = 0; \quad G \ll 1, \quad |\tau| \gg \frac{D^2 b}{n^2 g^3 a_0}. \tag{5.8.26}$$

Both these conditions restrict the applicability of Landau theory in the critical region, though each in its own way. We shall dispose of these restrictions separately. First we discard the condition $\varepsilon_N = 0$, enabling us to apply Landau theory equations to the critical region including the critical point itself. This attempt leads to fundamental complications which can only be overcome by rejecting the second restriction.

5.8.5 Extrapolation of Landau Theory into the Critical Region

It follows from (5.8.15, 23) that $\chi(0,0)$, r_c tends to infinity at $T \to T_c$ $(a \to 0)$. At the critical point, however, the quantities χ and r_c can actually tend to infinity only in the thermodynamic limit. Indeed, in a system of a finite volume the correlation radius of fluctuations of polarization cannot exceed the dimensions of the system.

In the critical region with respect to parameter ε_N, the order parameter is $\eta \sim \delta\eta$. In order to extend the above formulae of Landau theory into the critical region, we substitute in them:

$$\eta^2 \to \langle x_{k=0}^2(t)\rangle = \langle x^2(t)\rangle, \quad x(t) = \int x(R, t) \frac{dR}{V}. \tag{5.8.27}$$

Doing this we in fact give up the definition of the order parameter as a quasimean [see (5.8.10, 11)]. The criterion of the phase transition is now be given by the behavior of the temporal correlation of random function $x(t)$ or the corresponding spectral density $(x^2)_\omega$.

Like in Sect. 5.7 (5.7.16), we introduce a universal approximation for susceptibility $\tilde\chi$

$$\tilde\chi(0,0) = \frac{1}{a + 3b\langle x^2(t)\rangle}. \tag{5.8.28}$$

The equation for $\langle x^2(t)\rangle$ [with the condition $(x^2)_{k=0} \to (x_{k=0})3$] follows from (5.8.5) and has the form [cf. (5.6.5)]

$$a\langle x^2\rangle + b\langle x^2\rangle^2 = D/N. \tag{5.8.29}$$

From (5.8.28, 29), in the critical point

$$\tilde\chi \sim \sqrt{\frac{N}{Db}} \propto V^{1/2}, \quad \tilde r_c = \sqrt{g\tilde\chi} \propto V^{1/4}. \tag{5.8.30}$$

From the second expression it follows that the relative volume in the critical point is

$$\frac{\tilde{r}_c^3}{V} \propto V^{-1/4}, \tag{5.8.31}$$

which in the thermodynamic limit therefore becomes zero. In contrast, the corresponding value of the Ginzburg parameter

$$\tilde{G} \sim \frac{V}{\tilde{r}_c^3} \frac{\langle (\delta \eta)^2 \rangle}{\langle x^2 \rangle} \sim \frac{V}{r_c^3} \propto V^{1/4} \tag{5.8.32}$$

in the thermodynamic limit tends to inifinity.

Natural as it may seem, generalization of Landau theory produces unsatisfying results (5.8.31, 32). It is noteworthy that the contradiction (5.8.31) disappears in four-dimensional space, which illustrates a well-known fact that four-dimensional space plays an outstanding role in the theory of mean field.

5.8.6 Critical Indices

From a contemporary point of view the dependence of the order parameter η, susceptibility χ, and correlation radius r_c on τ in the critical region is, as in Landau theory, of the power type,

$$\chi \propto |\tau|^\beta, \quad \chi \propto |\tau|^{-\gamma}, \quad r_c \propto |\tau|^{-\nu}. \tag{5.8.33}$$

The actual values of the critical indices, however, differ significantly from those accepted in Landau theory.

Eight critical indices are introduced in all $\alpha, \beta, \gamma, \delta, \nu, \varepsilon, \mu, \xi$ [5.12, 14, 15, 18] connected by five general relations,

$$2\beta + \gamma = 2 - \alpha, \quad \beta\delta = \beta + \gamma, \quad \varepsilon(\beta + \gamma) = 2.$$

$$\mu(\beta + \gamma) = \nu, \quad \nu(2 - \xi) = \gamma. \tag{5.8.34a}$$

A sixth relation can be obtained from the hypothesis of scale invariance proposed by *Kadanoff* [5.13], and *Patashinsky* and *Pokrovsky* [5.14]. It has the form

$$\nu d = 2 - \alpha \quad \text{(or} \quad \nu d = \gamma + 2\beta). \tag{5.8.34b}$$

By dint of these relations all eight critical indices can be reduced to two independent ones. The calculations of critical indices carried out by *Wilson* [5.15a] and *Fisher* [5.15b] by the so-called method of ε = expansion show that two (α and ξ) of the eight indices are much smaller than the rest. Approximating $\alpha = 0$ and $\xi = 0$ from the above six relations (5.8.34), one can obtain the values of the rest of the indices, which for three-dimensional space are

$$\beta = \tfrac{1}{3}, \quad \gamma = \tfrac{4}{3}, \quad \nu = \tfrac{2}{3}, \quad \delta = 5, \quad \varepsilon = 0, \quad \mu = \tfrac{2}{5}. \tag{5.8.35a}$$

Compare these with the critical indices of Landau theory,

$$\alpha_L = 0, \quad \xi_L = 0, \quad \beta_L = \tfrac{1}{2}, \quad \gamma_L = 1, \quad \nu_L = \tfrac{1}{2},$$

$$\delta_L = 3, \quad \varepsilon_L = 0, \quad \mu_L = \tfrac{1}{3}. \tag{5.8.35b}$$

The dependence of the parameters upon $|\tau|$ given above does not hold for the critical point itself, when $|\tau| = 0$, since the correlation radius, for example, becomes infinite while it cannot exceed the size of the system. It is clear that the dependent parameters (5.8.33) are only true when the thermodynamic limit transition has already been made.

In order to obtain qualitative results which are also true for the critical point itself, we shall coordinate two limit transitions, $T \to T_c$ and the thermodynamic limit transition. It will be shown that the requirement that these two transitions be coordinated in the critical point itself leads to renormalization of parameters and to introduction of critical indices which characterize dependence on V (or N). Moving away from the critical point with respect to parameter $\varepsilon \sim 1/N$ (5.8.20), the power-type dependence upon N becomes a power-type dependence upon $|\tau|$ of type (5.8.33), with the critical indices in (5.8.35a). To determine small critical indices as well, the coordination conditions of the limit transitions should be generalized.

5.8.7 Coordination of Limit Transitions

We shall proceed from two conditions [5.19].
1) At the critical point in the transition to the thermodynamic limit the correlative volume increases proportional to V, i.e.,

$$r_c^{*3}/V \propto V^0. \tag{5.8.36}$$

The asterisk marks renormalized quantities.
2) At the critical point the renormalized Ginzburg parameter remains finite at the thermodynamic limit, i.e.,

$$G^* \propto V^0 \propto N^0. \tag{5.8.37}$$

At first we shall assume that the relation between renormalized parameters r_c^* and χ^* retains its form,

$$r_c^* \propto \sqrt{g\chi^*}. \tag{5.8.38}$$

Later we shall discuss a more general case when the equation $\xi = 0$ imposed by (5.8.38) is rejected.

The first of these two conditions enables us to determine the critical indices. Indeed, according to (5.8.36), one can write the following relation for dimensionless quantities

$$n r_c^{*3} \propto (n \tilde{r}_c^3)^{2\nu} \sim V. \tag{5.8.39}$$

Later we shall realize the convenience of this method of definiting index ν. Using (5.8.30) for $\tilde{r}_c$ and equalizing the powers of N on both sides, we get

$$\nu = \tfrac{2}{3} \tag{5.8.40}$$

and therefore

$$r_c^* \sim n^{1/9} \tilde{r}^{4/3} . \tag{5.8.41}$$

Let us now employ the second condition. First we present the parameter G (5.8.25) as a product of three dimensionless parameters,

$$G \sim \frac{1}{n r_c^3} \frac{\chi \sqrt{Db}}{\eta^2 / \sqrt{Db}} . \tag{5.8.42}$$

Turning to quantities $\tilde{r}_c$, $\tilde{\chi}$, $\langle x^2 \rangle$ (5.8.28) we define the renormalized Ginzburg parameter as

$$G^* \sim \left(\frac{1}{n \tilde{r}_c^3} \right)^{2\nu} \frac{(\chi \sqrt{Db})^\gamma}{(\langle x^2 \rangle / \sqrt{D/b})^{2\beta}} , \tag{5.8.43}$$

where we have introduced two new critical indices γ and β.

In accordance with the second condition about the finiteness of the parameter G^* at the critical point in thermodynamic limit transition we find one more relation between indices ν, γ, β,

$$3\nu = \gamma + 2\beta . \tag{5.8.44}$$

At $d = 3$ this equation coincides with the second one in (5.8.34b) and therefore corresponds to the condition of scale invariance.

Finally, the relation

$$2\nu = \gamma \tag{5.8.45}$$

follows from (5.8.38).

From the three relations (5.8.40, 44, 45) we obtain for β, γ, ν the same values as in (5.8.35a). Thus, the renormalized quantities r^*, χ^*, $\langle x^2 \rangle^*$ (provided the dimension of each is retained after renormalization) are linked with $\tilde{r}_c$, $\tilde{\chi}$, $\langle x^2 \rangle$ by relations

$$r_c^* \sim n^{1/9} \tilde{r}^{4/3}, \quad \chi^* \sim (Db)^{1/6} (\tilde{\chi})^{4/3}, \quad \langle x^2 \rangle^* \sim \left(\frac{D}{b} \right)^{1/6} \langle x^2 \rangle^{2/3} . \tag{5.8.46}$$

These are the values of indices which correspond to the requirement that the small indices be equal to zero. The relation (5.8.44) also remains intact for a more general case, while (5.8.40, 45) are replaced by more general relations,

$$3\nu = 2 - \alpha, \quad (2 - \xi)\nu = \gamma . \tag{5.8.47}$$

In order to obtain, e.g., the second of these, one has to replace (5.8.37) by a more general condition about the finiteness of spatial correlations at the critical point for every value of r within the region $r < r_c$. In order to take advantage of this condition we make the following substitution in (5.8.22):

$$\frac{V\langle(\delta\eta)^2\rangle}{r_c^2 r} = \frac{nD\chi}{r_c^2 r} \rightarrow \frac{nD\chi^*}{(r_c^*)^{2-\xi} r^{1+\xi}} \tag{5.8.48}$$

where a new index ξ is introduced, which characterizes the deviation from dependence of types $1/r$. Then we set $\chi^* \sim \tilde{\chi}^\gamma$ and $r^* \sim (\tilde{r}_c)^{2\nu}$, and use (5.8.30) which gives the dependence on N. Then the second equation in (5.8.47) results from the condition about the finiteness of correlations.

If α and ξ are nonzero, the critical indices depend on the particularities of the model, e.g., on the dimension of the "spin" value. Our qualitative treatment then becomes inadequate.

Let us now discuss these relations for various regions of values of parameters.

5.8.8 The Critical Point

The temperature region where (5.8.46) is true can be divided into two subregions. The first includes the critical point and its immediate neighborhood, where the parameter ε_N is not small. For the critical point we get from (5.8.46) and employing (5.8.29, 30)

$$r^* \sim n^{1/9}\left(\frac{g^2 N}{Db}\right)^{1/3}, \quad \chi^* \sim \frac{N^{2/3}}{\sqrt{Db}}, \quad \langle x^2\rangle^* \sim \sqrt{\frac{D}{b}}\, N^{-1/3}. \tag{5.8.49}$$

The corresponding expression for the Ginzburg parameter has the form

$$G^* \sim \frac{Db}{g^2 n^{4/3}}, \tag{5.8.50}$$

which reveals total independence of N. Consequently, this equation defines a new dimensionless parameter which characterizes correlations in the critical point.

5.8.9 The Region of Scale Invariance

Consider the region far enough from the critical point that the parameter

$$\varepsilon_N = \frac{Db}{Na^2} \ll 1 \quad \text{or} \quad \tau^2 \gg \frac{Db}{Na_0^2}. \tag{5.8.51}$$

The parameter G, however, is not small.

According to (5.8.29) in zero-order approximation with respect to ε_N, the quantity $\langle x^2 \rangle = \eta^2$, and therefore the expressions for $\tilde{\chi}$ and $\tilde{r}_c$ coincide with the corresponding formulae in Landau theory [see (5.8.15, 23)]. Thus when $a < 0$,

$$\tilde{\chi} = \chi = \frac{1}{2a_0|\tau|}, \quad \tilde{r}_c = r_c = \sqrt{\frac{g}{2a_0|\tau|}}, \quad \langle x^2 \rangle = \eta^2 = \frac{|a|}{b}, \tag{5.8.52}$$

and (5.8.46) takes the form

$$\chi^* \sim \frac{(Db)^{1/6}}{(2a_0)^{4/3}} |\tau|^{-4/3}, \quad \eta^* \sim \left(\frac{a_0}{b}\right)^{1/3} \left(\frac{D}{b}\right)^{1/12} |\tau|^{1/3},$$

$$r_c^* \sim n^{1/9} \left(\frac{g}{2a_0}\right)^{2/3} |\tau|^{-2/3} \tag{5.8.53}$$

where we recognize power-type dependence (5.8.33) with power indices (5.8.35a).

For the region $a > 0$, one should employ, in place of (5.8.52), the expressions

$$\tilde{\chi} = \frac{1}{a}, \quad \tilde{r}_c = \sqrt{\frac{g}{a}}, \quad \langle x^2 \rangle = \frac{D}{Na}. \tag{5.8.54}$$

As a result (5.8.46) becomes

$$\chi^* \propto \tau^{-4/3}, \quad r_c^* \propto \tau^{-2/3}. \tag{5.8.55}$$

We see that the values of critical indices above and below the critical point coincide. This result is in agreement with general conclusions from the theory of phase transitions.

The Ginzburg parameter G^* for this subregion, following from (5.8.43, 52), is again determined by (5.8.50).

5.8.10 The Transition to the Results of Landau Theory

A further increase in the temperature difference $|T - T_c|$ brings us to the transition stage between the region of scale invariance and the domain of Landau theory.

Since Landau theory corresponds to the approximation of a self-consistent field when correlations are negligible, it would be natural to characterize the intermediate stage by a parameter connected in some way with the Ginzburg parameter. We call it the Wilson parameter ε_W and define it as

$$\varepsilon_W = \frac{G}{1 + G}. \tag{5.8.56}$$

Since the parameter G depends on $|\tau|$ and on approaching the critical point increases as $|\tau|^{-1/2}$, the Wilson parameter is confined within the limits

$$1 \geqslant \varepsilon_W \geqslant 0. \tag{5.8.57}$$

Those values of ε_W which are close to 1 correspond to large values of G (the region of scale invariance), and very small values of ε_W correspond to small values of G (the domain of Landau theory).

The ε-expansion method of calculating critical indices is based on the values of indices derived for a four-dimensional space ($d = 4$), which are the same as in Landau theory. The unknown indices are calculated in a space of $d = 4 - \varepsilon_W$ dimensions with $\varepsilon_W \ll 1$, and presented in the form of series in ε_W. Landau theory therefore corresponds to a zero-order approximation.

In the final results, ε_W is taken to be equal to one, thus causing the formal transition to conventional three-dimensional space since $d = 4 - \varepsilon_W = 3$ if $\varepsilon_W = 1$.

Now we attempt to give a qualitative notion of the transition from the region of scale invariance to the domain where Landau theory is applicable. We have already noted that in four-dimensional space there are no grounds to single out the region of scale invariance. This fact by itself is not yet sufficient to justify Wilson's ε-expansion; however it prompts the way of finding the connection between critical indices in three-dimensional space ν, γ, β with the corresponding critical indices ν_L, γ_L, β_L of Landau theory. For this purpose d is replaced by $4 - \varepsilon_W$ in (5.8.34b) (at $a = 0$). The critical indices can then be presented in the form

$$\nu = \frac{2}{4 - \varepsilon_W}, \quad \gamma = \frac{4}{4 - \varepsilon_W}, \quad \beta = \frac{2 - \varepsilon_W}{4 - \varepsilon_W}. \tag{5.8.58}$$

At $\varepsilon_W = 1$ they coincide with the critical indices for the region of scale invariance, and at $\varepsilon_W = 0$ with the critical indices of Landau theory. The intermediate values of indices correspond to the transition region.

Of course, these formulae only given a simplified qualitative notion of changes of the critical indices after departing the critical point. More precise results are only obtainable by solving the initial equation (5.8.1).

5.8.11 Correlation Times and Spectrum Widths at a Phase Transition [5.19]

The definition of the order parameter is based on the concept of quasi-means and therefore implies that the choice of positive or negative direction along a given axis (so far we have been concerned only with one-dimensional polarization) is determined by an external field. Of course, a phase transition in a system may occur in the absence of external field. The choice of direction in an ensemble of systems occurs then at random, so that the mean value $\langle x \rangle = 0$; the occurrence of a phase transition can be detected by the character of time correlations.

Bearing this in mind we consider two correlation times τ_η and τ_x. The former reflects the statistical behavior of fluctuations of the order parameter $\delta\eta$ at different moments in time, and the latter the statistical connection of functions

$$x(t) = \int x(R, t)\frac{dR}{V} = x_{k=0}(t) .$$

We shall see that at above-critical temperatures the time of correlations τ_x is finite for all values of N and thus also in the thermodynamical limit. In contrast, at below-critical temperatures the correlations depend upon N, such that in the thermodynamic limit $\tau_x \to \infty$. This behavior of correlation time τ_x in the vicinity of the critical point indicates the phase transition.

Let us designate by $\Delta\omega_\eta$, $\Delta\omega_x$ the corresponding spectrum widths of fluctuations. From the aforesaid it follows that the spectrum width $\Delta\omega_x$ for below-critical temperatures tends to zero in the thermodynamic limit. The state with infinitesimal spectrum width corresponds to the presence of the order parameter in the system.

First we find the quantities τ_η and $\Delta\omega_\eta$ by employing the equation for $\delta\eta$ in the linear approximation. As a result we get the following expressions for spatial and temporal and for spatial spectral densities, see (5.8.16, 17). The functions $\chi_\eta(\omega, k)$ and $\chi_\eta(0, k)$ entering these equations are defined by (5.8.15). It follows that the quantities τ_η and $\Delta\omega_\eta$ are expressible through the statical susceptibility,

$$\tau_\eta(k) = \frac{1}{\Delta\omega_\eta} = \chi_\eta(0, k) . \tag{5.8.59}$$

From this and (5.8.46) we find the correlation time of the order parameter $\delta\eta_{k=0}$, taking heed of renormalization,

$$\tau_\eta \sim \chi^* \sim (Db)^{1/6}(\tilde{\chi})^{4/3} \tag{5.8.60}$$

where $\tilde{\chi}$ is determined by (5.8.28). As a result we get the following expression for the critical point ($a = 0$) and for the region of scale invariance

$$\tau_\eta \sim \frac{N^{2/3}}{\sqrt{Db}} \quad \text{at} \quad a = 0 ,$$

$$\tau_\eta \sim \frac{(Db)^{1/6}}{(2a_0)^{4/3}}|\tau|^{-4/3} . \tag{5.8.61}$$

We notice that at the critical point the correlation time $\tau_\eta \propto N^{2/3}$. In the region of scale invariance the time τ_η decreases as $|\tau|^{-4/3}$ with an increase of $|\tau|$, while in the domain where Landau theory is applicable it decreases as $|\tau|^{-1}$ [5.20].

In order to find the quantities τ_x and $\Delta\omega_x$ we employ the equation for function $\langle x(t)x(t-\tau)\rangle$, which we write in the same approximation as (5.8.29) for the one-time correlator $\langle x^2(t)\rangle$ [cf. (5.6.4)],

$$\frac{d}{d\tau} \langle x(t)x(t-\tau)\rangle + [a + b\langle x^2(t)\rangle]\langle x(t)x(t-\tau)\rangle = 0 . \tag{5.8.62}$$

For the initial condition (at $t = 0$) we take the solution of (5.8.29).

It follows (so far without renormalization, i.e., without replacing $\langle x^2 \rangle$ by $\langle x^2 \rangle^*$) that the sought quantities are described by

$$\tau_x = \frac{1}{\Delta\omega_x} = \frac{1}{a + b\langle x^2(t)\rangle} = \frac{\langle x^2(t)\rangle}{D} N . \tag{5.8.63}$$

By dint of (5.8.29) both of the definitions in this equation are equivalent, though the second of them is more practical since it explicitly depends on the intensity of the random source D/N.

In the domain where Landau theory is applicable, when $\langle x^2\rangle = |a|/b$, it follows from (5.8.63) that $\tau_x \propto N$, and that therfore in the thermodynamic limit the correlation time tends to infinity while the spectrum width $\Delta\omega_x$ tends to zero. This also serves to indicate a phase transition.

In order to describe the behavior of τ_x in the critical region we replace $\langle x^2\rangle$ by $\langle x^2\rangle^*$ in (5.8.63). As a result [cf. (5.6.8)],

$$\tau_x \sim \langle x^2\rangle^* N/D . \tag{5.8.64}$$

The quantity $\langle x^2\rangle^*$ is determined by the last part of (5.8.46). According to (5.8.49), at the critical point

$$\tau_x \sim N^{2/3}/\sqrt{Db} , \tag{5.8.65}$$

i.e., is of the same order as τ_η, see (5.8.61).

Employing the function $\langle x(t)x(t-\tau)\rangle$ enables us to describe a phase transition while avoiding involving the order parameter as a quasi-mean quantity (5.8.10). One can show that the susceptibility which characterizes the fluctuations $\delta x^2 = x^2 - \langle x^2\rangle$ behaves qualitatively like the susceptibility χ_η. A corresponding similarity also exists for the dependence of correlation times τ_{x^2}, τ_η on the values of N, T.

We must point out once more that everything said here only gives a qualitative notion about the behavior of various characteristics in the critical region in the whole temperature range and with a finite number of particles. A more detailed description is only possible by accurately solving the Ginzburg–Landau equation.

For the model described by (5.8.1) the fluctuation characteristics, such as the correlation radius r_c, depend on the diffusion coefficient g. In the next chapter we shall discuss the example of a phase transition in a system of atoms and field which may take place without any connection through spatial diffusion ($g = 0$). The domain where Landau theory is applicable is then much wider.

In recent years a number of authors have drawn attention to the existing similarity between phase transitions and generation thresholds in classical and quantum generators – autooscillating systems. We shall discuss this analogy in the following sections as well as in Chaps. 12, 13. This analogy happens to be rather instructive both for the theory of phase transitions and for the theory of fluctuations in self-oscillatory systems. The juxtaposition of these two classes also reveals their fundamental dissimilarities.

The models of phase transitions discussed here are based on the Langevin equations with nondissipative nonlinearity. The dissipative terms are linear, although dissipation is large (the characteristic frequencies $\omega \ll \gamma$). The situation is reversed for autooscillating systems with near-harmonic oscillations: the dissipative terms are nonlinear, and dissipation is small ($\gamma \ll \omega$). In order to understand these peculiarities better we shall now discuss the theory of fluctuations in systems with dissipative nonlinearity.

5.9 Dissipative Nonlinearity

Consider the Langevin equations for an oscillator with nonlinear friction,

$$\frac{dx}{dt} = v , \quad \frac{dp}{dt} + F(v)p + m\omega_0^2 x = \sqrt{D(v)}\, y(t) , \tag{5.9.1}$$

where $F(v)$ is an even function of velocity, e. g.,

$$F(v) = \gamma + \delta v^2 . \tag{5.9.2}$$

For a system with nonlinear friction, Einstein's formula is no longer valid since the intensity of pushes, whose influence on the system is analogous to thermal motion, is now a function of velocity. It is convenient to include this yet unknown function in the Langevin equations, while defining the correlation of the random source as

$$\langle y(t)\rangle = 0 , \quad \langle y(t)y(t')\rangle = 2\delta(t-t') . \tag{5.9.3}$$

A large number of works are dedicated to the study of fluctuations in passive systems with nonlinear friction [5.7, 21, 22]. Here we shall confine ourselves to a special case: we shall find the function $D(v)$ – the intensity of a Gaussian δ-correlated random source for the nonlinear system in question.

Employing the method described in Sect. 5.2 we find the Fokker–Planck equation which corresponds to the Langevin equations (5.9.1). For that purpose we first write the equation which corresponds to (5.2.5). It has the form

$$\frac{\partial f}{\partial t} + v\frac{\partial f}{\partial x} - m\omega_0^2 r\frac{\partial f}{\partial p} = \frac{\partial}{\partial p}[F(v)pf]$$

$$-\frac{1}{n}\frac{\partial}{\partial p}[\sqrt{D(v)}\langle y(t)\delta N(x,p,t)\rangle] . \tag{5.9.4}$$

We also write the equation in δN similar to (5.2.8),

$$\frac{\partial \delta N_{(x,p,t)}}{\partial t} = -\frac{\partial}{\partial p}\left[n\sqrt{D(v)}\,y\,f(x,p,t)\right],\tag{5.9.5}$$

whence

$$\delta N(x,p,t) = -\frac{\partial}{\partial p}\int_0^\infty \sqrt{D(v)}\,y(t-\tau)\,f(x,p,t-\tau)d\tau.\tag{5.9.6}$$

We substitute this solution into the second term in the right-hand side of (5.9.4) and use (5.9.3), getting as a result

$$-\frac{1}{n}\frac{\partial}{\partial p}\left[\sqrt{D(v)}\langle y(t)\,\delta N(x,p,t)\rangle\right]$$

$$=\frac{\partial}{\partial p}\left[\sqrt{D(v)}\frac{\partial}{\partial p}\sqrt{D(v)}\,f(x,v,t)\right].\tag{5.9.7}$$

If D is constant, this expression coincides with (5.2.9). Taking advantage of the equation

$$\frac{\partial}{\partial p}\left[\sqrt{D(v)}\frac{\partial}{\partial p}\sqrt{D(v)}\,f\right] = \frac{\partial}{\partial p}\left[D(v)\frac{\partial f}{\partial p}\right] + \frac{\partial}{\partial p}\left(\frac{1}{2}\frac{\partial D}{\partial p}f\right),\tag{5.9.8}$$

(5.9.4) can be written in the form

$$\frac{\partial f}{\partial t} + v\frac{\partial f}{\partial x} - m\omega_0^2 x\frac{\partial f}{\partial p} = \frac{\partial}{\partial p}\left[D(v)\frac{\partial f}{\partial p}\right]$$

$$+\frac{\partial}{\partial p}\left[F(v)p + \frac{1}{2}\frac{\partial D(v)}{\partial p}\right]f.\tag{5.9.9}$$

The equilibrium solution does not depend on the type of dissipation and therefore must coincide with the equilibrium solution of the Fokker–Planck equation (5.4.3) for an oscillator with linear friction, i.e., with (5.4.4). If this solution is substituted into (5.9.9), the left-hand side becomes zero in the state of equilibrium. The right-hand side equals zero when the unknown function $D(v)$ satisfies the equation

$$\frac{1}{2}\frac{dD}{dp} - \frac{Dv}{\varkappa T} + F(v)p = 0.\tag{5.9.10}$$

For the case with linear friction, when $F(v) = \gamma$, the solution of this equation

$$D = \gamma m \mathscr{K} T \tag{5.9.11}$$

coincides with Einstein's formula. For an even function and with the condition

$$D(p) \exp\left(-\frac{p^2}{m \mathscr{K} T}\right) \to 0 \quad \text{at} \quad p \to \infty, \tag{5.9.12}$$

the solution of (5.9.10) can be presented in the form

$$D(p) = -2e^{p^2/m\mathscr{K}T} \int_{\infty}^{P} F(v') e^{-p'^2/m\mathscr{K}T} p' \, dp' . \tag{5.9.13}$$

If the friction is linear, (5.9.11) follows, while for a function of type (5.9.2),

$$D(p) = m\gamma \mathscr{K} T \left(1 + \frac{\delta}{\gamma} \frac{\mathscr{K}T}{m}\right) + \delta \frac{\mathscr{K}T}{m} p^2 . \tag{5.9.14}$$

This solution can also be used at $\gamma < 0$ provided

$$|\gamma| < \delta \frac{\mathscr{K}T}{m} . \tag{5.9.15}$$

We see that if the intensity of noise in (5.9.1) is determined by (5.9.13), the equilibrium solution of the Fokker – Planck equation (5.9.9) has the form of (5.4.3) and, consequently, in the state of equilibrium the mean values $\langle x \rangle$ and $\langle p \rangle$ equal zero. This consequence means that in the equivalent electric circuit ($mv \leftrightarrow LJ$) the mean current $\langle J \rangle$ is zero no matter which function $F(v)$ enters the initial equation (5.9.1).

From the very beginning we have benefitted from the suggestion that the random source $y(t)$ in the Langevin equation (5.9.1) is Gaussian and δ correlated. Thanks to this the kinetic equation (5.9.9) contains no higher derivatives of p. A more accurate way of constructing the kinetic equation for a system with dissipative nonlinearity, employing methods of the nonlinear thermodynamics of irreversible processes, leads to a kinetic equation with higher derivatives (see references to Chap. 13).

5.10 The Langevin Equations for a Self-Oscillatory System. The Fokker – Planck Equation

Consider once more the Langevin equations (5.9.1). If a random source is not present and $\gamma < 0$, these equations describe an self-oscillatory process and are equivalent to the Van der Pol equation, which is employed to describe self-oscillatory processes in an extensive number of physical (lasers), as well as chemical and biological systems [5.7, 21 – 27].

When taking random effects into consideration one should be aware of the fact that the statistical characteristics of random sources in autooscillatory systems are generally determined by nonequilibrium processes. For example, in an electronic generator the random source is determined not only by thermal motion, but also by the shot noise in the electronic tube. The processes in lasers are affected by nonequilibrium fluctuations of the polarization of the active medium. For that reason the intensity of the random source at a given dissipative nonlinearity cannot be found by solving (5.9.10). Indeed, this equation was obtained by assuming that the collision integral in the Fokker – Planck equation becomes zero if the equilibrium distribution with respect to momenta (the Maxwell distribution) is substituted into it. The stationary states of autooscillating systems are nonequilibrium, so that this method of determining the noise intensity with a given dissipative nonlinearity does not work here. Determining the intensity of a random source becomes a separate problem.

In Sects. 4.8, 9 we have shown that the field equation for a system of free charged particles and a field includes two independent random sources. In a system of atoms and a field there are also two independent random sources. One of them is spontaneous emission by atoms in the active medium of a laser, and the other thermal radiation of the system of field oscillators. The combined activity of these sources at certain conditions (Sects. 12.3, 4) can be interpreted as a δ-correlated noise of a given intensity, i.e.,

$$\langle y(t) \rangle = 0 , \quad \langle y(t)y(t') \rangle = 2D\delta(t-t') . \tag{5.10.1}$$

We assume that the intensity of a random source in the Langevin equations for an autooscillating system is set and equals D. The value of D, however, this time is not connected with temperature via Einstein's formula.

Let us present the Langevin equations in a different notation, which is more suitable for comparing the results of calculations of the fluctuation processes in autooscillating systems and in systems with a phase transition,

$$\frac{dx}{dt} = v , \quad \frac{dv}{dt} + \left(a + \frac{2}{3}bv^2 \right)v + \omega_0^2 x = y(t) . \tag{5.10.2}$$

The inclusion of the factor $\frac{2}{3}$ further simplifies the formulae to be obtained.

Since the intensity of the random source in these equations is set, one can follow the scheme proposed in Sect. 5.2 and go from the Langevin equations (5.10.2) to a corresponding Fokker – Planck equation for the distribution function $f(x, v, t)$, which has the form

$$\frac{\partial f}{\partial t} + v\frac{\partial f}{\partial x} - \omega_0^2 x\frac{\partial f}{\partial v} = D\frac{\partial^2 f}{\partial v^2} + \frac{\partial}{\partial v}\left[\left(a + \frac{2}{3}bv^2 \right)vf \right] . \tag{5.10.3}$$

The right-hand side of this equation does not become zero upon substitution of an equilibrium distribution (5.4.4), due to the fact that the fluctuations included in the source are nonequilibrium.

The exact solution of equations describing self-oscillations has not been obtained even for the dynamic mode, when the random source is absent. The equations can be significantly simplified if relaxation times are much greater than the period of oscillations, which opens the possibility of averaging over a period.

In order to carry out this averaging it is convenient to switch over to new variables,

$$x = \tilde{x} \cos \omega_0 t + \frac{\tilde{v}}{\omega_0} \sin \omega_0 t ,$$

$$v = \tilde{v} \cos \omega_0 t - \omega_0 \tilde{x} \sin \omega_0 t . \tag{5.10.4}$$

The equations for functions $\tilde{x}$ and $\tilde{v}$ follow from (5.10.2) and have the form

$$\frac{d\tilde{x}}{dt} = \frac{1}{\omega_0} \left[\left(a + \frac{2}{3} b v^2 \right) v - y(t) \right] \sin \omega_0 t , \tag{5.10.5}$$

$$\frac{d\tilde{v}}{dt} = - \left[\left(a + \frac{2}{3} b v^2 \right) v - y(t) \right] \cos \omega_0 t . \tag{5.10.6}$$

These equations are exact. After averaging over a period they take the form

$$\frac{d\tilde{x}}{dt} + \frac{1}{2} (a + bE) \tilde{x} = y_x(t) ,$$

$$\frac{d\tilde{v}}{dt} + \frac{1}{2} (a + bE) \tilde{v} = y_0(t) , \tag{5.10.7}$$

with the designation for specific energy (energy divided by mass)

$$E = \tilde{v}^2/2 + \omega_0^2 \tilde{x}^2/2 , \tag{5.10.8}$$

and for new sources

$$y_x(t) = - \frac{1}{\omega_0} y(t) \sin \omega_0 t , \quad y_v(t) = y(t) \cos \omega_0 t . \tag{5.10.9}$$

The correlations of these sources can be found from (5.10.1, 9),

$$\omega_0^2 \langle y_x(t) y_x(t') \rangle = \langle y_v(t) y_v(t') \rangle = D \delta(t - t') ,$$

$$\langle y_x(t) y_v(t') \rangle = \langle y_v(t) y_x(t') \rangle = 0 . \tag{5.10.10}$$

In getting these formulae we have dropped those terms which disappear if averaged over the period.

The Fokker–Planck equation for the distribution function $f(\tilde{x}, \tilde{v}, t)$ can be obtained according to the pattern demonstrated in Sect. 5.2; it has the form (omitting the sign $\sim$)

$$\frac{\partial f}{\partial t} = \frac{D}{2\omega_0^2} \frac{\partial^2 f}{\partial x^2} + \frac{D}{2} \frac{\partial^2 f}{\partial v^2} + \frac{1}{2} \frac{\partial}{\partial x} [(a + bE)x f]$$

$$+ \frac{1}{2} \frac{\partial}{\partial v} [(a + bE)v f]. \tag{5.10.11}$$

The corresponding equation for the distribution function of energy

$$f(E, t) = \int \delta(E - \tfrac{1}{2}(v^2 + \omega_0^2 x^2)) f(x, v, t) \, dx \, dv \tag{5.10.12}$$

has the form

$$\frac{\partial f(E, t)}{\partial t} = D \frac{\partial}{\partial E} \left(E \frac{\partial f}{\partial E} \right) + \frac{\partial}{\partial E} [(a + bE)E f]. \tag{5.10.13}$$

Consider the equations for moments of energy. By dint of (5.10.13) we find

$$\frac{d\langle E^n \rangle}{dt} = n^2 D \langle E^{n-1} \rangle - n(a\langle E^n \rangle + b\langle E^{n+1} \rangle). \tag{5.10.14}$$

Hence at $n = 1$ the equation for mean energy follows,

$$\frac{d\langle E \rangle}{dt} + (a\langle E \rangle + b\langle E^2 \rangle) = D. \tag{5.10.15}$$

In the absence of noise it has a stationary solution

$$\langle E \rangle_0 = |a|/b. \tag{5.10.16}$$

The Langevin equation corresponding to (5.10.13) has the form

$$\frac{dE}{dt} + (a + bE)E = y_E(t) \tag{5.10.17}$$

with random source

$$y_E = \omega_0^2 x y_x(t) + v y_v(t), \quad \langle y_E \rangle = D. \tag{5.10.18}$$

By virtue of (5.10.15, 17) we find the equation for the fluctuations of energy. In linear approximation, it is

$$\frac{d\delta E}{dt} + (a + 2b\langle E\rangle)\,\delta E = \delta y_E(t) \equiv y_E - \langle y_E\rangle\,. \tag{5.10.19}$$

To determine the correlation δy_E of a random source we use (5.10.18, 19) and single out the δ-type source. As a result we get

$$\langle \delta y_E(t)\,\delta y_E(t')\rangle = 2D\langle E\rangle\delta(t-t')\,. \tag{5.10.20}$$

The intensity of the random source is proportional not only to the intensity of pushes D, but to the mean energy of self-oscillations, as well.

5.11 The Stationary Distribution of the Energy of Oscillations

The stationary solution of (5.10.13) has the form

$$f(E) = c\exp\left(-\frac{aE + bE^2/2}{D}\right),\quad \int_0^\infty f(E)\,dE = 1\,. \tag{5.11.1}$$

The constant of integration is found from the normalization condition

$$\frac{1}{c} = \sqrt{\frac{\pi D}{2b}}\left[1 - \Phi\left(\frac{a}{\sqrt{2bD}}\right)\right]\exp\left(\frac{a^2}{2bD}\right) \tag{5.11.2}$$

where

$$\Phi(x) = \frac{2}{\sqrt{\pi}}\int_0^\infty e^{-t^2}dt = -\Phi(-x)\,. \tag{5.11.3}$$

Distribution (5.11.1) and condition (5.11.2) can be rewritten in the form

$$f(E) = \sqrt{\frac{2b}{\pi D}}\exp\left[-\frac{\left(E - \frac{|a|}{b}\right)^2}{2D/b}\right]\left[1 + \Phi\left(\frac{|a|}{\sqrt{2bD}}\right)\right]^{-1}\,. \tag{5.11.4}$$

Consider three extreme cases.

1) The state of equilibrium ($a > 0$, $b = 0$):

$$f(E) = \frac{D}{a}\exp\left(-\frac{E}{D/a}\right),\quad \langle E^n\rangle = n!\left(\frac{D}{a}\right)^n,$$

$$D = a\,\varkappa T/m\,. \tag{5.11.5}$$

2) The generation threshold ($a = 0$):

$$f(E) = \sqrt{\frac{2b}{\pi D}}\, \exp\left(-\frac{bE^2}{2D}\right),$$

$$\langle E^n \rangle = \frac{\Gamma\left(\dfrac{n+1}{2}\right)}{\sqrt{\pi}} \left(\frac{2D}{b}\right)^{n/2}. \tag{5.11.6}$$

It follows in particular that

$$\langle E \rangle = \frac{2}{\sqrt{\pi}} \sqrt{\frac{D}{2b}}, \quad \langle E^2 \rangle = \frac{D}{b}, \quad \langle (\delta E)^2 \rangle = \left(1 - \frac{2}{\pi}\right)\frac{D}{b}. \tag{5.11.7}$$

3) Greatly exceeding the generation threshold (zero-order approximation with respect to parameter $Db/a^2 = \varepsilon$):

$$f(E) = \delta\left(E - \frac{|a|}{b}\right), \quad \langle E^n \rangle = \left(\frac{|a|}{b}\right)^n. \tag{5.11.8}$$

At $n = 1$, (5.10.16) can be derived.

To calculate the energy fluctuations, we first use the approximate equation (5.10.19) to find the expression for the spectral density of energy,

$$(\delta E)^2_\omega = \frac{2D\langle E \rangle}{\omega^2 + (\Delta\omega_E)^2}, \quad \Delta\omega_E = a + 2b\langle E \rangle. \tag{5.11.9}$$

The expression for the dispersion of energy fluctuations follows,

$$\langle (\delta E)^2 \rangle = \frac{1}{2\pi} \int_{-\infty}^{\infty} (\delta E)^2_\omega d\omega = D\langle E \rangle / (a + 2b\langle E \rangle). \tag{5.11.10}$$

Again we consider three special cases.

1) The state of equilibrium at $\eta = 0$:

$$\Delta\omega_E = a, \quad \langle E \rangle = \mathscr{K}T/m, \quad \langle (\delta E)^2 \rangle = (\mathscr{K}T/m)^2. \tag{5.11.11}$$

2) The generation threshold [compare with the exact result (5.11.7) for $\langle (\delta E)^2 \rangle$]:

$$\Delta\omega_E = 2b\langle E \rangle, \quad \langle (\delta E)^2 \rangle = D/2b. \tag{5.11.12}$$

3) The regime of well-developed generation:

$$\Delta\omega_E = |a|, \quad \langle (\delta E)^2 \rangle = \frac{D}{b} = \frac{Db}{a^2}\langle E \rangle^2 \equiv \varepsilon\langle E \rangle^2. \tag{5.11.13}$$

5.12 Fluctuations of Amplitude. Diffusion of a Phase

Let us obtain the Fokker–Planck equation for the distribution function of amplitude A and phase φ. We use the relations between variables x, v, A, and φ,

$$x = \frac{A}{\omega_0} \cos \varphi, \quad v = -A \sin \varphi, \quad E = A^2/2, \tag{5.12.1}$$

and of the definition of the distribution function

$$f(A, \varphi, t) = \int \delta(A - \sqrt{v^2 + \omega_0^2 x^2}) \, \delta(\varphi + \arctan(v/\omega_0 x)) \, f(x, v, t) \, dx \, dv,$$

$$\int_0^\infty \int_0^{2\pi} f(A, \varphi, t) \, dA \, d\varphi = 1. \tag{5.12.2}$$

Using (5.10.11) for the distribution function we obtain the sought equation,

$$\frac{\partial f}{\partial t} = \frac{D}{2} \frac{\partial^2 f}{\partial A^2} + \frac{D}{2A^2} \frac{\partial^2 f}{\partial \varphi^2} + \frac{\partial}{\partial A} \left\{ \left[\frac{1}{2} \left(a + b \frac{A^2}{2} \right) A - \frac{D}{2A} \right] f \right\}. \tag{5.12.3}$$

We need the equation for the mean value of amplitude. With the aid of (5.12.3) we find

$$\frac{d\langle A \rangle}{dt} = \frac{D}{2} \left\langle \frac{1}{A} \right\rangle - \frac{1}{2} \left(a\langle A \rangle + b \frac{\langle A^2 \rangle}{2} \right). \tag{5.12.4}$$

The Langevin equations corresponding to the Fokker–Planck equation (5.12.3) can be written in the form

$$\frac{dA}{dt} + \frac{1}{2} \left(a + b \frac{A^2}{2} \right) A = y_a(t),$$

$$\frac{d\varphi}{dt} = \frac{1}{A} y_\varphi(t) \tag{5.12.5}$$

with the designations

$$y_A(t) = \omega_0 y_x \cos \varphi - y_v \sin \varphi,$$

$$y_\varphi(t) = -\omega_0 y_x \sin \varphi - y_v \cos \varphi. \tag{5.12.6}$$

The mean value of the source y_A is nonzero and can be found from (5.12.4)

$$\langle y_A \rangle = \frac{D}{2} \left\langle \frac{1}{A} \right\rangle. \tag{5.12.7}$$

Let us now calculate the fluctuations of amplitude. From (5.12.4) in the absence of noise ($D = 0$) we find

$$\langle A \rangle_0 = \sqrt{2|a|/b} \ . \tag{5.12.8}$$

In (5.12.5) we set

$$A = \langle A \rangle_0 + \delta A \tag{5.12.9}$$

and write the equation for fluctuations in linear approximation

$$\frac{d\delta A}{dt} + \frac{1}{2} b \langle A \rangle_0^2 \delta A = \delta y_A(t) ,$$

$$\langle \delta y_A(t) \delta y_A(t') \rangle = D \delta(t - t') . \tag{5.12.10}$$

Hence follow the expressions for spectral density and dispersion of amplitude fluctuations

$$(\delta A)_\omega^2 = \frac{D}{\omega^2 + (\Delta \omega_A)^2} , \quad \langle (\delta A)^2 \rangle = \frac{D}{b \langle A \rangle_0^2} = \frac{D}{2|a|} \tag{5.12.11}$$

with a designation for the halfwidth of the spectral line

$$\Delta \omega_A = \tfrac{1}{2} b \langle A \rangle_0^2 = |a| . \tag{5.12.12}$$

From (5.12.8, 11) it follows that

$$\langle (\delta A)^2 \rangle / \langle A \rangle_0^2 = Db/4a^2 \equiv \varepsilon/4 \tag{5.12.13}$$

where we have introduced a designation for a dimensionless parameter [cf. (5.5.17)],

$$\varepsilon = Db/a^2 . \tag{5.12.14}$$

For well-developed generation, $\varepsilon \ll 1$ and therefore fluctuations of amplitude are small.

In zero-order approximation with respect to fluctuations of amplitude the Langevin equation for phase becomes

$$\frac{d\varphi}{dt} = \frac{1}{\langle A \rangle} y_\varphi(t) , \quad \langle y_\varphi(t) y_\varphi(t') \rangle = D \delta(t - t') . \tag{5.12.15}$$

The corresponding equation for the distribution function of phase is [cf. (5.3.7, 12)]

$$\frac{\partial f(\varphi, t)}{dt} = D_\varphi \frac{\partial^2 f}{\partial \varphi^2} , \quad D_\varphi = \frac{D}{2 \langle A \rangle^2} . \tag{5.12.16}$$

From this [cf. (5.3.10)] we find the two-time distribution

$$f(\varphi, t; \varphi', t-\tau) = \frac{1}{\sqrt{4\pi D_\varphi \tau}} \exp\left(-\frac{\Delta\varphi_\tau^2}{4D_\varphi\tau}\right); \quad \Delta\varphi_\tau = \varphi(t) - \varphi(t-\tau),$$

$$(5.12.17)$$

which describes the process of phase diffusion. In the next section we shall show that phase diffusion plays an important role in the process of establishing the distribution of autooscillation energy over frequency.

5.13 Spectral Distribution of the Energy of Autooscillations

In calculating the distribution of oscillation energy over frequency for a stationary state, it is convenient to use the following designation for the half-sum of two-time correlations of slow-changing coordinates and velocities (sign $\sim$ omitted over x, v) [cf. (5.6.1)]:

$$\langle E(\tau)\rangle = \tfrac{1}{2}[\omega_0^2\langle x(t)x(t-\tau)\rangle + \langle v(t)v(t-\tau)\rangle]. \tag{5.13.1}$$

At $\tau = 0$ this expression is connected with the mean energy

$$\langle E(0)\rangle \equiv \langle E\rangle = \frac{1}{2\pi} \int_{-\infty}^{\infty} E(\omega)d\omega. \tag{5.13.2}$$

The function $E(\omega)$ describes the distribution of the mean energy over frequencies.

In order to find the equation for the function (5.13.1), consider the Fokker – Planck equation as an equation for the two-time distribution function $f(x, v, t; x', v', t')$, and utilize it to obtain the desired equation

$$\frac{d\langle E(\tau)\rangle}{d\tau} + \frac{1}{2}\langle(a + bE)E(\tau)\rangle = 0. \tag{5.13.3}$$

This equation is not closed since it includes, besides $\langle E(\tau)\rangle$, a higher moment

$$\langle EE(\tau)\rangle = \langle E\rangle\langle E(\tau)\rangle + \langle \delta E\delta E(\tau)\rangle. \tag{5.13.4}$$

We start by calculating the spectral energy distribution for a regime of well-developed generation, when $\varepsilon \ll 1$ (5.2.14). We have seen that in zero-order approximation the fluctuations of amplitude can be neglected, see (5.12.13). Then, taking (5.12.1) into account, we can present (5.13.1) in the form

$$\langle E(\tau)\rangle = \tfrac{1}{2}\langle A(t)A(t-\tau)\cos[\varphi(t) - \varphi(t-\tau)]\rangle$$

$$= \tfrac{1}{2}\langle A\rangle_0^2\langle\cos[\varphi(t) - \varphi(t-\tau)]\rangle. \tag{5.13.5}$$

Then in zero-order approximation in ε, the calculation of the spectral distribution of energy becomes the calculation of the mean value of the cosine of phase gain. Using the distribution function (5.12.17) we find

$$\langle E(\tau)\rangle = \frac{\langle A\rangle_0^2}{2} e^{-D_\varphi|\tau|}, \quad E(\omega) = \frac{2D_\varphi\langle E\rangle_0}{\omega^2 - D_\varphi^2}. \tag{5.13.6}$$

It follows that the halfwidth of the energy distribution spectrum is determined by the coefficient of phase diffusion,

$$\Delta\omega = D_\varphi = \frac{D}{2\langle A\rangle_0^2} = \frac{D}{4\langle E\rangle_0} \tag{5.13.7}$$

with D_φ defined by (5.12.16).

Thus for the regime of well-developed generation one can introduce two characteristic parameters of time,

$$\tau_A \sim \frac{1}{|a|}, \quad \tau_\varphi = \frac{1}{D_\varphi} \sim \frac{|a|}{Db}; \quad \frac{\tau_A}{\tau_\varphi} \sim \frac{Db}{a^2} = \varepsilon \ll 1. \tag{5.13.8}$$

The first of these (τ_A) determines the relaxation time for the amplitude fluctuations and the second, the time of phase diffusion.

We have found that in zero-order approximation in ε the distribution of energy over the spectrum has the form of a Lorentz line with halfwidth $\Delta\omega = D_\varphi$. In first-order approximation in ε the spectrum consists of two Lorentz lines, one wide and one narrow [5.28, 22]. Taking the amplitude fluctuations in the first approximation in ε into account leads to a broadening of the narrow line proportional to εD_φ. The width of the narrow line is $D_\varphi(1 + 3\varepsilon/4)$, and of the wide line, $4D_\varphi/\varepsilon$.

To estimate the upper and lower limits of the linewidth [5.28] we assume that the spectrum can be presented as a sum of Lorentz lines, which corresponds to presenting the function $\langle E(\tau)\rangle$ in the form of a sum of exponential distributions,

$$\langle E(\tau)\rangle = \sum_i \langle E\rangle_i e^{-\lambda_i|\tau|} \equiv \sum_i \langle E\rangle_i e^{-\lambda|\tau|} = \langle E\rangle e^{-\lambda|\tau|}. \tag{5.13.9}$$

We have replaced the sum of exponential distributions by a single one whose area is equal to mean energy $\langle E\rangle$, $\lambda = \Delta\omega$, where $\Delta\omega$ is the halfwidth of the resulting Lorentz line. In order to determine this λ we substitute (5.13.9) into (5.13.3). As a result we get

$$\lambda = \frac{1}{2\langle E\rangle}\langle(a + bE)E(\tau)\rangle e^{\lambda|\tau|}. \tag{5.13.10}$$

The right-hand side of this equation does not depend on τ if

$$\langle(a + bE)E(\tau)\rangle = Be^{-\lambda|\tau|}. \tag{5.13.11a}$$

We shall find the constant B for two extreme cases.

The first corresponds to zero-order approximation with respect to ε when τ_A is assumed to be equal to zero. This means that $\tau \gg \tau_A$ in (5.13.11). Then in order to find B one should take the moment of time when the amplitude fluctuations have just faded out for the initial one ($\tau = 0$). In other words, one substitutes $A(t-\tau)$ for $\langle \rangle$ in the left-hand side of (5.13.11a), and then sets $\tau = 0$. The value of B will then be determined by

$$B = \left\langle \left(a + b\frac{A^2}{2}\right)\frac{A}{2}\right\rangle \langle A \rangle \, ; \tag{5.13.11b}$$

this can be further transformed with the aid of (5.12.4), assuming that the state is stationary. As a result we get the following expression for B:

$$B = \frac{D}{2}\left\langle \frac{1}{A}\right\rangle \langle A \rangle \approx \frac{D}{2} \, , \tag{5.13.12}$$

and hence, with (5.13.7),

$$\lambda = \Delta\omega = \frac{D}{4\langle E\rangle_0} = D_\varphi \, . \tag{5.13.13}$$

As ought to be expected, in zero-order approximation with respect to amplitude fluctuations (zero-order approximation in ε), the halfwidth of the spectral line coincides with the magnitude of the coefficient of phase diffusion D_φ.

Consider now another extreme case, when all terms in the series (5.13.9) are taken into consideration. Then in order to determine B we simply assume $\tau = 0$ in (5.13.11a). As a result we get the following expression for B:

$$B = \langle (a + bE)E\rangle \, . \tag{5.13.14}$$

In place of (5.12.4) for mean amplitude we now employ the equation for mean energy $\langle E\rangle$ (5.10.15), resulting in

$$B = D \, . \tag{5.13.15}$$

Now B is twice as big as in (5.13.12), and therefore instead of (5.13.13) we have

$$\lambda \equiv \Delta\omega = \frac{D}{2\langle E\rangle_0} = 2D_\varphi \, . \tag{5.13.16}$$

Depending on the circumstances, the halfwidth of the spectral line is confined within the limits

$$D_\varphi \leqslant \Delta\omega \leqslant 2D_\varphi \, . \tag{5.13.17}$$

Here we have obtained the expressions for the halfwidth of the line in the regime of well-developed generation. Based on this one can immediately write the general approximate formulae, which are true for any value of the parameters [5.28]

$$\Delta\omega = \frac{D}{2\langle A\rangle}\left\langle\frac{1}{A}\right\rangle, \quad \Delta\omega = \frac{D}{2\langle E\rangle}. \tag{5.13.18}$$

These equations define the limiting curves between which the true values lie. In the state of equilibrium (far below the generation threshold, when one may assume $b = 0$) they give identical results since

$$\langle E\rangle = \langle A\rangle\Big/\left\langle\frac{1}{A}\right\rangle = \frac{\varkappa T}{m}, \quad D = a\frac{\varkappa T}{m}, \quad \Delta\omega = \frac{a}{2}. \tag{5.13.19}$$

At the generation threshold their results are close to one another, and in the regime of well-developed generation the linewidths differ by a factor of two.

The linewidth can be calculated for a self-consistent field approximation, similar to how it was done in Sects. 5.5, 6 for the model of a phase transition. In a self-consistent field approximation the equations for the stationary state (5.10.15, 13.3) become

$$\frac{d\langle E(\tau)\rangle}{d\tau} + \frac{1}{2}(a + b\langle E\rangle)\langle E(\tau)\rangle = 0, \tag{5.13.20}$$

$$(a + b\langle E\rangle)\langle E\rangle = D. \tag{5.13.21}$$

Like in Sect. 5.5, the latter equations defines the mean energy value $\langle E^*\rangle$, renormalized to take fluctuations into account. It can be easily proven to be connected with the exact value of the mean energy through the equation

$$\langle E^*\rangle = \langle E\rangle + \langle(\delta E)^2\rangle/\langle E\rangle, \tag{5.13.22}$$

which coincides with (5.5.30). The asterisk is omitted in (5.13.20, 21) and further on.

With the aid of the equation for $\langle E\rangle$ we simplify (5.13.20) [cf. (5.6.4, 8)],

$$\frac{d\langle E(\tau)\rangle}{d\tau} + \lambda\langle E(\tau)\rangle = 0, \quad \lambda = D/2\langle E\rangle. \tag{5.13.23}$$

From this at the initial condition $\langle E(0)\rangle = \langle E\rangle$ we find

$$\langle E(\tau)\rangle = \langle E\rangle e^{-\lambda|\tau|}, \quad E(\omega) = \frac{D}{\omega^2 + (\Delta\omega)^2}, \tag{5.13.24}$$

where

$$\Delta\omega = \lambda = D/2\langle E\rangle \tag{5.13.25}$$

is the halfwidth of the mean energy distribution over frequency. This expression coincides with the second part of (5.13.18) with a higher order of accuracy in ε, and consequently in the regime of well-developed generation, $\Delta\omega = 2D_\varphi$ (5.13.25).

Let us use the solution of (5.13.21) to find the halfwidth of the spectral line at any value of the parameters. The values for the three special cases above are:

1) $\Delta\omega = \dfrac{a}{2}$;

2) $\Delta\omega = \frac{1}{2}\sqrt{Db}$;

3) $\Delta\omega = 2D_\varphi$. (5.13.26)

In a self-consistent field approximation one can obtain the curve of the lower values of the linewidth as well [the first formula in (5.13.18) was obtained neglecting the amplitude fluctuations in the nonlinear term in going from (5.13.11a to b)]. To do this we replace $\langle E \rangle$ with $\frac{1}{2}\langle A \rangle^2$ in the nonlinear term of (5.13.20), which corresponds to neglecting the amplitude fluctuations, and use, instead of (5.13.21), the corresponding equation derived from (5.12.4) for the stationary state. As a result we get

$$\frac{d\langle E(\tau) \rangle}{d\tau} + \frac{1}{2}\left(a + b\frac{\langle A \rangle^2}{2}\right)\langle E(\tau) \rangle = 0 ,$$

$$\frac{1}{2}\left(a + b\frac{\langle A \rangle^2}{2}\right)\langle A \rangle = \frac{D}{2}\left\langle \frac{1}{A} \right\rangle \qquad (5.13.27)$$

in place of (5.13.20, 21). From this set we find

$$\frac{d\langle E(\tau) \rangle}{d\tau} + \lambda_A\langle E(\tau) \rangle = 0 , \quad \lambda_A = \frac{D}{2\langle A \rangle}\left\langle \frac{1}{A} \right\rangle , \qquad (5.13.28)$$

and the linewidth is therefore determined by the first equation of (5.13.18).

A method of measuring natural fluctuations in Thomson generators has been developed by *Bershtein* [5.29], see also [5.7].

5.14 The Response to Resonant Force

Two "response functions" can be defined in the region above the generation threshold for the response to a tuned-in external force, like in the model of a phase transition in Sect. 5.7.

If the generator is under the infuence of a resonant force

$$F(t) = X\cos\omega_0 t + Y\sin\omega_0 t , \qquad (5.14.1)$$

the Langevin equations (5.10.7) become

$$\frac{dx}{dt} + \frac{1}{2}(a + bE)x = -\frac{1}{2\omega_0}Y + y_x(t), \quad E = \frac{1}{2}(v^2 + \omega_0^2 x^2)$$

$$\frac{dv}{dx} = \frac{1}{2}(a + bE)v = \frac{1}{2}X + y_v(t) \tag{5.14.2}$$

where X and Y are either constants, or slowly changing (with frequencies within the synchronization band) functions of time. Naturally, the force (5.14.1) may be set in a simpler form, e.g., assuming $Y = 0$.

The equations for the mean values $\langle x \rangle$ and $\langle v \rangle$ are from (5.14.2)

$$\frac{d\langle x \rangle}{dt} + \frac{1}{2}\langle (a + bE)x \rangle = -\frac{1}{2\omega_0}Y,$$

$$\frac{d\langle v \rangle}{dt} + \frac{1}{2}\langle (a + bE)v \rangle = \frac{1}{2}X. \tag{5.14.3}$$

Similar to the treatment of phase transition model in Sect. 5.7, we can introduce a number of susceptibilities which determine the responses to different exertions. Consider the susceptibilities χ and χ_E. By definition,

$$\langle E \rangle_F - \langle E \rangle = \chi_E(\overline{\langle v \rangle_F F})^{2\pi/\omega_0} = \chi_E(-\omega_0\langle x \rangle Y + \langle v \rangle X)/2. \tag{5.14.4}$$

With the aid of (5.14.3) we find

$$\chi_E = \frac{1}{a + 2b\langle E \rangle} = \frac{1}{\Delta \omega_E}. \tag{5.14.5}$$

In order to determine the dynamic susceptibility we consider the response to an alternating force; then

$$\chi_E = \frac{1}{2} \frac{1}{-i\omega + \Delta \omega_E/2}. \tag{5.14.6}$$

The function χ_E is connected with fast fluctuations of energy δE, while $\Delta \omega_E$ is the halfwidth of the spectrum of these fluctuations. In order to obtain the expression for the susceptibility, determined by the spectral linewidth $\Delta \omega$ of slow fluctuations, one may employ the stationary solution of the Fokker – Planck equation accounting for a resonant force. Such an equation includes an additional term [cf. (5.10.11)]

$$\frac{1}{2}\left[\frac{\partial}{\partial x}\left(\frac{Y}{\omega_0}f \right) - \frac{\partial}{\partial v}(Xf) \right]; \tag{5.14.7}$$

its stationary solution has the form

$$f(x, v) = c \exp\left(-\frac{aE + bE^2/2 + \omega_0 x Y - vX}{D}\right).$$ (5.14.8)

In the absence of external force this distribution coincides with (5.11.1).

The mean values $\langle x \rangle$ and $\langle v \rangle$, obtained by dint of the distribution (5.14.8), can be presented in linear approximation in the form

$$\omega_0 \langle x \rangle = -\chi_x Y, \quad \langle v \rangle = \chi_v X, \quad \chi = \tfrac{1}{2}(\chi_x + \chi_v).$$ (5.14.9)

The response χ introduced here is determined by the expressions

$$\chi = \frac{1}{2}\left(\frac{\omega_0^2 \langle x^2 \rangle}{D} + \frac{\langle v^2 \rangle}{D}\right) = \frac{\langle E \rangle}{D} = \frac{1}{2\Delta\omega},$$ (5.14.10)

where $\langle E \rangle$ is the mean value of energy in the absence of external force. It can be found with the aid of the distribution (5.11.1). In the last equation of (5.14.10) we have used (5.13.25) for the halfwidth of the line.

Consider the ratio of the two quantities χ_E and χ. From (5.14.5, 10),

$$\frac{\chi_E}{\chi} = \frac{D}{\langle E \rangle (a + 2b\langle E \rangle)}.$$ (5.14.11)

Let us investigate this ratio for each of the three cases discussed above.

1) The state far below the generation threshold ($a > 0$, $b = 0$),

$$\frac{\chi_E}{\chi} = \frac{D}{a\langle E \rangle} = 1.$$ (5.14.12)

The values of susceptibilities χ_E and χ coincide here.

2) The generation threshold ($a = 0$), taking (5.11.7) into account,

$$\frac{\chi_E}{\chi} = \frac{\pi}{4}, \quad \text{i.e.,} \quad \chi_E \ll \chi.$$ (5.14.13)

3) The regime of well-developed generation,

$$\frac{\chi_E}{\chi} = \frac{Db}{2a^2} = \frac{\varepsilon}{2} \ll 1.$$ (5.14.14)

The ratio χ_E/χ depends on the proximity to the generation threshold in the same manner as for a phase transition (Sect. 5.7).

For the susceptibility χ_E, the dependence on the degree to which the generation threshold is exceeded far from the threshold is determined by

$$\chi_E = \frac{1}{|a|} \quad \text{at} \quad a > 0 \, ; \qquad \chi_E = \frac{1}{|a|} \quad \text{at} \quad a < 0 \, . \tag{5.14.15}$$

They correspond to formulae expressing Curie's law for phase transitions. Indeed, susceptibility increases approaching the generation threshold from both above and below, and crossing the generation threshold plays the role of temperature difference.

The expression for susceptibility χ_E can be also obtained via the distribution function (5.14.8).

To calculate the susceptibilities (5.14.5, 10) one can employ the self-consistent field approximation. The mean energy $\langle E \rangle$ will then be determined by the solution of (5.13.21). Let us use (5.14.3) and find the expression for the dynamic response $\chi(\omega)$ in the self-consistent field approximation. It has the form

$$\chi(\omega) = \frac{1}{2} \, \frac{1}{-i\omega + \Delta\omega} \, , \quad \Delta\omega = D/2\langle E \rangle \, . \tag{5.14.16}$$

The static susceptibility can also be found through the equation for dispersion,

$$\chi(0) = \frac{2}{\pi} \int_0^\infty \frac{\text{Im}\{\chi(\omega)\}}{\omega} \, d\omega = \frac{1}{2\Delta\omega} = \frac{\langle E \rangle}{D} \, . \tag{5.14.17}$$

We see that the dynamic response $\chi(\omega)$ is nonzero within the limits of the line $\Delta\omega$, and $\chi_E(\omega)$ is nonzero within the limits of the line $\Delta\omega_E$. In the regime of well-developed generation $\Delta\omega/\Delta\omega_E \sim \varepsilon \ll 1$, and therefore the former region is much narrower than the latter.

The distribution of the mean energy over frequencies $E(\omega)$ can also be found by using the function $\chi(\omega)$ if one uses an expression similar to the fluctuation dissipation theorem

$$E(\omega) = \frac{2\,\text{Im}\{\chi(\omega)\}}{\omega} \quad D = \frac{D}{\omega^2 + (\Delta\omega)^2} \, . \tag{5.14.18}$$

This expression coincides with (5.13.24).

We see that in the regime of well-developed generation the susceptibility χ is much greater than χ_E. The function $\text{Re}\{\chi(\omega)\}$ however, is only nonzero in a narrow band $\Delta\omega \ll \Delta\omega_E$.

In this case we also notice an analogy with phase transitions. For states far below the critical point (which correspond to a regime of well-developed generation), $\chi_\eta/\chi \sim \varepsilon \ll 1$. At phase transitions it is hard to discern the response determined by susceptibility $\chi(\omega)$ because it is only nonzero in a narrow band $\Delta\omega$ around $\omega = 0$. One therefore usually notices the response determined by χ_η

(see the end of Sect. 5.8). In order to be able to study the response determined by $\chi(\omega)$ at phase transitions, the influence of excess noise has to be minimized.

So far we have only discussed linear response, i.e., the response to weak influence. Let us now consider an example of the response to a strong force. Let $X = 0$ in (5.14.1, 8). With the aid of (5.12.1) we can rewrite (5.14.9) in the form

$$\langle x \rangle = \left\langle \frac{A}{\omega_0} \cos \varphi \right\rangle = \chi_x Y/\chi_0 . \tag{5.14.19}$$

Consider the regime of well-developed generation. In zero-order approximation with respect to fluctuation amplitude $A = A_0 = \sqrt{2|a|/b}$, and the distribution over phases follows from (5.14.8) and has the form

$$f(\varphi) = c \exp\left(-\frac{A_0 Y}{D} \cos \varphi\right), \quad \frac{1}{2\pi} \int_0^{2\pi} f(\varphi) d\varphi = 1 . \tag{5.14.20}$$

With the help of this distribution we find the equation similar to the Langevin formula in the theory of paramagnetics,

$$\langle \cos \varphi \rangle = I_1(d)/I_0(d) , \quad d = A_0 Y/D . \tag{5.14.21}$$

For two extreme cases of very strong and very weak influence we find

$$\langle \cos \varphi \rangle = d/2 \quad |d| \ll 1 , \quad \langle \cos \varphi \rangle = \pm 1 \quad |d| \gg 1 . \tag{5.14.22}$$

A number of works have been dedicated to the experimental and theoretical study of spatial correlations in lasers. It has been discovered that there are correlations with characteristic lengths described by formulae similar to those for the correlation radius of the order parameter at phase transitions. This once more proves the similarities between processes occurring at phase transitions and in self-oscillatory systems (Sect. 12.6, 7).

5.15 Kinetic Theory of Fluctuations in Brownian Motion

At the end of Sect. 5.2 we pointed out that the Fokker–Planck equations (5.2.10) were obtained under the neglect of large–scale fluctuations (fluctuations with $\tau_{cor} > 1/\gamma$). For that reason the distribution function in the Fokker–Planck equation is deterministic (nonrandom). The theory of large-scale kinetic fluctuations in Brownian motion can be constructed analogous to the kinetic theory of fluctuations in a plasma (Sect. 4.8) [5.30].

The equation for the smoothed phase density of charged particles $\tilde{N}$ [see (4.8.1)] may serve as the initial equation,

$$\frac{\partial \tilde{N}}{\partial t} + v \frac{\partial \tilde{N}}{\partial r} + \tilde{F} \frac{\partial \tilde{N}}{\partial p} = D \frac{\partial^2 \tilde{N}}{\partial p^2} + \frac{\partial}{\partial p} (\gamma p \tilde{N}) , \qquad (5.15.1)$$

where $\tilde{F}$ is a random force with $\tau_{cor} \gtrsim 1/\gamma$ and $D = \gamma m \varkappa T$.

We average this equation and use (5.2.4), getting the equation for a one-particle distribution function,

$$\frac{\partial f}{\partial t} + v \frac{\partial f}{\partial r} + F \frac{\partial f}{\partial p} = I + \tilde{I} \qquad (5.15.2)$$

where I is the collision integral in the Fokker – Planck equation, see (5.2.10), and

$$\tilde{I} = - \frac{1}{n} \frac{\overline{\partial \delta \tilde{N} \delta \tilde{F}}}{\partial p} \qquad (5.15.3)$$

is an additional collision integral which is determined by large-scale fluctuations; F is the mean force.

The equation for δN can be written similar to (4.8.15),

$$\left(\frac{\partial}{\partial t} + v \frac{\partial}{\partial r} + F \frac{\partial}{\partial p} + \delta \hat{I} \right) \delta N + \delta F \frac{\partial nf}{\partial p} = y(x, t) . \qquad (5.15.4)$$

The operator $\delta \hat{I}$ is determined by the equation

$$\delta \hat{I} = - D \frac{\partial^2}{\partial p^2} - \frac{\partial}{\partial p} (\gamma p \dots) , \qquad (5.15.5)$$

and the Langevin source correlation is given by

$$\overline{(yy)}_{x,t;x',t'} = A(x, x', t) \, \delta(t - t') . \qquad (5.15.6)$$

Like in Sect. 4.8, see (4.8.20), the intensity of the Langevin source can be presented as a sum of two parts,

$$A(x, x', t) = \tilde{A} + A^{F-P} , \qquad (5.15.7)$$

the first term being described by the collision integral $\tilde{I}$ [cf. (4.8.21)],

$$\tilde{A} = n \left\{ \delta(x - x') \tilde{I}(x, t) - \frac{1}{V} [\tilde{I}(x, t) f(x', t) + \tilde{I}(x', t) f(x, t)] \right\} . \qquad (5.15.8)$$

The expression for the second term follows from (4.8.22) by substituting (5.15.5) it,

$$A^{F-P} = 2D\delta(r-r')\frac{\partial^2}{\partial p \partial p'} n\delta(p-p')f(r,p,t).$$ (5.15.9)

Equations (5.15.2, 4) form a closed set of equations for the one-particle distribution function and large-scale fluctuations. The Fokker–Planck equation follows from this set of equations in first moments approximation when an additional integral $\bar{I}$ in (5.15.2) is equal to zero.

The characteristic relaxation times in (5.15.2, 4) are, in general, of the same order of magnitude. This makes it impossible to express the collision integral through f in an explicit form and thus obtain a kinetic equation accounting for large-scale fluctuations without further concessions (Sect. 4.8).

Equations (5.15.2, 4) can be united in a single one,

$$\frac{\partial \tilde{N}}{\partial t} + v\frac{\partial \tilde{N}}{\partial r} + F\frac{\partial \tilde{N}}{\partial p} + \delta\tilde{F}\frac{\partial nf}{\partial p} = D\frac{\partial^2 \tilde{N}}{\partial p^2} + \frac{\partial(\gamma p\tilde{N})}{\partial p} + \tilde{I} + y(x,t).$$ (5.15.10)

If averaged, this equation coincides with (5.15.2); if written for δN, with (5.15.4).

With the aid of both of these equations one can also obtain, e.g., the equation for density fluctuations which account for large-scale fluctuations of the force acting on a Brownian particle. In spatial diffusion approximation this equation has the form

$$\frac{\partial}{\partial t}\delta n - D\Delta_r\delta n + \frac{1}{m\gamma}n\frac{\partial}{\partial r}\delta\tilde{F} = \eta(r,t), \quad D = \frac{\mathcal{K}T}{m\gamma}.$$ (5.15.11)

For a spatially uniform distribution of Brownian particles the Langevin source correlator η is determined by the expression

$$(\eta\eta)_{r,t;r',t'} = 2D\delta(t-t')\frac{\partial^2}{\partial r \partial r'}n\delta(r-r').$$ (5.15.12)

Let us also cite the expression for spectral density of fluctuations $(\delta n\,\delta n)_{\omega,k}$,

$$(\delta n\,\delta n)_{\omega,k} = \frac{nk^2}{\omega^2 + D^2k^4}\left[2D + \frac{n}{m^2\gamma^2}(\delta\tilde{F}\delta\tilde{F})_{\omega,k}\right].$$ (5.15.13)

The second term here describes the contribution of large-scale fluctuations of forces acting on Brownian particles. In a turbulent medium this can be the dominant contribution.

To conclude this chapter two more problems connected with Brownian motion which are of principal interest will be discussed. The results obtained in Sect. 5.3 regarding the diffusion of Brownian particles are true for a limitless system. In a limited system, however, the nature of the diffusion of "Brownian particles" of any kind is quite different. At long times (low frequencies), when

$\sqrt{D\tau}$ (or $\sqrt{D/\omega}$) is much greater than the dimensions of the system, additional time correlations appear. Correspondingly, the spectrum $1/\omega$ is born − natural flicker noise. The minimal frequency of the flicker noise is restricted only by the length of observation. These problems are discussed in [5.31].

The formulae for the correlator of the sources of hydrodynamic fluctuations (5.1.2) which determine the random force in the Langevin equations (5.1.3) are true, as noted in Sect. 5.1, only for the state of local equilibrium. At large deviations from equilibrium, additional terms appear in the expressions for the strengths of the sources of hydrodynamic fluctuations; they are proportional to squares of the derivatives of velocity and temperature. Correspondingly, one can then introduce the nonlinear coefficients of viscosity and thermal conductivity. Nonlinear (with respect to gradients of velocity and temperature) dissipative terms then appear in the hydrodynamic equations, and can be important for both turbulent and laminar flows. These problems are treated in [5.32].

6. Kinetic Equations for an Atom – Field System

Two approaches are developed to describe kinetic processes in an atom – field system. The first is based on the polarization approximation, already employed in Chap. 4. It is convenient for taking into account the collective effects in a rarefied gas. The second is based on the distinction made between fast (intraatomic) movements and the motion of an atom as a whole. The equation for the polarization vector is employed as an example to demonstrate the influence of an atom's correlations in a system of arbitrary density on the effective field (Lorentz field), frequency shift, and damping. The theory of large-scale (kinetic) fluctuations is developed and employed to describe a phase transition in a system of atoms and a field.

6.1 Electromagnetic Fluctuations in a Gas

The kinetic theory of a gas which accounts for the inner degrees of freedom is naturally far more complicated than that of a totally ionized plasma (see [Ref. 6.1, Chap. 14]). We shall start with a relatively simple case, i.e., when the concentration of atoms is so small that one may take a zero-order approximation with respect to each of the density parameters ε, ε_d, ε_{em} (Sect. 4.2). The dissipative processes are then determined by the spontaneous emission of radiation by oscillating atoms, and the principal temporal parameters are determined by (4.2.16).

The kinetic equations can be obtained if we start from the set of equations for the phase density of atoms N_{ab} and the microscopic field strengths. First we shall consider the dipole approximation. For the function N_{ab} (3.2.8) can be employed, in which we set $-\partial \Phi/\partial r = \mu \omega_0^2 r$ for a system of oscillating atoms. We employ the method used in Sect. 4.4, presented in greater detail in [Ref. 6.1, Chap. 7], to construct the kinetic equation.

Before averaging (3.2.8), we replace the variables r and p by r_0 and p_0 according to [cf. (5.10.4)]

$$r = r_0 \cos \omega_0 t + \frac{p_0}{m \omega_0} \sin \omega_0 t , \quad p = p_0 \cos \omega_0 t - m \omega_0 r \sin \omega_0 t , \qquad (6.1.1)$$

and the function $N_{ab}(r, p, R, P, t)$ by a new function

$$N_{ab}(r_0, p_0, R, P, t) \equiv N_{ab}\left(r_0 \cos \omega_0 t + \frac{p_0}{m \omega_0} \sin \omega_0 t, \right.$$
$$\left. p_0 \cos \omega_0 t - m \omega_0 r_0 \sin \omega_0 t, R, P, t\right). \qquad (6.1.2)$$

For the new functions we retain the same designations and distinguish them by their arguments.

The equation for the distribution function in new variables $X_0 = (r_0, p_0, R, P)$ follows from (3.2.8) after averaging and has the form

$$\left(\frac{\partial}{\partial t} + v \frac{\partial}{\partial R} \right) f(x_0, t) + e \, \mathscr{E}(R, t) \left(\cos \omega_0 t \frac{\partial}{\partial p_0} - \frac{1}{m \omega_0} \sin \omega_0 t \frac{\partial}{\partial r_0} \right) f$$

$$\equiv \hat{L}_{X_0} f = I(X_0, t), \quad X_0 = (x_0, R, P) \tag{6.1.3}$$

with the Lorentz force

$$e \, \mathscr{E} = eE + \frac{e}{c}[v \times B] \tag{6.1.4}$$

and the collision integral

$$I = -\frac{e}{n} \left(\cos \omega_0 t \frac{\partial}{\partial p_0} - \frac{1}{m \omega_0} \sin \omega_0 t \frac{\partial}{\partial r_0} \right) \overline{(\delta N \delta \mathscr{E})}_{X_0, R, t}, \tag{6.1.5}$$

where $\delta \mathscr{E}$ are Lorentz force fluctuations. The collision integral is expressed here, as in the case of a plasma, by a correlation of fluctuations δN and $\delta \mathscr{E}$.

To calculate the fluctuations we employ the same approximation as in Sect. 4.13; the equation for δN in variables r_0, p_0, R, P thus is similar in form to (4.4.10),

$$\hat{L}_{X_0} (\delta N - \delta N^{\text{source}})_{X_0, t}$$

$$= - e \delta \mathscr{E}(R, t) \left(\cos \omega_0 t \frac{\partial}{\partial p_0} - \frac{1}{m \omega_0} \sin \omega_0 t \frac{\partial}{\partial r_0} \right) n f(X_0, t). \tag{6.1.6}$$

Here we have used the operator introduced in (6.1.3).

The correlation of source fluctuations can be found from the equation

$$\hat{L}_{X_0} \overline{(\delta N \delta N)}^{\text{source}}_{X_0 t; X_0', t'} = 0, \tag{6.1.7}$$

which is solved at the initial $(t = t')$ condition,

$$\overline{(\delta N \delta N)}^{\text{source}}_{X_0, X_0', t} = n \delta(X_0 - X_0') f(X_0, t). \tag{6.1.8}$$

We presume that the mean field has no influence upon a collision itself; thus $\mathscr{E}$ may be set equal to zero in the equation for fluctuations.

From (6.1.3) it follows that $f(X_0, t)$ is a slowly changing function at $\gamma \ll \omega_0$ (γ is radiative friction). In zero-order approximation with respect to γ / ω_0, the expression for correlation is, from (6.1.7),

$$\overline{(\delta N \delta N)}^{\text{source}}_{X_0, t; X_0', t'} = n \delta(R - V(t - t') - R') \delta(P - P') \delta(x_0 - x_0') f(X_0, t), \tag{6.1.9a}$$

where $x_0 = (r_0, p_0)$. This leads to the corresponding equation for spectral density,

$$(\delta N \delta N)_{\omega, k, R, t, x_0, x_0'} = n 2\pi \delta(\omega - kV) \delta(P - P') \delta(x_0 - x_0') f(X_0, t). \quad (6.1.9b)$$

With the aid of these results we can find expressions for the correlations and spectral densities of the source of polarization and field. Let us use the expression for the polarization vector fluctuations,

$$\delta P(R, t) = e \int \left(r_0 \cos \omega_0 t + \frac{p_0}{m \omega_0} \sin \omega_0 t \right) \delta N(x_0, R, P, t) dx_0 dP. \quad (6.1.10)$$

With the aid of (5.1.9a) we find

$$\overline{(\delta P \delta P)}_{k, t; R; t-\tau}^{\text{source}}$$

$$= e^2 n \int \delta(R - V\tau - R') \frac{v_0^2 + \omega_0^2 r_0^2}{2\omega_0} f(x_0, R, P, t) dx_0 dP \cos \omega_0 \tau. \quad (6.1.11a)$$

Hence the spectral density is described by the expression

$$(\delta P \delta P)_{\omega, k, R, t}^{\text{source}} = e^2 n \int 2\pi \frac{\delta(\omega - kV - \omega_0) + \delta(\omega - kV + \omega_0)}{2}$$

$$\times \frac{v_0^2 + \omega_0^2 r_0^2}{2\omega_0^2} f(x_0, R, P, t) dx_0 dP. \quad (6.1.11b)$$

For the state of equilibrium this expression can be presented in the form

$$(\delta P \delta P)_{\omega, k}^{\text{source}} = \frac{1}{2\pi \omega} \text{Im}\{\varepsilon(\omega, k)\} 3 \mathcal{K}T, \quad (6.1.12)$$

where $\varepsilon(\omega, k)$ is the relevant dielectric permittivity.

In order to find the dielectric permittivity we employ the solution of (6.1.6). In zero-order approximation with respect to γ/ω_0 it has the form

$$\delta N(X_0, t) = \delta N^{\text{source}}(X_0, t) - e \int_0^\infty e^{-\Delta \tau} \delta \mathcal{E}(R - V\tau, V, t - \tau)$$

$$\times \left[\cos \omega_0(t - \tau) \frac{\partial}{\partial p_0} - \frac{1}{m \omega_0} \sin \omega_0(t - \tau) \frac{\partial}{\partial r_0} \right] n f(x_0, R, P, t - \tau) d\tau. \quad (6.1.13)$$

Here we again encounter the problem of separating the fluctuations into small-scale and large-scale ones. Small-scale (quick) fluctuations with $\tau_{\text{cor}} \sim \tau_{\text{ph}}^{\text{sp}}$ (4.2.16) determine the collision integral, while slow fluctuations are themselves determined by the collision integral, i.e., by the kinetic equation. In (6.1.13) $\Delta \sim 1/\tau_{\text{ph}}^{\text{sp}}(\gamma \ll \Delta \ll \omega_0)$. From (4.2.16) it is clear that too low a concentration of atoms prevents singling out fast fluctuations (and hence the construction of a corresponding kinetic equation) since $\tau_{ph}^{sp} \to \infty$ at $n \to 0$. The minimal concentration $n_{min} \sim 1/\varepsilon_{em}^3$ at $\gamma \sim 10^8 \, s^{-1}$, $v \sim 10^5 \, cm \, s^{-1}$ is $n_{min} \sim 10^9 \, cm^{-3}$.

The expression for the polarization vector fluctuation in zero-order approximation in $\partial f/\partial t$ is, from (6.1.13),

$$\delta P(R, P, t)$$
$$= \delta P^{source} + \frac{e^2 n}{m \omega_0} \int_0^\infty e^{-\Delta \tau} \sin \omega_0 \tau \, \delta \mathscr{E} \, (R - V\tau, V, t - \tau) \, d\tau \, f(R, P, t) \quad (6.1.14)$$

where $f(R, P, t)$ is the distribution over coordinates and momenta. Having carried out the Fourier transform with respect to t and R, we find from (6.1.14)

$$\delta P(\omega, k, V) = \delta P^{source} + \alpha(\omega, k, V) \delta \mathscr{H}(\omega, k, V) . \quad (6.1.15)$$

Here we use the designation for electric polarizability for a group of particles with momentum $P = MV$,

$$\alpha(\omega, k, V) = \frac{e^2 n}{m} f(R, P, t) \frac{1}{\omega_0^2 - (\omega - kV + i\Delta)^2} . \quad (6.1.16)$$

Hence the permittivity is described by

$$\varepsilon(\omega, k) = 1 + \frac{4\pi e^2 n}{m} \int \frac{f(R, P, t)}{\omega_0^2 - (\omega - kV + i\Delta)^2} \, dP \quad (\Delta \to 0) , \quad (6.1.17)$$

from which the expression for the imaginary part of permittivity follows,

$$Im\{\varepsilon(\omega, k)\} = \frac{2\pi^2 e^2 n}{m \omega_0} \int [\delta(\omega - \omega_0 - kV) - \delta(\omega + \omega_0 - kV)] \, f \, dP . \quad (6.1.18)$$

In the case of a Maxwell distribution we find from (6.1.17)

$$\varepsilon(\omega, k) = 1 + \frac{2\pi e^2 n}{\omega_0 \sqrt{2\pi k v_T}} (I_- - I_+) , \quad (6.1.19)$$

with the designations for integrals

$$I_{\mp} = \int_{-\infty}^\infty \frac{e^{-t^2}}{t - Z_\pm - i\Delta} \, dt = i\pi e^{-Z_\mp^2} - 2\sqrt{\pi} e^{-Z_\mp^2} \int_0^{Z_\mp} e^{\pm t^2} dt . \quad (6.1.20)$$

In these formulae,

$$v_T = \sqrt{\frac{\mathscr{K} T}{M}} , \quad Z_\mp = \frac{\omega \mp \omega_0}{\sqrt{2} k v_T} . \quad (6.1.21)$$

At $Z_- \ll 1$, $Z_+ \gg 1$,

$$\mathrm{Re}\{\varepsilon\} = 1 - \frac{4\pi e^2 n}{m\omega_0\sqrt{2kv_T}} \frac{\omega - \omega_0}{\sqrt{2kv_T}},$$

$$\mathrm{Im}\{\varepsilon\} = \sqrt{\pi}\, \frac{2\pi e^2 n}{m\omega_0\sqrt{2kv_T}} e^{-Z_-^2}. \tag{6.1.22}$$

Notice that (6.1.19 – 22) are true provided that $\Delta/kv_T \sim (\tau_{\mathrm{ph}}^{\mathrm{sp}} kv_T)^{-1} \ll 1$ when the Doppler linewidth kv_T is much greater than $1/\tau_{\mathrm{ph}}^{\mathrm{sp}}$. Hence

$$kv_T \gg 1/\tau_{\mathrm{ph}}^{\mathrm{sp}} \gg \gamma. \tag{6.1.23}$$

These conditions can only be satisfied for nonuniform broadening when $kv_T \gg \gamma$ and in a certain range of concentration values which can be determined from the definition for $\tau_{\mathrm{ph}}^{\mathrm{sp}}$ (4.2.16). Provided

$$\omega_0 > 1/\tau_{\mathrm{ph}}^{\mathrm{sp}} \gg kv_T, \gamma \tag{6.1.24}$$

in zero-order approximation with respect to $kv_T\tau_{\mathrm{ph}}^{\mathrm{sp}}$, the Maxwell distribution function is replaced by $\delta(P)$ and therefore

$$\varepsilon(\omega, k) = 1 + \frac{4\pi e^2 n}{m} \frac{1}{\omega_0^2 - (\omega + i\Delta)^2}. \tag{6.1.25}$$

Conditions (6.1.24) can be satisfied both at uniform and nonuniform broadening.

The spectral density of the fluctuations of an electromagnetic field can be calculated on the basis of (4.4.14, 5.1) by using the expressions

$$(\delta E^{\parallel} \delta E^{\parallel})_{\omega,k}^{\mathrm{source}} = (4\pi)^2 (\delta P^{\parallel} \delta P^{\parallel})_{\omega,k}^{\mathrm{source}},$$

$$(\delta E^{\perp} \delta E^{\perp})_{\omega,k}^{\mathrm{source}} = (4\pi)^2 (\delta P^{\perp} \delta P^{\perp})_{\omega,k}^{\mathrm{source}}. \tag{6.1.26}$$

The spectral densities of fluctuations $\delta P^{\parallel}$ and $\delta P^{\perp}$ are, in turn, determined by (6.1.11 b) where $v^2 + \omega_0 r^2$ is replaced,

$$v^2 + \omega_0^2 r^2 \to \frac{(k \cdot v)^2}{k^2} + \omega_0^2 \frac{(k \cdot r)^2}{k^2}, \quad \frac{[k \times v]^2}{k^2} + \frac{\omega_0^2 [k \times r]^2}{k^2}. \tag{6.1.27}$$

For the state of equilibrium we once more obtain (4.4.14, 19, 7.16, 17) with dielectric permittivity given by (6.1.17). Thus in the state of equilibrium the spectral density of electromagnetic field fluctuations can be described by

$$(\delta E^{\perp} \delta E^{\perp})_{\omega,k} = \frac{\omega^4 \dfrac{16\pi}{\omega} \left[\mathrm{Im}\{\varepsilon(\omega,k)\} + \dfrac{\gamma_k}{\omega} \right]}{|\omega^2 \varepsilon(\omega,k) + i\omega\gamma_k - c^2 k^2|^2} \mathscr{K}T. \tag{6.1.28}$$

This expression, like (4.7.17), accounts for the losses γ_k, which are not represented by the imaginary part of permittivity of a system of atoms.

Let us consider the region of transparency, and obtain the Rayleigh–Jeans distribution using (6.1.28). In the region of transparency it is possible to carry out a limit transition with attenuation tending to zero. From (6.1.28) we find

$$(\delta E^\perp \delta E^\perp)_{\omega,k} = 16\pi^2 |\omega| \delta(\omega^2 \text{Re}\{\varepsilon\} - c^2 k^2) \mathcal{H}T$$

$$= \frac{8\pi^2}{\text{Re}\{\varepsilon\}} \left[\delta\left(\omega - \frac{ck}{\sqrt{\text{Re}\{\varepsilon\}}} \right) + \delta\left(\omega + \frac{ck}{\sqrt{\text{Re}\{\varepsilon\}}} \right) \right] \mathcal{H}T. \tag{6.1.29}$$

Integrating over k leads to [Ref. 6.2, Eq. (9.11)]

$$(\delta E^\perp \delta E^\perp)_\omega = 4 \frac{\sqrt{\text{Re}\{\varepsilon\}}\,\omega^2}{c^3} \mathcal{H}T \tag{6.1.30}$$

for the distribution over frequencies. This depends on the index of refraction of the medium $n = \sqrt{\text{Re}\{\varepsilon\}}$. At $n = 1$, (6.1.30) yields the Rayleigh–Jeans distribution,

$$\rho_\omega = \frac{1}{4\pi^2} (\delta E^\perp \delta E^\perp)_\omega = \frac{\omega^2}{\pi^2 c^3} \mathcal{H}T. \tag{6.1.31}$$

We employ (6.1.30) in studying a phase transition in an atom–field system.

6.2 The Kinetic Equation. The Collision Integral

In the accepted approximation the forces acting on atoms are not taken into account. The distribution of atoms over velocities therefore remains unchanged in the process of relaxation and is assumed to be fixed and described, e. g., by a Maxwell distribution. This approximation does not provide for a consistent account of the influence of atoms' motion on the dissipative processes described by the collision integral. For that reason we shall only retain zero-order terms with respect to kv_T (Doppler broadening) over ω_0 in our calculation of the spectral densities of fluctuations. In addition, in order to single out the dissipative contribution to the collision integral due to radiative friction, we shall only take into consideration the transverse field fluctuations. The role of longitudinal field fluctuations will be considered in the next section.

Substituting (6.1.13) into the expression for the collisions integral (6.1.5), the latter can be presented as a sum of two parts,

$$I(x_0, \boldsymbol{R}, \boldsymbol{P}, t) = I^{\text{ind}} + I^{\text{source}}. \tag{6.2.1}$$

The first term is called the "induced" part since it is proportional to the correlation of field fluctuations,

$$I^{ind} = e^2 \int_0^\infty d\tau (\delta E_i \delta E_j)_{R - V\tau, t - \tau, R, t}([\,]_{t - \tau})_i([\,]_t)_j f(x_0, R, P, t) .$$ (6.2.2)

Here we have used the designation for operator

$$[\,]_t = \cos \omega_0 t \frac{\partial}{\partial p_0} - \frac{1}{m \omega_0} \sin \omega_0 t \frac{\partial}{\partial r_0} .$$ (6.2.3)

Hereafter we shall omit the subscript "0" in r_0 and p_0. The second term in (6.2.1) is proportional to δN^{source} and is determined by

$$I^{source} = - \frac{e}{n} [\,]_t (\delta N^{source} \delta E)_{x_0, R, t} .$$ (6.2.4)

From (6.1.3) it follows that the harmonics of function f are of the order γ/ω_0, so that the main contribution to (6.2.1) is provided by a slowly changing part of the distribution function.

The expression for I^{ind} in zero-order approximation with respect to lagging [see (6.2.9)] as well as with respect to $k v_T/\omega_0$ can be presented in the form

$$I^{ind} = \frac{e^2}{4(2\pi)^3} \int dk (\delta E_i \delta E_j)_{\omega_0, k} \left(\frac{\partial^2 f}{\partial p_i \partial p_j} + \frac{1}{m^2 \omega_0^2} \frac{\partial^2 f}{\partial r_i \partial r_j} \right) .$$ (6.2.5)

The tensor of the spectral density of field fluctuations has the structure of (4.5.5), so the integral

$$\frac{1}{(2\pi)^3} \int (\delta E_i \delta E_j)_{\omega, k} \, dk = \frac{1}{3} \delta_{ij} (\delta E \delta E)_\omega .$$ (6.2.6)

As a result (6.2.5) takes the form [6.3]

$$I^{ind} = \frac{D}{2} \left(\frac{\partial^2 f}{\partial p^2} + \frac{1}{m^2 \omega_0^2} \frac{\partial^2 f}{\partial r^2} \right) ,$$ (6.2.7)

with the coefficient of diffusion designated

$$D = \frac{e^2}{6} (\delta E \delta E)_\omega .$$ (6.2.8)

Retaining the first-order term for lagging in the transition from (6.2.1) to (6.2.5), results in [6.3]

$$D = \frac{e^2}{6} \left[(\delta E \delta E)_{\omega_0} + \frac{m}{4\pi e^2 n} \frac{1}{2\pi} \int \frac{\partial \omega (\text{Re}\{\varepsilon\} - 1)}{\partial \omega} \frac{\partial (\delta E \delta E)_\omega}{\partial t} d\omega \right] ,$$ (6.2.9)

instead of (6.2.8).

Looking for the expression for I^{source}, we find, from the wave equation for transverse field and using (6.1.15 – 17), the connection between fluctuations δE, $(\delta P^{\perp})^{source}$,

$$\delta E(\omega, k) = - \frac{4 \pi \omega^2}{\omega^2 \varepsilon(\omega, k) - c^2 k^2} [\delta P^{\perp}(\omega, k)]^{source} . \tag{6.2.10}$$

Now we substitute this equation into (6.2.4) and employ the formula (6.1.9b) for spectral density of fluctuations δN^{source}. Retaining only zero-order terms with respect to $k v_T / \omega_0$, we present the expression for I^{source} in the form

$$I^{source} = \frac{1}{2} \left[\frac{\partial}{\partial p} (\gamma p f) + \frac{\partial}{\partial r} (\gamma r f) \right] + \frac{1}{2} \left[m \omega_0 \Delta \, \omega r \frac{\partial f}{\partial p} - \frac{\Delta \omega p}{m \omega_0} \frac{\partial f}{\partial r} \right]. \tag{6.2.11}$$

Here the designations for the dissipative constant γ and frequency shift $\Delta \omega$ are introduced. Consider first the expression for γ,

$$\gamma = \frac{2e^2}{3m} \frac{1}{(2\pi)^3} \int dk \frac{4 \pi \omega_0^3 \text{Im}\{\varepsilon(\omega_0)\}}{|\omega_0^2 \varepsilon(\omega_0) - c^2 k^2|^2} , \tag{6.2.12}$$

where the expressions

$$\frac{[k \times p]^2}{k^2} ; \quad \frac{[k \times r]^2}{k^2} \rightarrow \frac{2}{3} p^2 ; \quad \frac{2}{3} r^2 \tag{6.2.13}$$

have already been taken into account in integrating over angles. For the transparency region integration over k is carried out in the same way as in the transition from (6.1.28) to (6.1.30), resulting in a well-known expression for the coefficient of radiative friction,

$$\gamma = \frac{2}{3} \frac{e^2 \omega_0^2}{mc^3} n(\omega_0) , \tag{6.2.14}$$

where $n(\omega_0) = \sqrt{\text{Re}\{\varepsilon(\omega_0)\}}$ is the index of refraction at resonant frequency. For the range of values of parameters where the real part of permittivity is given by (6.1.22, 25), the index of refraction is $n(\omega_0) = 1$, and

$$\gamma = \frac{2e^2 \omega_0^2}{3mc^3} . \tag{6.2.15}$$

The frequency shift $\Delta \omega$ is given by the expression

$$\Delta \omega = \frac{2e^2}{3m} \frac{1}{(2\pi)^3} \int dk \frac{4 \pi \omega_0^3 \text{Re}\{\varepsilon(\omega_0)\}}{|\omega_0^2 \varepsilon(\omega_0) - c^2 k^2|^2} , \tag{6.2.16}$$

and by substituting (6.2.13). This equation only differs from (6.2.12) by having the real part of permittivity where (6.2.12) has the imaginary one. The range of integration in (6.2.16) must exclude the side of big values of k (small distances), since the dipole approximation accepted here does not hold for small distances.

Using expressions (5.2.7, 11) we can write the kinetic equation for the slowly changing part of the distribution function in zero-order approximation with respect to

$$\frac{\partial f}{\partial t} - \frac{1}{2}\left(\frac{eE^s - \Delta\omega p}{m\omega_0}\right)\frac{\partial f}{\partial r} + \frac{1}{2}(eE^c - \Delta\omega m\omega_0 r)\frac{\partial f}{\partial p}$$

$$= \frac{D}{2}\left[\frac{\partial^2 f}{\partial p^2} + \frac{1}{m^2\omega_0^2}\frac{\partial^2 f}{\partial r^2}\right] + \frac{1}{2}\left[\frac{\partial(\gamma p f)}{\partial p} + \frac{\partial(\gamma r f)}{\partial r}\right]. \tag{6.2.17}$$

The mean field strength has the form

$$E = E^c\cos\omega_0 t + E^s\sin\omega_0 t. \tag{6.2.18}$$

The quantities D, γ, $\Delta\omega$ are determined by (6.2.9, 14, 16).

The right-hand side of (6.2.17) is uniform with the right-hand side of the Fokker–Planck equation (5.10.11) at $b = 0$. There is, however, a significant difference. Indeed, in the Fokker–Planck equation the diffusion coefficient is fixed, and at $a > 0$, $b = 0$ is determined by Einstein's formula (5.4.5). In the kinetic equation discussed here the diffusion coefficient is connected with the spectral density of field functions, which in nonequilibrium itself depends on the type of distribution function. Equation (6.2.17) is therefore nonlinear and non-closed, and has to be supplemented by the equation for spectral density of field fluctuations. We find this equation from that for the energy balance,

$$\frac{\partial}{\partial t}\left[\frac{(\delta E\,\delta E)_\omega + (\delta B\,\delta B)_\omega}{8\pi}\right] = -\operatorname{Re}\{\delta j\,\delta E\}_\omega, \tag{6.2.19}$$

which follows from the equations for field fluctuations (2.11.8). The spectral density $\operatorname{Re}\{\delta j\,\delta E\}_\omega$ is calculated similar to the spectral density determining the collision integral. As a result we get the field energy balance equation at frequency ω (see [Ref. 6.3, 1.60]),

$$\frac{1}{8\pi}\frac{\partial}{\partial t}\left[\frac{\partial\omega\operatorname{Re}\{\varepsilon\}}{\partial\omega}(\delta E\,\delta E)_\omega + (\delta B\,\delta B)_\omega\right]$$

$$= -\frac{\omega}{4\pi}\operatorname{Im}\left\{\varepsilon\left(\omega,\frac{\omega}{c}n(\omega)\right)\right\}\left[(\delta E\,\delta E)_\omega - \frac{4}{3}\frac{n(\omega)}{c^3}\omega^2\frac{\overline{p^2}}{m}\right]. \tag{6.2.20}$$

In the state of equilibrium the expression in brackets on the right-hand side must become zero, which brings us to (6.1.30). The distribution function for the state of equilibrium is Gaussian,

$$f = C \exp\left(-\frac{p^2/2m + m\omega_0^2 r^2/2}{\mathscr{K}T}\right). \tag{6.2.21}$$

If the system of atoms is in a thermostat, i.e., if the spectral density of field fluctuations is determined by the equilibirum expression (6.1.30), then the diffusion coefficient is

$$D = \frac{2}{3}\frac{e^2\omega^2}{c^3}n(\omega)\,\mathscr{K}T = \gamma m\,\mathscr{K}T, \tag{6.2.22}$$

leading once more to Einstein's formula.

Turning to the equation for the polarization vector, we set, like in (6.2.18),

$$P = P^c \cos \omega_0 t + P^s \sin \omega_0 t. \tag{6.2.23}$$

The components P^c and P^s are linked to the distribution function f by

$$P^c(R, t) = en\int rf(x, R, P, t)\,dx\,dP, \quad P^s = en\int \frac{p}{m\omega_0}f\,dx\,dP. \tag{6.2.24}$$

With the kinetic equation (6.2.17) we find the equations for slow-changing components of the polarization vector,

$$\frac{\partial P^c}{\partial t} + \frac{1}{2}\gamma P^c - \frac{1}{2}\varDelta\omega P^s = -\frac{e^2n}{2m\omega_0}E^s,$$

$$\frac{\partial P^s}{\partial t} + \frac{1}{2}\gamma P^s + \frac{1}{2}\varDelta\omega P^c = \frac{e^2n}{2m\omega_0}E^c. \tag{6.2.25}$$

This set of equations can be replaced by a single equivalent equation for the polarization vector,

$$\frac{\partial^2 P}{\partial t^2} + \gamma\frac{\partial P}{\partial t} + \left(\omega_0 + \frac{1}{2}\varDelta\omega\right)^2 P = \frac{e^2n}{m}E. \tag{6.2.26}$$

When carrying out the transition from this to the previous two equations it is sufficient to retain just first-order terms with respect to parameters γ/ω_0, $\varDelta\omega/\omega_0$.

The kinetic equation obtained here and the corresponding equation for the polarization vector only account for dissipative processes due to radiative friction.

The dissipative term in the equation for polarization vector can also be obtained from the equation for the balance of energy of an atom oscillator. Conventional calculation [Ref. 6.4, Sect. 75] yields the force of radiative friction due to the radiation of a dipole,

$$f_{\text{rad}} = \frac{2e}{3c^3} \overset{...}{d} ,$$ (6.2.27)

where d is the dipole moment of an atom oscillator. Because of radiative friction, the dissipative term in the equation for the polarization vector is proportional to a third time derivative and equals

$$-\frac{2e^2}{3mc^3} \frac{\partial^3 p}{\partial t^3} .$$ (6.2.28)

Provided that $\gamma \ll \omega_0$, this term differs but little from the dissipative term in (6.2.26). The difference may, however, become important, for instance, in the study of a frequency-dependent coefficient of extinction (attentuation) of light traveling through a substance, and hence in the study of light diffusion (Sect. 11.12).

6.3 The Equation for the Polarization Vector

We shall now consider a more general equation for the polarization vector, which would account for the interaction of atoms. The microscopic polarization vector is again labeled $P^m(R, t)$; its equation in dipole approximation is

$$\frac{\partial^2 P^m}{\partial t^2} + \omega_0^2 P^m = \frac{e^2}{\mu} \sum_i \delta[R - R_i(t)] E^m(R, t) .$$ (6.3.1)

Substituting the solution of field equations (3.6.4) into the right-hand side of this equation and taking (3.6.5) into account, we obtain a closed equation for the polarization vector (3.6.6).

The immediate task is to obtain the equation for the averaged polarization vector $\bar{P}^m$ which accounts for dissipation due to small-scale fluctuations. In obtaining the kinetic equations for the distribution functions of charged particles (see Sects. 4.3, 4, 6), we divided a field acting on charged particles into two parts: the mean field E and the fluctuational field δE. The dissipation in the kinetic equations was then determined by small-scale fluctuations with $\tau_{\text{cor}} < \tau_{\text{ph}}$ and $r_{\text{cor}} < l_{\text{ph}}$. The collision integrals are determined by correlations of these very fluctuations. Here we adopt a similar approach.

The magnitude of the physically infinitesimal volume l_{ph}^3 for a system of dipole atoms is determined by (4.2.13). Let us use the stationary solution of the field equations for atoms within the volume centered around the chosen atom (3.6.4). The influence of all the other atoms, like in (6.3.1), will be accounted for by the field. Having carried out this separation we obtain the following equation for the polarization vector ($P = \bar{P}^{m*}$):

$$\frac{\partial^2 P}{\partial t^2} + \gamma \frac{\partial P}{\partial t} + \omega_0^2 P = \frac{e^2}{\mu} \frac{1}{V_{\text{ph}}} \int \text{rot}_R \, \text{rot}_R \frac{1}{|R - R'|} \overline{(nP^m)}_{R, t, R'; t - (|R - R'|/c)} dR'$$

$$- 4\pi \overline{(nP^m)}_{k, t} + \frac{e^2 n}{\mu} E'(R, t) \tag{6.3.2}$$

[the operator $\text{rot}_R \, \text{rot}_R$ does not act upon the argument of the function $n(R, t)$]. Here $E'(R, t)$ is the field created by atoms outside the volume V_{ph}. Its connection to the mean microscopic field follows from (3.6.4, 5)

$$E(R, t) = \left[\int_{V_{\text{ph}}} \text{rot}_R \, \text{rot}_R \frac{1}{|R - R'|} P\left(R' t - \frac{|R - R'|}{c}\right) dR' - 4\pi P(R, t) \right]$$

$$+ E'(R, t) . \tag{6.3.3}$$

The first term in the right-hand side describes the field of the atoms within V_{ph}, and the second the field from the rest of the atoms. It is noteworthy that this equation, unlike (6.3.2), includes only first moments. With this equation we exclude the field E' from (6.3.2). Using the identity

$$\overline{nP^m} = nP + \overline{\delta n \, \delta P} ,$$

we obtain the following equation:

$$\frac{\partial^2 P}{\partial t^2} + \gamma \frac{\partial P}{\partial t} + \omega_0^2 P = \frac{e^2 n}{\mu} E$$

$$+ \frac{e^2}{\mu} \left[\int_{V_{\text{ph}}} \text{rot}_R \, \text{rot}_R \frac{1}{|R - R'|} \overline{(\delta n \, \delta P)}_{R, t, R'; t - (|R - R'|/c)} dR' \right.$$

$$\left. - 4\pi \overline{(\delta n \, \delta P)}_{R, t} \right] . \tag{6.3.4}$$

Equations (6.3.2, 4) already account for dissipation from radiative friction [see (6.2.28)] which is therefore not connection with particle interaction. It would be possible to take the corresponding frequency shift into consideration from the very beginning; it is more practical, however, to consider the expression for $\Delta \omega$ as a whole.

The second term in the right-hand side of (6.3.4) contains both dissipative and nondissipative contributions. This is clear, e.g., from the fact that there are terms with both odd- and even-order time derivatives in the expansion in lagging, the former determining the dissipative terms. The direction of time was determined by using the stationary solution (lagging potentials) for the field of particles within the volume V_{ph}.

Equation (6.3.4) was obtained without distinguishing between microscopic and local fields. To consider this distinction we shall use the connection of the second moment of fluctuations of phase density δN_{ab} with the corresponding correlation function [cf. (2.12.6),

$$\overline{(\delta N_{ab} \delta N_{ab})}_{X, X', t} = n^2 g(X, X', t) + n\delta(X - X') f_{ab}(X, t) . \tag{6.3.5}$$

Recall that $X = (r, p, R, P)$ is the complete set of variables for an atom. The subscript ab of the function g_{ab} is omitted.

The introduction of local field E_l^m, instead of E^m, for atoms within the volume V_{ph} leads to replacing the second moment of fluctuations δN_{ab} by the correlation function g_{ab}. As a result in (6.3.4)

$$(\overline{\delta n \, \delta P})_{R, t, R'; t'} \rightarrow en^2 \int r' g(X, X', t, t') dr \, dp \, dP \, dr' \, dp' \, dP'$$
$$= en^2 \int r' g(R, t, r', R', t) dr' , \tag{6.3.6}$$

and the equation for the polarization vector becomes

$$\frac{\partial^2 P}{\partial t^2} + \gamma \frac{\partial P}{\partial t} + \omega_0^2 P = \frac{e^2 n}{\mu} E$$

$$+ \frac{e^2 n}{\mu} \left[n \int \text{rot}_R \, \text{rot}_R \frac{1}{|R - R'|} er' g \left(R, t, r', R', t - \frac{|R - R'|}{c} \right) \right.$$

$$\left. \times dr' \, dR' - 4\pi en \int r g(R, r, R, t) dr \right] . \tag{6.3.7}$$

Now let us substantiate the definition of the quantity V_{ph}. For a dipole – dipole interaction the quantity l_{ph} and the number of particles in the volume V_{ph} are determined by (4.2.13), and for atom – atom interactions (atoms considered as rigid spheres) the corresponding quantities are given by (4.2.1). The physically infinitesimal volume in (6.3.7) is determined in the sense of a dipole – dipole interaction, so that according to (4.2.13),

$$V_{ph} \sim \varepsilon_d^{3/2} l_d^3 \sim \frac{\rho_w^3}{\varepsilon_d^{3/2}} \gg \frac{1}{n} . \tag{6.3.8}$$

since $\varepsilon_d = n\rho_w^3$ and $\varepsilon_d \ll 1$.

From (4.2.9) it follows that $r_0 \lesssim r_{av}$; then

$$V_{ph} \gg r_0^3 , \qquad l_{ph} \gg r_0 , \tag{6.3.9}$$

and hence the physically infinitesimal length element in (6.3.7) is much greater than the correlation radius of the positions of atoms considered as rigid spheres. This enables us to simplify the equation for the polarization vector.

The two-particle distribution function corresponding to the correlation function $g(R, r', R', t)$ entering (6.3.7) is

$$f_2(R, r', R', t) = f_1(R, t) f_1(R', r', t) + g(R, r', R', t) . \tag{6.3.10}$$

The function $g(R, r', R', t)$ describes two types of correlations: first, correlations of atoms' positions with respect to variables R and R' with the corresponding correlation radius $r_{cor} \sim r_0 \lesssim r_{av}$; secondly, correlations due to dipole – dipole interaction with correlation radius $r_{cor} \sim \rho_W$. According to (6.3.8), however, $\rho_W \gg l_{ph}$ for dipole – dipole interaction, so these can be considered large-scale correlations and neglected in the analysis of the second term in the right-hand side of (6.3.7).

This gives us reason to consider (6.3.7) as only taking correlations of atoms' positions into account, i.e., the correlation $g(R, R')$. This function can be assumed to be set in advance since we do not take into consideration those forces whose distribution varies over atomic variables R and P. This justifies using, instead of (6.3.10), the approximate expression

$$f_2(R, r', R', t) = f_1(R, t) f_1(r', R', t) [1 + g(R, R')] \tag{6.3.11}$$

which corresponds to the following substitution in (6.3.7):

$$g(R, t, r', R', t') \rightarrow f_1(r', R', t') g(R, R'), \quad f(R, t) = 1. \tag{6.3.12}$$

This condition is a consequence of the constancy of atoms' concentration.

As a result the equation for the polarization vector (6.3.7) takes the form

$$\frac{\partial^2 P}{\partial t^2} + \gamma \frac{\partial P}{\partial t} + \omega_0^2 P = \frac{e^2 n}{\mu} E(R, t)$$

$$+ \frac{e^2 n}{\mu} \left[\int \mathrm{rot}_R \, \mathrm{rot}_R \left(\frac{P\left(R', t - \dfrac{|R - R'|}{c}\right)}{|R - R'|} \right) g(R, R') \, dR' \right.$$

$$\left. - 4\pi P(R, t) g(R, R) \vphantom{\int} \right]. \tag{6.3.13}$$

Employing the definition of the polarization vector and taking heed of the fact that the correlation radius r_{cor} of the function $g(R, R')$ is of the order $r_0 \ll l_{ph}$, integration over R' can be extended to the whole volume.

Later we shall be concerned with those polarization processes whose wavelength λ is much greater than the correlation radius r_{cor} of the function $g(R, R')$, i.e.,

$$\lambda \gg r_0. \tag{6.3.14}$$

This condition enables one to carry out expansion in (6.3.13) with respect to lagging, which corresponds to development in r_0/λ. Let us now investigate the physical meaning of the separate members of this expansion, starting with zero-order terms.

6.4 The Effective Lorentz Field

In the zero-order approximation in lagging and in $(R - R')\partial/\partial R$ (in both cases expansion is carried out with respect to r_0/λ),

$$P\left(R', t - \frac{|R - R'|}{c}\right) = P(R, t) .\tag{6.4.1}$$

We label a unit vector along P with n and introduce the designation $r = R - R'$. Then with (6.4.1) the terms in brackets in (6.3.13) can be written in the form

$$|P(R, t)|\int g(|r|) \operatorname{rot}_r \operatorname{rot}_r \left(\frac{n}{|r|}\right) dr - 4\pi P(R, t) g(|R - R'|) .\tag{6.4.2}$$

This takes account of the fact that at $r_0 \ll \rho_w$ the equilibrium over variables R and R' is attained much sooner than the equilibrium with respect to internal variables, and hence

$$g(R, R', t) \to g(|R - R'|) .\tag{6.4.3}$$

The integration in (6.4.2) is carried out over the whole volume since $l_{\text{ph}} \gg r_0$.

The identity (2.2.10) and equation (2.2.8) can be used to simplify (6.4.2), giving

$$|P(R, t)|\int g(|r|) \operatorname{grad}_r \operatorname{div}_r \frac{n}{|r|} dr \equiv I_0 .\tag{6.4.4}$$

The integrand displays a δ-type singularity at $r = 0$. To prove this we shall expand the integrand in the Fourier integral. Using

$$\frac{1}{|r|} = \frac{1}{(2\pi)^3} \int \frac{4\pi}{k^2} e^{ikr} dk ,$$

$$g(|r|) = \frac{1}{(2\pi)^3} \int g(|k|) e^{ikr} dk ,\tag{6.4.5}$$

we get

$$I_0 = - |P(R, t)|\frac{4\pi}{(2\pi)^3} \int \frac{k(kn)}{k^2} g(|k|) dk = - \frac{4\pi}{3} \frac{1}{(2\pi)^3} \int g(|k|) dk\, P(R, t) .$$

Here we have taken advantage of the fact that the function g only depends on the absolute value of k. Since the integral

$$\frac{1}{(2\pi)^3} \int g(|k|) dk = g \quad (r = 0) ,$$

the final result has the form

$$I_0 = -\frac{4\pi}{3} g(|\boldsymbol{R} - \boldsymbol{R}|) P(\boldsymbol{R}, t), \qquad (6.4.6)$$

which proves the presence of δ-type singularity at $r = 0$ in the integrand in (6.4.4).

Thus, in zero-order approximation with respect to r_0/λ the polarization vector change in (6.3.13) is conditioned by the effective field,

$$E_{\text{eff}}(\boldsymbol{R}, t) = E(\boldsymbol{R}, t) - \frac{4\pi}{3} g(|\boldsymbol{R} - \boldsymbol{R}|) P(\boldsymbol{R}, t). \qquad (6.4.7)$$

If the correlation function is such that

$$g(|\boldsymbol{R} - \boldsymbol{R}|) = -1, \qquad (6.4.8)$$

i.e., if the probability of finding two particles (two atoms) at a distance $r = 0$ is zero, then (6.4.7) becomes

$$E_{\text{eff}}(\boldsymbol{R}, t) = E(\boldsymbol{R}, t) + \frac{4\pi}{3} P(\boldsymbol{R}, t), \qquad (6.4.9)$$

which is the effective Lorentz field.

The result (6.4.7) was obtained under two assumptions: 1) zero-order approximation in r_0/λ, and 2) condition (6.3.12), which allows neglecting the correlations between internal motion in atoms and the motion of an atom as a whole. Finally, the transition from (6.4.7) to the Lorentz equation (6.4.9) can be performed provided (6.4.8) is satisfied. This is the situation with the model of a gas composed of rigid spheres, as well as with rigid bodies, although the reason for the latter is different. In this case the atoms oscillate slightly around the equilibrium positions, and hence the probability that the coordinates of two or more atom's centers coincide is practically zero. In general the result is governed by correlations at small distances. These undergo radical changes in the transition from atoms to free charged particles, i.e., to a plasma. The effective field in a plasma is therefore quite different. A general result true for both bound and free charged particles can only be obtained within the framework of quantum theory. We shall discuss this matter again in Chap. 11; here it suffices to point out that the above-mentioned singularity of the integrand is observed in zero-order approximation with respect to r_0/λ when lagging is neglected and only the simultaneous interactions of atoms are taken into account. We shall now analyze the role of higher terms of expansion in r_0/λ.

6.5 Dissipative Processes Due to Close Correlations

The first-order approximation with respect to r_0/λ is, instead of (6.4.1),

$$P\left(R', t - \frac{|R - R'|}{c}\right) = \left[1 - (R - R')\frac{\partial}{\partial R} - \frac{|R - R'|}{c}\frac{\partial}{\partial t}\right]P(R, t) .$$
$$(6.5.1)$$

The contribution of the term containing $\partial/\partial t$ is zero since the operand of $\mathrm{rot}_R \mathrm{rot}_R$ is a constant in this approximation. The contribution of the term with $\partial/\partial R$ is also zero. Indeed, the integral $(R - R = r)$

$$I_1 = -n \int \left(r\frac{\partial}{\partial R}\right) P(R, t) g(|r|) \mathrm{rot}_R \mathrm{rot}_R \frac{n}{|r|} dr = 0 \tag{6.5.2}$$

equals zero since the integrand changes its sign when $-r$ is substituted for r. First-order terms in r_0/λ, therefore, do not contribute to the right-hand side of (6.3.13).

The second-order terms determine nondissipative processes and will be discussed in the next section. Here we shall treat third-order terms. Four such terms appear in the expansion of the function $P(R', t - |R - R'|/c)$. The contribution of terms with $(r \, \mathrm{grad}_R)^3$ and $(r \, \mathrm{grad}_R)$ is zero for the same reason that (6.5.2) is. The term with $(r \, \mathrm{grad}_R)^2 \partial/\partial t$ also contributes zero, so that only the term with $\partial^3/\partial t^3$ is left. As a result the expression for the integral J_3, which determines the contribution of third-order terms, is

$$I_3 = -\frac{1}{6c^3}\frac{\partial^3 P}{\partial t^3}\int g(|r|)(n \, \mathrm{rot}_r \, \mathrm{rot}_r (n \cdot r^2)) dr . \tag{6.5.3}$$

With the equation

$$n \, \mathrm{rot}_r \, \mathrm{rot}_r (n \times r^2) = n \, \mathrm{rot}_r [\mathrm{grad}_r (r^2 \cdot n)] = 2n \, \mathrm{rot}_r [r \times n] = -4 ,$$

the final result has the form

$$I_3 = \frac{2}{3c^3}\int g(|r|) dr \frac{\partial^3 P}{\partial t^3} . \tag{6.5.4}$$

This term is also determined by the form of correlation function, although unlike (6.4.6) it involves the total contribution of correlations at all distances. Equation (6.5.4) can be rewritten in a more comprehensible form by employing the connection between the integral of the correlation function $g(|r|)$ and the mean-square value of the fluctuations in the number of particles in the volume $\Delta V \gg r_{\mathrm{cor}}^3$,

$$n \int g(|r|) dr = \frac{\overline{(\delta N_{\Delta V})^2}}{\bar{N}_{\Delta V}} - 1 , \tag{6.5.5}$$

which follows from (2.12.6). The volume ΔV can be taken to be equal to V_{ph} since $l_{\text{ph}} \gg r_0 \sim r_{\text{cor}}$.

From (6.5.4, 5) it follows that the dissipative contribution into the right-hand side of (6.3.13) is given by the expression

$$\frac{2e^2}{3\mu c^3} \left(\frac{\overline{(\delta N_{V_{\text{ph}}})^2}}{\bar{N}_{V_{\text{ph}}}} - 1 \right) \frac{\partial^3 P}{\partial t^3} \tag{6.5.6}$$

and is therefore proportional to the third derivative of the polarization vector. Combining this equation with the dissipative term in the left-hand side of (6.3.13) [employing a more general presentation (6.2.28)], we find the final form

$$-\frac{2e^2}{3mc^3} \frac{\overline{(\delta N_{v_{\text{ph}}})^2}}{\bar{N}_{v_{\text{ph}}}} \frac{\partial^3 P}{\partial t^3} \equiv -\frac{2e^2}{3mc^3} [1 + n \int g(r) dr] \frac{\partial^3 P}{\partial t^3} . \tag{6.5.7}$$

This result was first obtained by *Klimontovich* and *Fursov* [6.5]. It describes the radiative friction effect taking into account the correlations of the positions of the atoms surrounding the chosen one within the physically infinitesimal volume V_{ph}.

To describe atom oscillations with frequencies close to ω_0, (6.5.7) can be rewritten in the form

$$\gamma [1 + n \int g(r) dr] \frac{\partial P}{\partial t} \equiv \gamma_{\text{eff}} \frac{\partial P}{\partial t} . \tag{6.5.8a}$$

For an ideal gas, this expression is equal to $\gamma \partial P / \partial t$) and is equal to $g = 0$. In the other extreme case (an ideal lattice), the density fluctuations are zero; consequently (6.5.7, 8a) also equal zero. This means that the field of surrounding atoms compensates for the radiative friction of the specified atom oscillator. The consequence of this result, obtained by *Mandelshtam* [6.6], is the absence of light scattering in an ideally ordered medium: a medium devoid of fluctuations in number of particles.

The quantity γ_{eff} can be related to thermodynamic functions by employing the connection between the isothermal compressibility and the correlation function,

$$\beta_T = \frac{1}{n} \left(\frac{\partial n}{\partial p} \right)_T = \frac{1}{n \mathscr{K} T} [1 + n \int g(r) dr] . \tag{6.5.9}$$

The expression for γ_{eff} can then be presented as

$$\gamma_{\text{eff}} = \gamma \mathscr{K} T \left(\frac{\partial n}{\partial p} \right)_T = \gamma \beta_T n \mathscr{K} T . \tag{6.5.8b}$$

It is important to note again that (6.5.7, 8 b) describe the effective extinction due both to the radiative friction of the atom oscillator and to the random (fluctuating) fields of the surrounding atoms. This was achieved by singling out the atom field within the physically infinitesimal volume V_{ph}. This technique corresponds to singling out small-scale field fluctuation with $r_{cor} < l_{ph}$ and $\tau_{cor} < \tau_{ph}$. Large-scale fluctuations are described on the basis of dissipative equations obtained earlier (cf. Sects. 4.8, 9).

Recall also that all the calculations have been carried out in dipole approximation, when the distribution of atoms over variables R and P are taken to be set in advance. For that reason all the effects of pressure in the equations for γ_{eff} are expressed through the equilibrium correlation function g. The effects of nonequilibrium with respect to atoms' variables can be taken into account when considering the contribution of large-scale fluctuations.

Note further that (6.5.8 a) is more general than (6.5.8 b) while (6.5.9) is only true when the density fluctuations comply with the Gaussian distribution.

If a system of N atoms occupies a volume with linear dimensions $l \ll \lambda$, then the influence of the mean field makes a cooperative effect possible because of the interaction of atoms via their common field [6.7].

6.6 The Equation for the Polarization Vector

The oscillator frequency shift due to interaction is determined by two causes, one of which — that connected with the effective Lorentz field (6.4.7, 9) — has already been discussed. The second contribution is proportional to the second time derivative and, therefore, can be treated either as the influence of interaction on the mass ("effective mass"), or as the eigenfrequency shift of an oscillator. We label the corresponding contribution (factor $e^2 n / \mu$ included) by I_2. From (6.3.13) we find

$$I_2 = - \frac{8 \pi e^2 n}{3 \mu c^3} \int_0^\infty g(r) r \, dr \, \frac{\partial^2 P}{\partial t^2} , \tag{6.6.1}$$

where a new combination with the function g appears.

The contribution of the second mixed partial derivative is zero, leaving the contribution which is proportional to the second derivative over coordinates. The corresponding integral is designated $I_{2(R)}$; it is given by the expression

$$I_{2(R)} = \frac{e^2 n}{2 \mu} \int g(|r|) (r \, \mathrm{grad}_R)^2 \frac{3 r (Pr) - r^2 P}{r^5} \, dr . \tag{6.6.2}$$

Consider the special case when

$$\mathrm{grad}_R |P| \perp P . \tag{6.6.3}$$

After integration over angles we get

$$I_{2(R)} = -\frac{8\pi}{15}\frac{e^2 n}{\mu}\int_0^\infty g(r)r\,dr\frac{\partial^2 P}{\partial R^2} \equiv b\frac{\partial^2 P}{\partial R^2}. \tag{6.6.4}$$

The sign of b is determined by the sign of the integral of the function g. Thus, $b > 0$ for a model of rigid spheres where the integral is negative.

Combining all the expressions which describe the different contributions to interaction, we get the following equation for the polarization vector [with designation (6.5.8a)]:

$$\frac{\partial^2 P}{\partial t^2} + \gamma_{\text{eff}}\frac{\partial P}{\partial t} + \omega_{\text{eff}}^2 P - b\frac{\partial^2 P}{\partial R^2} = \frac{e^2 n}{\mu}E_{\text{eff}}. \tag{6.6.5}$$

Recall that the quantity γ_{eff} is determined by (6.5.8a, b). The effective frequency square is found from (6.6.1),

$$\omega_{\text{eff}}^2 = \omega_0^2\left[1 - \frac{8\pi e^2 n}{3\mu c^2}\int_0^\infty g(r)r\,dr\right]. \tag{6.6.6}$$

The constant b is determined by (6.6.4), and the effective field by (6.4.7) or by (6.4.9) at $g(0) = -1$.

Estimating the magnitude of the frequency shift according to (6.6.6), we find for a model of rigid spheres

$$\frac{|\omega_{\text{eff}} - \omega_0|}{\omega_0} \sim \frac{\gamma}{\omega_0}n\lambda^3\left(\frac{r_0}{\lambda}\right)^2. \tag{6.6.7}$$

Thus, the relative frequency shift is determined by the product of three parameters, one of which is the optical density $n\lambda^3$.

The factor $b^{1/2}$ gives the dimension of speed. The dimensionless quantity

$$\frac{b}{c^2} \sim \frac{\gamma}{\omega_0}n\lambda^3\left(\frac{r_0}{\lambda}\right)^2 \tag{6.6.8}$$

is therefore of the same order of magnitude as the relative frequency shift. The sign of the frequency shift and that of constant b are determined by one and the same integral.

The Lorentz correction $4\pi P/3$ also determines the frequency shift. The relative magnitude of the shift is then determined by the parameter χ (4.2.23),

$$\chi \sim \frac{\gamma}{\omega_0}n\lambda^3. \tag{6.6.9}$$

From a comparison with (6.6.7) we see that the small parameter $(r_0/\lambda)^2$ is not present here, so that this contribution into the frequency shift is thus the principal one.

6.7 The Dielectric Permittivity. The Lorentz–Lorenz Formula. The Equation of Dispersion

The estimate given above only allows us to keep the main contribution to the frequency shift in the equation for the polarization vector, leaving out the term with the spatial derivative which determines spatial dispersion. Instead of (6.6.5) we then have a simpler equation,

$$\frac{\partial^2 P}{\partial t^2} + \gamma_{\text{eff}} \frac{\partial P}{\partial t} + \omega_0^2 P = \frac{e^2 n}{\mu} \left[E - \frac{4\pi}{3} g(0) P \right] \equiv \frac{e^2 n}{\mu} E_{\text{eff}} . \tag{6.7.1}$$

Based on this equation we can introduce two functions which describe the polarizability of the medium. One of these (α_{eff}) gives the response of the system of atoms to the effective field E_{eff}, and the other (α) the response to the mean field E.

The polarizability $\alpha(\omega)$ is determined by the expression

$$\alpha_{\text{eff}}(\omega) = \frac{e^2 n}{\mu} \frac{1}{\omega_0^2 - \omega^2 - i\omega\gamma_{\text{eff}}} . \tag{6.7.2}$$

In order to find the expression for the polarizability we shift the term with the Lorentz correction from the right-hand to the left-hand side, and employ the designation

$$\tilde{a} = \omega_0^2 + \frac{4\pi e^2 n}{\mu} g(0) \equiv \tilde{\omega}_0^2 \qquad \tilde{a} > 0 \tag{6.7.3}$$

for the square of the frequency with the Lorentz correction taken into account. Then, using the relation between polarizability and permittivity, we get

$$\varepsilon(\omega) = 1 + 4\pi\alpha(\omega) = 1 + \frac{4\pi e^2 n}{\mu} \frac{1}{\tilde{\omega}_0^2 - \omega^2 - i\omega\gamma_{\text{eff}}} . \tag{6.7.4}$$

Naturally, the functions $\alpha(\omega)$ and α_{eff} [or $\varepsilon(\omega)$ and α_{eff}] are related to one another, the relationship following from (6.7.1). It can be found by substituting the expression for the polarization vector $P = (\varepsilon - 1)E/4\pi$ into both sides of (6.7.1), resulting in the formula

$$\frac{(\varepsilon - 1)}{3 - g(0)(\varepsilon - 1)} = \frac{4\pi}{3} \alpha_{\text{eff}} . \tag{6.7.5}$$

At $g(0) = -1$ this expression has the form

$$\frac{\varepsilon - 1}{\varepsilon + 2} = \frac{4\pi}{3}\alpha_{\text{eff}} . \tag{6.7.6}$$

This is the well-known Lorentz – Lorenz formula except that the damping decrement γ_{eff} in the expression for α_{eff} is determined by (6.5.7, 8) and therefore not only depends on the concentration but also on the compressibility β_t.

From (6.7.2, 4) it follows that the difference between these functions stems from replacing the eigenfrequency ω_0 [in (6.7.2)] with the renormalized (taking the Lorentz correction into account) frequency $\tilde{\omega}_0$. The correction is, according to (6.6.9), proportional to the optical density. At the density value

$$\frac{4\pi e^2 n_c}{\mu \omega_0^2} |g(0)| = 1 , \tag{6.7.7}$$

the renormalized frequency (6.7.3) becomes zero and a further increase in density makes $\tilde{a}$ negative. Thus, a soft mode can appear in a system with a sufficiently high concentration of atoms.

Note that the Lorentz correction is proportional to the concentration of atoms, that is, pairs of bound charged particles. The number of these pairs may depend on temperature, increasing as the latter decreases. Such dependence is characteristic of all substances which may be thought of as partially ionized plasmas. The concentration may depend on temperature also due to thermal expansion.

The concept of a critical temperature be can introduced, at which the renormalized frequency is reduced to zero due to the Lorentz correction, i. e., at which a soft mode appears. The appearance of the soft mode, as will be shown, leads to a phase transition which results in a state with ordered polarization. Such a model of phase transitions is widely employed in the theory of ferroelectrics [6.8, 9]. This is, however, not the only model [6.10].

Consider now the dispersional properties of the medium using the dispersional equation for the electromagnetic field,

$$\omega^2 \varepsilon(\omega) - c^2 k^2 = 0 . \tag{6.7.8}$$

In the first of two extreme cases, the medium is assumed to be ideally uniform and isotropic, and the fluctuations of density are zero. Consequently, the damping decrement γ_{eff} is also zero (6.5.8). Then the dispersional equation takes the form

$$\omega^2 \left(1 + \frac{\omega_L^2}{\tilde{\omega}^2 - \omega^2} \right) - c^2 k^2 = 0 \tag{6.7.9}$$

with the designation

$$\omega_L^2 = \frac{4\pi e^2 n}{\mu}. \tag{6.7.10}$$

Equation (6.7.9) has two solutions for ω^2, which at small values of $\tilde{\omega}^2$ are

$$\omega_1^2 = \omega_L^2 + c^2 k^2 + \frac{\omega_L^2}{\omega_L^2 + c^2 k^2}\tilde{\omega}^2, \quad \omega_2^2 = \tilde{\omega}^2 \frac{c^2 k^2}{\omega_L^2 + c^2 k^2}. \tag{6.7.11}$$

With the appearance of the soft mode the second solution turns into zero.

At below-critical ($\tilde{a} < 0$) temperatures instability arises, and the linear equation for the polarization vector is no longer sufficient. The development of instability is limited by nonlinear terms which are determined by the anharmonicity of oscillations. The nonlinear term taken into account, (6.7.1) becomes

$$\frac{\partial^2 P}{\partial t^2} + \gamma_{\text{eff}} \frac{\partial P}{\partial t} + (\tilde{a} + \tilde{b}P^2)P = \frac{e^2 n}{\mu} E, \tag{6.7.12}$$

where $\tilde{b}$ is the nonlinearity constant, and $\tilde{a}$ is given by (6.7.3).

The solution of the dispersional equation (6.7.11) was obtained for $\gamma_{\text{eff}} = 0$. Let us now consider the low-frequency branch of oscillations at $\gamma_{\text{eff}} \neq 0$, assuming the fulfillment of conditions [1],

$$|\omega| \ll \gamma_{\text{eff}}, \quad |\omega| \lesssim \frac{\tilde{a}}{\gamma_{\text{eff}}} \ll \sqrt{|\tilde{a}|} \quad \sqrt{|\tilde{a}|} \ll \gamma_{\text{eff}}. \tag{6.7.13}$$

For the low-frequency region the second derivative in the equation for the polarization vector can be dropped; the latter can then be written in the form

$$\frac{\partial P}{\partial t} + (a + bP^2)P = \frac{e^2 n}{\mu \gamma_{\text{eff}}} E \tag{6.7.14}$$

with

$$a = \frac{\tilde{a}}{\gamma_{\text{eff}}}, \quad b = \frac{\tilde{b}}{\gamma_{\text{eff}}}, \tag{6.7.15}$$

similar to (5.5.9). At the critical point, $a = 0$.

In linear approximation the polarizability at low frequencies is described by the expression

$$a(\omega) = \frac{e^2 n}{\mu \gamma_{\text{eff}}} \frac{1}{-i\omega + a}. \tag{6.7.16}$$

---

[1] Here we employ the model where the dissipation is determined by (6.5.8a), and where γ_{eff} is frequency independent.

Hence it follows that the static polarizability

$$\alpha(0) = \frac{e^2 n}{\mu \gamma_{\text{eff}} a} \tag{6.7.17}$$

tends to infinity as the critical temperature is approached.

6.8 Fluctuations of Polarization and Field at Above-Critical Temperatures

In Sect. 6.1 we discussed fluctuations of polarization and field in connection with the collision integral. These fluctuations are small scale; they determine the dissipative processes resulting from radiational friction. In Sect. 6.5 we have developed the theory of small-scale field fluctuations, which determine the dissipative processes with the interaction of particles within a physically infinitesimal volume around the chosen particle taken into account. Now, starting from the dissipational equation for the polarization vector, we can develop the theory of large-scale (kinetic) fluctuations, with $\tau_{\text{cor}} \gtrsim 1/\gamma_{\text{eff}}$, employing the method developed in Sect. 4.8.

By analogy with (4.8.11) we write the equation for the polarization vector. We start with equilibrium fluctuations, for which $\langle P \rangle = 0$ and $\langle E \rangle = 0$, and so $\delta P = P$ and $\delta E = E$. The equation for fluctuations of polarization can be written (cf. Sect. 4.8)

$$\left(\frac{\partial^2}{\partial t^2} + \gamma_{\text{eff}} \frac{\partial}{\partial t} + \tilde{\omega}^2 \right) (P - P^{\text{source}}) = \frac{e^2 n}{\mu} E . \tag{6.8.1}$$

The spectral density of source fluctuations can be calculated either according to the scheme set forth in Sect. 4.8, or by direct application of the equations for correlation P^{source}. By analogy with (4.6.13), these equations can be replaced by a first-order set of equations,

$$\frac{\partial (\overline{PP})^{\text{source}}_{R, t; R', t'}}{\partial t} = (\overline{JP})^{\text{source}}_{R, t; R', t'},$$

$$\left(\frac{\partial}{\partial t} + \gamma_{\text{eff}} \right) (\overline{JP})^{\text{source}}_{R, t; R', t'} + \tilde{\omega}^2 (\overline{PP})^{\text{source}}_{R, t; R', t'} = 0 . \tag{6.8.2}$$

To these we add the initial conditions the values of correlations at $t = t'$. Here we take advantage of the fact that the distribution over the internal variables r and p is Gaussian, and that the radii of large-scale fluctuations $r_{\text{cor}} \gtrsim l_{\text{ph}} \gg r_0$ and are therefore much greater than the correlation radius of the function g. Then,

$$\overline{(PP)}_{R,R'}^{\text{source}} = \frac{3e^2 n}{\mu \tilde{\omega}^2} \delta(R - R') \mathcal{K}T , \quad \overline{(JP)}_{R,R'}^{\text{source}} = 0 . \tag{6.8.3}$$

Here it has been taken into account that the mean value $\langle r^2 \rangle = 3 \mathcal{K}T/\mu \tilde{\omega}^2$ for an oscillator with eigenfrequency $\tilde{\omega}$.

From (6.8.2, 3) the desired expression for the spectral density of fluctuations P^{source} can be presented in the form

$$(PP)_{\omega,k}^{\text{source}} = \frac{1}{2\pi\omega} \text{Im}\{\varepsilon(\omega)\} 3 \mathcal{K}T . \tag{6.8.4}$$

The expression for the imaginary part of permittivity follows from (6.7.4),

$$\text{Im}\{\varepsilon(\omega)\} = \omega_L^2 \frac{\omega \gamma_{\text{eff}}}{(\omega^2 - \tilde{\omega}^2) + \omega^2 \gamma_{\text{eff}}^2} . \tag{6.8.5}$$

The fluctuations $P^{\perp}$ and $P^{\parallel}$ enter the equations of the field as sources. Their spectral densities are expressed through (6.8.4),

$$(P^{\perp}P^{\perp})_{\omega,k}^{\text{source}} = 2(P^{\parallel}P^{\parallel})_{\omega,k}^{\text{source}} = \tfrac{2}{3}(PP)_{\omega,k}^{\text{source}} . \tag{6.8.6}$$

The expression for the spectral density of fluctuations of a transverse field can be written, by analogy with (4.7.16),

$$(E^{\perp}E^{\perp})_{\omega,k} = \frac{16\pi}{\omega} \frac{\omega^4 [\text{Im}\{\varepsilon(\omega)\} + \gamma_k/\omega]}{|\omega^2 \varepsilon(\omega) + i\omega \gamma_k - c^2 k^2|^2} \mathcal{K}T , \tag{6.8.7}$$

where γ_k is the damping decrement of the field for attenuation of the borders of the region.

The spectral density of longitudinal field fluctuations can be written, by analogy with (4.4.14, 19), as

$$(E^{\parallel}E^{\parallel})_{\omega,k} = \frac{8\pi}{\omega} \frac{\text{Im}\{\varepsilon(\omega)\}}{|\varepsilon(\omega)|^2} \mathcal{K}T . \tag{6.8.8}$$

Hence, the spatial correlation of fluctuations of the longitudinal field is given by

$$(E^{\parallel}E^{\parallel})_{\omega,r} = \frac{8\pi}{\omega} \frac{\text{Im}\{\varepsilon(\omega)\}}{|\varepsilon(\omega)|^2} \delta(r) . \tag{6.8.9}$$

Because of the lack of spatial dispersion [which is described by the term with the second spatial derivative in (6.6.5)], the correlation radius of longitudinal field fluctuations is zero.

Using (6.8.7), the spatial correlation of transverse field fluctuations [Ref. 6.2, Eq. (90.24)] can be expressed as

$$(E^\perp E^\perp)_{\omega,r} = \frac{2\omega}{c^2 i} \frac{1}{r} \left(\exp\left[-\frac{\omega}{c} \sqrt{-(\varepsilon + i\gamma_k/\omega)}\, r \right] \right.$$

$$\left. - \exp\left[-\frac{\omega}{c} \sqrt{-(\varepsilon^* - i\gamma_k/\omega)}\, r \right] \right) \mathscr{K} T. \tag{6.8.10}$$

In contrast to (6.8.9) this expression is finite at $r = 0$,

$$(E^\perp E^\perp)_\omega = \frac{2\omega^2}{c^3} \left(\sqrt{\varepsilon + i\gamma_k/\omega} + \sqrt{\varepsilon^* - i\gamma_k/\omega} \right) \mathscr{K} T. \tag{6.8.11}$$

In the transparency region, where $\mathrm{Im}\{\varepsilon\}$, $\gamma_k = 0$, we find from (6.8.11),

$$(E^\perp E^\perp)_\omega = \frac{4\omega^2}{c^3} \sqrt{\mathrm{Re}\{\varepsilon\}}\, \mathscr{K} T. \tag{6.8.12}$$

This equation coincides with the form of (6.1.30). Now, however, it describes large-scale fluctuations, and the real part of permittivity is determined by (6.7.4). The contribution of the longitudinal field fluctuations in the transparency region is negligibly small, and (6.8.12) completely describes the spectral density of field fluctuations. The superscript $\perp$ in (6.8.12) may thus be omitted. Equations (6.8.9 – 12) correspond to those given in [Ref. 6.2, Eq. (90.24, 91.1)], differing by the factor 2π due to a different definition of Fourier components.

Substituting the expression for dielectric permittivity (6.7.4) into (6.8.12), it is easy to prove that, approaching the critical point (when the soft mode appears: $\tilde{\omega} \to 0$), the intensity of fluctuations at low frequencies $|\omega| < \tilde{\omega}^2/\gamma_{\mathrm{eff}}$ increases as $1/\tilde{\omega} \sim (T - T_c)^{1/2}$. For this frequency range in (6.8.12),

$$\mathrm{Re}\{\varepsilon(\omega)\} \to 4\pi\alpha(0) = \frac{4\pi e^2 n}{\mu\tilde{\omega}^2} \equiv \frac{\omega_L^2}{\tilde{\omega}^2} \tag{6.8.13}$$

and therefore

$$(E \cdot E)_\omega = \frac{4\omega^2}{c^3} \frac{\omega_L}{\tilde{\omega}} \mathscr{K} T. \tag{6.8.14}$$

Thus, approaching the critical point from the side $T > T_c$ ("symmetric phase"), an increase of the intensity of fluctuations is registered at low frequencies.

To investigate the behavior of fluctuations crossing the critical point, we employ a method similar to that in Sects. 5.5 – 8. First, however, we must make one more note which will be helpful for the future discussion. From the Langevin equation (5.4.1) for a one-dimensional oscillator it follows that the spectral density of the random force $y(t)$ is given by the expression

$$(y \cdot y)_\omega = 2D = 2m\gamma \mathscr{K} T. \tag{6.8.15}$$

Our equation for the polarization vector corresponds to the Langevin equation for a three-dimensional oscillator; the Langevin source plays the role of the random force eE. Hence the corresponding spectral density of the force (reckoned for one dimension) is given by [employing also (6.8.12)]

$$(y \cdot y)_\omega = \frac{1}{3} e^2 (E \cdot E)_\omega = 2m \frac{2e^2 \omega^2}{3 m c^3} \sqrt{\mathrm{Re}\{\varepsilon\}} \, \mathscr{K} T . \tag{6.8.16}$$

Comparing (6.8.15, 16) it is apparent that the decrement γ in this case is defined by

$$\gamma = \frac{2}{3} \frac{e^2 \omega^2}{m c^3} \sqrt{\mathrm{Re}\{\varepsilon\}} , \tag{6.8.17}$$

which, as expected, coincides with the equation for radiational friction in a medium with refraction index $\sqrt{\mathrm{Re}\{\varepsilon\}}$ [cf. (6.2.28)]. Approaching the critical point, the radiational friction coefficient increases as $(T - T_c)^{-1/2}$, from (6.8.13, 17).

6.9 Phase Transition in an Atom – Field System

6.9.1 Initial Equations

Now we shall attempt to conceive a phase transition in an atom – field system in its entirety. This means that fluctuations of polarization and field at any temperature including the critical point fall within our scope.

Recall that in Sect. 5.8 we discussed on a qualitative level a phase transition in a system of atom oscillators, interacting through spatial diffusion. We saw that the theory of phase transitions elaborated by Landau cannot be applied to such a system in the critical region even after a thermodynamic limiting transition has been carried out. Today we known many examples which show that the region of applicability of Landau theory (or the theory of the mean field) depends significantly on the character of the interaction of particles in the system.

All systems can be roughly divided into two classes. In the first the systems display a rather wide temperature range, in which the critical indices differ markedly from those of Landau theory (see Sect. 5.8). This is a so-called region of scale invariance. In particular, the systems described by the Ginzburg – Landau equation with a random source – (5.8.1) – fall into this class.

In systems of the second type, Landau theory holds almost up to the critical point – "almost" since in the thermodynamical limit the width of the region where Landau theory is not true is reduced to zero. Among this class is a system described by the Bardin – Cooper – Schriffer Hamiltonian (BCSch) [6.11]. Indeed, *Bogolyubov*, *Zubarev*, and *Tserkovnikov* [6.12, 13] have shown that such an initial Hamiltonian for such a system can be replaced by an approximative Hamiltonian. Such a replacement corresponds to the mean field approximation. The solution obtained with the approximative Hamiltonian is asymptotically exact. This means that after carrying out the thermodynamical limiting transition the solution obtained in this manner coincides with the exact one. The results of more recent work on the construction of approximative Hamiltonians can be found in [6.14].

The BCSch model is, however, an idealized one. More practical models of superconductivity always display a region in the neighborhood of the critical point where Landau theory is inapplicable after carrying out the thermodynamic limit transition. This is a region of so-called scale invariance. For superconductors it is narrow yet finite. Another example of a system whose region of scale invariance is extremely narrow is a uniaxial ferroelectric, where dipole – dipole interaction plays an important role [6.15].

Take an example of an atom – field system where direct interaction between atoms is absent, and all the interactions take place through fluctuational electromagnetic radiation. This does not mean, of course, that direct interaction between particles is not important. It determines the values of renormalized parameters in the initial equation for the polarization vector, i.e., the effective damping decrement γ_{eff}, the frequency square of the soft mode $\tilde{a}$ resulting from the Lorentz field.

The initial equation for the polarization vector in the low-frequency region, the most important at phase transitions, can be written employing (6.7.15) with a random source for the polarization vector,

$$\frac{\partial P}{\partial t} + (a + bP^2)P = \frac{e^2 n}{m \gamma_{\text{eff}}} E + y(R, t) . \tag{6.9.1}$$

The presence of a random source y is, as we know, a consequence of the atomic structure of the system in question. The moments of a random source are given by [cf. (5.8.3)]

$$\langle y \rangle = 0 , \quad \langle y(R, t) y(R', t') \rangle = 2D e^2 n \delta(R - R') \delta(t - t')$$

$$D = \frac{3 \mathcal{K} T}{m \gamma_{\text{eff}}} . \tag{6.9.2}$$

Equation (6.9.1) differs significantly from the Ginzburg – Landau equation (5.8.1). Indeed, the diffusional term is absent here ($g = 0$). This means that the correlation radius (5.8.23) is zero. Consequently, the mechanism which gives rise to coherent state here is entirely different.

We shall see that spontaneous polarization is due to the ordering action of large-scale fluctuations of an electromagnetic field in (6.9.1). In this sense, the phase transition is an induced one. It is induced, however, by equilibrium field fluctuations, as we shall see later. Such a process is made possible at temperatures $T < T_c$ by the appearance of the soft mode in an atomic subsystem.

Attention should be paid to the fact that in this system the damping decrement γ_{eff} is constant and does not depend on frequency. This attenuation, therefore, cannot be interpreted as radiational friction (6.8.17, 28). Radiational friction is proportional to ω^2 and thus plays no important role at phase transitions. As a consequence, the spectral density of the random source in (6.9.1) is not described by (6.8.16).

This does not mean, however, that field fluctuations are not important in (6.9.1). On the contrary, it will be shown that fluctuations of polarization induced by the field are responsible for the phase transition. In order to describe this process we supplement (6.9.1) with field equations.

The role of the order parameter is here played by the quantity [cf. (5.8.10)]

$$\eta = \langle P_{k=0}(t) \rangle_{E=0} = \langle P(t) \rangle_{E=0}. \tag{6.9.3}$$

In the absence of field and polarization fluctuations it follows from (6.9.1) that

$$\eta = \sqrt{|a|/b}. \tag{6.9.4}$$

For fluctuations of the order parameter [cf. (5.8.12)],

$$\delta\eta_k(t) \equiv \delta P_k(t) = P_k(t) - \delta\eta_{k=0}, \tag{6.9.5}$$

we obtain from (6.9.1)

$$\frac{\partial \delta P_k}{\partial t} + (a + 3b\eta^2)\,\delta P_k = \frac{e^2 n}{m\,\gamma_{\text{eff}}}\,\delta E_k(t) + y_k(t). \tag{6.9.6}$$

Hence, the function δP can be presented as a sum of two parts,

$$\delta P = \delta P^{\text{ind}} + \delta P^{\text{source}}. \tag{6.9.7}$$

6.9.2 Fluctuations of the "Source" δP^{source}

The second term is called a "source." It is determined by the solution of (6.9.6) at $\delta E = 0$ and is, therefore, connected with spontaneous fluctuations of the polarization vector. In contrast, the first term in the right-hand side of (6.9.7) is called "induced." It is determined by the solution of the nonuniform equation (6.9.6) at $y = 0$.

From (6.9.6) the spectral densities of fluctuations δP^{source} are

$$(\delta P \, \delta P)^{\text{source}}_{\omega, k} = \frac{2e^2 nD}{\omega^2 + (a + 3b\eta^2)^2}, \tag{6.9.8}$$

$$(\delta P \, \delta P)_k^{\text{source}} = \frac{e^2 n D}{a + 3 b \eta^2} \, . \tag{6.9.8}$$

With the second equation we can find the expression for the spatial correlation of fluctuations,

$$(\delta P \, \delta P)_{R,R'}^{\text{source}} = \frac{e^2 n}{a + 3 b \eta^2} \, \delta(R - R') \, . \tag{6.9.9}$$

Approaching the critical point the intensity of the source fluctuations grows as $[(T - T_c)/T_c]^{-1}$. The fluctuations remain, however, δ correlated in space; the correlation radius is therefore zero.

The susceptibility which connects fluctuations δP^{source} and y is given by

$$\chi(\omega) = \frac{1}{-i\omega + a + 3 b \eta^2} \, , \quad \delta P^{\text{source}}(\omega) = \chi(\omega) y(\omega) \, . \tag{6.9.10}$$

Let us designate with τ_{source} the correlation time of fluctuations δP^{source}. From the spatial – temporal spectral density (6.9.8) it follows that the correlation time τ_{source} is described by

$$\tau_{\text{source}} \sim \frac{1}{a + 3 b \eta^2} = \chi(0) \, . \tag{6.9.11}$$

The correlation time, consequently, reacts to changes in temperature in a similar manner as the statical susceptibility $\chi(0)$ [cf. (5.8.59) for the correlation time of fluctuations τ_η]. This means that the correlation time decreases on departure from the critical point according to Curie's law,

$$\tau_{\text{source}} \sim \frac{1}{a}(T > T_c) \, , \quad \tau_{\text{source}} \sim \frac{1}{2|a|} \quad \text{at} \quad T < T_c \, . \tag{6.9.12}$$

Later we shall see that at the critical point itself $(a = 0)$,

$$\tau_{\text{source}} \propto V^{1/2} \, . \tag{6.9.13}$$

Thus the fluctuations of polarization δP^{source} cannot give rise to a coherent state with permanent polarization of the whole system.

6.9.3 Induced Fluctuations of the Polarization Vector

In order to obtain results which hold for the critical point as well, we shall proceed in a way similar to the approach adopted in Sect. 5.8 (5.8.27). We substitute

$$\eta^2 \to \langle P_{k=0}^2(t) \rangle = \langle P^2(t) \rangle \, , \quad P(t) = \int P(R, t) \frac{dR}{V} \tag{6.9.14}$$

where P is the induced polarization (later we omit the superscript). Thus we do not employ the definition of the order parameter as a quasi-mean. The criterion for phase transition is the behavior of the temporal correlation of induced fluctuations of the polarization vector.

The equation for two-time correlators of induced fluctuations of the polarization vector is similar to (5.8.62) and has the form

$$\left[\frac{\partial}{\partial \tau} + (a + b\langle P^2 \rangle)\right] \langle PP \rangle_\tau = \frac{e^2 n}{m \gamma_{\text{eff}}} \langle EP \rangle_\tau, \quad \tau > 0. \tag{6.9.15}$$

Unlike in (5.8.62), the right-hand side is not equal to zero, but is determined by the temporal correlator of the field and polarization fluctuations; its calculation can be based on classical expressions. The fact that it is nonzero is of great importance for above critical temperatures as well as for the critical region when $\delta \eta \sim P(t)$. We shall therefore concern ourselves with fluctuations of the field which are determined by the fluctuations of the polarization vector δP^{source}.

The fluctuations of the longitudinal (with respect to the polarization vector) field are δ correlated in space (6.8.9) and therefore play no role in phase transition. Spatial correlations of transverse field fluctuations are determined by (6.8.10).

If

$$\text{Im}\{\varepsilon(\omega)\} \ll \text{Re}\{\varepsilon(\omega)\}, \tag{6.9.16}$$

which allows defining the transparency region, (6.8.10) can be presented in the form

$$(\delta E^\perp \delta E^\perp)_{\omega,r} = \frac{4\omega}{c^2} e^{-r/r_c} \frac{\sin\left[\frac{\omega}{c}\sqrt{\text{Re}\{\varepsilon(\omega)\}}r\right]}{r}, \tag{6.9.17}$$

with the following designation for the correlation radius of field fluctuations at frequency ω:

$$r_c(\omega) = \frac{c}{2\sqrt{\text{Re}\{\varepsilon\}}\,\omega} \frac{\text{Re}\{\varepsilon\}}{\text{Im}\{\varepsilon\}}. \tag{6.9.18}$$

The functions $\text{Re}\{\varepsilon\}$ and $\text{Im}\{\varepsilon\}$ in (6.9.17, 18) are determined by the polarizability (6.9.10),

$$\alpha(\omega) = \frac{e^2 n}{m \gamma_{\text{eff}}} \frac{1}{-i\omega + a + 3b\eta^2}. \tag{6.9.19}$$

The solution of the equation for the one-time correlator of the polarization vector must serve as the initial (at $\tau = 0$) condition; it is similar to (5.8.29), having the form

$$(a + b\langle P^2\rangle)\langle P^2\rangle = \frac{e^2 n}{m\,\gamma_{\text{eff}}}\langle EP\rangle . \tag{6.9.20}$$

The set of equations (6.9.15, 20) is not closed since the correlators of field and polarization fluctuations are not yet determined.

In place of this set of equations for correlators, one might use an equation for the vector of polarization induced by the field,

$$\frac{\partial P}{\partial t} + [a + b\langle P^2(t)\rangle]P = \frac{e^2 n}{m\,\gamma_{\text{eff}}}E . \tag{6.9.21}$$

This leads to the corresponding expression for polarizability,

$$\alpha(\omega) \equiv \frac{e^2 n}{m\,\gamma_{\text{eff}}}\chi_{\text{ind}}(\omega) = \frac{e^2 n}{m\,\gamma_{\text{eff}}}\frac{1}{-i\omega + a + b\langle P^2(t)\rangle} . \tag{6.9.22}$$

From (6.9.21, 22) it follows that the correlation time of induced fluctuations is given to an order of magnitude by

$$\tau_{\text{ind}} \sim \frac{1}{a + b\langle P^2(t)\rangle} . \tag{6.9.23}$$

Therefore the correlation time of induced fluctuations is infinitely great at $\langle P^2\rangle = \eta^2 = |a|/b$ (6.9.4).

From the above equations it is clear that the correlation radius of field fluctuations increases with a decrease in frequency ω. For the principal mode, when

$$k_{\text{min}} \sim \frac{\omega_{\text{min}}}{c}\sqrt{\text{Re}\{\varepsilon\}} \propto V^{1/3} , \tag{6.9.24}$$

the correlation radius

$$r_{\text{c}}(\omega_{\text{min}}) > V^{1/3} . \tag{6.9.25}$$

The correlation radius of field fluctuations can therefore exceed the dimensions of the system at low frequencies. Field fluctuations cause induced polarization fluctuations, which spread over the whole system. The conditions at which this process leads to a phase transition are determined by the behavior of the lifetime (correlation time) of the polarization fluctuations thus induced.

To solve this problem we turn to (6.9.20) for the one-time correlator $\langle P^2(t)\rangle$ since this correlator determines the sought-for correlation time (6.9.23). If the mean field is zero, the correlator $\langle EP\rangle$ entering the right-hand side of (6.9.20) can be presented in the form

$$\langle\delta E\,\delta P\rangle = \frac{1}{2\pi}\text{Re}\{\int\alpha(\omega)(\delta E\,\delta E)_\omega d\omega\} = 2\frac{\mathscr{K}T}{V} . \tag{6.9.26}$$

Here we have (6.8.12, 9.24) and retained only the main mode in the sum over k. This results in a closed equation for the function $\langle P^2 \rangle$,

$$(a + b\langle P^2 \rangle)\langle P^2 \rangle = \frac{2}{3} \frac{e^2 n}{V} D, \quad D = \frac{3 \mathcal{H} T}{m \gamma_{\text{eff}}} \tag{6.9.27}$$

which only differs from (5.8.29) in its designations.

Returning to (6.9.23), by using (6.9.27) we can express the correlation time τ_{ind} in the form

$$\tau_{\text{ind}} = \frac{3}{2} \frac{\langle P \rangle^2}{c^2 n D}. \tag{6.9.28}$$

From (6.9.11, 28) it follows that at above critical temperatures as well as at the critical point itself the times τ_{ind} and τ_{source} are of the same order of magnitude,

$$\tau_{\text{ind}} \sim \tau_{\text{source}} \sim \frac{1}{a} \quad T > T_c, \quad \tau_{\text{ind}} \sim \tau_{\text{source}} \propto V^{1/2} \quad T = T_c. \tag{6.9.29}$$

Finally, at below critical temperatures

$$\tau_{\text{ind}} \sim \frac{|a|}{e^2 n D b} V. \tag{6.9.30}$$

Consequently, when crossing the critical point (at $T < T_c$), the correlation time of fluctuations τ_{source} starts to decrease according to Curie's law, while the correlation time of induced fluctuations is proportional to the volume of the system, and in the thermodynamic limit tends to infinity. This points to a phase transition which results in a state with permanent polarization in time and space.

This example of a phase transition differs significantly from those described by the Ginzburg – Landau equation. In Chap. 13 we shall discuss a similar phase transition in a quantum system of atoms and field in detail. Then we shall also discuss an example of a nonequilibrium phase transition in an atom – field system which is also an induced one, but this time by nonequilibrium electromagnetic radiation, e.g., laser radiation. Naturally, nonequilibrium phase transitions are much more multifarious than equilibrium ones.

7. Microscopic Equations

In this chapter we shall discuss different forms of equations for the operator density matrices, which form the basis of the quantum kinetic theory.

7.1 A System of Free Charged Particles with Coulomb Interaction

A many-particle system in quantum mechanics can be described on the basis of equations for operator (quantized) wave functions $\Psi_a(r, t)$ and $\Psi_a^+(r, t)$ (a is the component subscript). These functions satisfy the following commutative relations [7.1, 2]:

$$\Psi_a(r, t)\, \Psi_b^+(r', t) \pm \Psi_b^+(r', t)\, \Psi_a(r, t) = \delta_{ab}\delta(r - r'),$$

$$\Psi_a(r, t)\, \Psi_b(r', t) \pm \Psi_b(r', t)\, \Psi_a(r, t) = 0,$$

$$\Psi_a^+(r, t)\, \Psi_b^+(r', t) \pm \Psi_b^+(r', t)\, \Psi_a^+(r, t) = 0. \tag{7.1.1}$$

The plus sign corresponds to a Fermi system, and minus to a Bose system. The integral

$$\int \Psi_a^+(r, t)\, \Psi_a(r, t)\, \Psi_a(r, t)\, dr = \hat{N}_a \tag{7.1.2}$$

determines the operator of the total number of particles of kind a.

In this section we shall discuss the equations for electron-ionic ($a = e, i$) completely ionized plasma. The ions are assumed to be singly charged ($e_i = |e_e|$), so $N_e = N_i = N$.

The operators Ψ and Ψ^+ can be presented in the form of a series over a complete system of eigenfunctions, e. g., plane waves. For instance, for a Bose system

$$\Psi(r, t) = \frac{1}{\sqrt{V}} \sum_k b_k e^{ikr}, \quad \Psi^+(r, t) = \frac{1}{\sqrt{V}} \sum_k b_k^+ e^{-ikr}, \tag{7.1.3}$$

where b_k and b_k^+ are the birth and annihilation operators of particles in the state with wave vector k. n_k is the corresponding occupation number. The action of birth and annihilation operators upon an arbitrary function of occupation numbers is determined by

$$b_k F(\ldots, n_k, \ldots) = \sqrt{n_k + 1}\, F(\ldots, n_k + 1, \ldots) \,,$$
$$b_k^+ F(\ldots, n_k, \ldots) = \sqrt{n_k}\, F(\ldots, n_k - 1, \ldots) \,. \tag{7.1.4}$$

Hence it follows that

$$b_k^+ b_k = n_k \,, \quad b_k b_k^+ = n_k + 1 \,. \tag{7.1.5}$$

The commutative relations for operators b_k and b_k^+ have the form

$$b_k b_{k'}^+ - b_{k'}^+ b_k = \delta_{kk'} \,,$$
$$b_k b_{k'} - b_{k'} b_k = 0 \,, \quad b_k^+ b_{k'}^+ - b_{k'}^+ b_k^+ = 0 \,. \tag{7.1.6}$$

Naturally, these relations follow from (7.1.1) and vice versa.

The theory of secondary quantization states [7.1, 2] that the operators which are additive, binary, etc., with respect to single-particle operators can be expressed with operator wave functions or with the corresponding birth and destruction operators of particles. For instance, the Hamiltonian operator of the system in question can be presented in the form

$$\hat{H} = \sum_a \left[-\frac{\hbar^2}{2m_a} \int \Psi_a^+(r, t) \Delta_r \Psi_a(r, t)\, dr \right]$$
$$+ \frac{1}{2} \sum_{ab} \int \Phi_{ab}(|r - r'|)\, \Psi_a^+(r, t)\, \Psi_b^+(r', t)\, \Psi_b(r', t)\, \Psi_a(r, t)\, dr\, dr' \,. \tag{7.1.7}$$

For a Coulomb plasma the potential of interaction is $\Phi_{ab} = e_a e_b / |r - r'|$.

In the Heisenberg presentation the quantized wave function satisfies the equation

$$i\hbar \frac{\partial \Psi_a}{\partial t} = \Psi_a \hat{H} - \hat{H} \Psi_a \tag{7.1.8}$$

since it is a time-dependent operator. For the Hamiltonian operator (7.1.7) this equation becomes

$$i\hbar \frac{\partial \Psi_a}{\partial t} = -\frac{\hbar^2}{2m_a} \Delta_r \Psi_a$$
$$+ \sum_b \int \Phi_{ab}(|r - r'|)\, \Psi_b^+(r', t)\, \Psi_b(r', t)\, dr'\; \Psi_a(r, t) \,, \tag{7.1.9}$$

which coincides in form with the equation for a one-particle wave function in a self-consistent field approximation, the Hartree equation. The form, however, is the only thing they have in common. In contrast to the approximate Hartree equation, (7.1.9) is exact.

We already know that a similar situation is encountered in the classical theory of plasma as well. Indeed, the equations for microscopic phase density $N_a(x, t)$ (2.3.7) coincide in form with equations for one-particle distribution functions f_a (Vlasov equations). In contrast to the latter, (2.3.7) is exact for the microscopic phase density $N_a(x, t)$ which is a random function, although it is not exact in respect to the distribution function.

In place of the equations for operator wave functions Ψ_a and Ψ_a^+, it is more convenient to employ the equation for a corresponding operator density matrix, which we define by

$$\frac{N_a}{V} \, \hat{\rho}_a(r', r'', t) = \Psi_a^+ (r'', t) \, \Psi_a(r', t) . \tag{7.1.10}$$

The operator matrix determined in this way is dimensionless. It follows from the equations for operators Ψ_a and Ψ_a^+, and for a system of particles with Coulomb interaction it can be written in the form

$$i\hbar \frac{\partial \rho_a}{\partial t} (r', r'', t)$$

$$= - \frac{\hbar^2}{2 m_a} (\Delta_{r'} - \Delta_{r''}) \, \hat{\rho}_a(r', r'', t) + e_a [\, \hat{\phi}(r', t) - \hat{\phi}(r'', t)] \, \hat{\rho}_a(r', r'', t) , \tag{7.1.11}$$

introducing the operator of electric potential. It is connected with the operator density matrix through the Poisson equation,

$$\Delta \, \hat{\phi}(r, t) = - 4 \pi \sum_a N_a e_a \, \hat{\rho}_a(r, r, t) , \quad N_a = N . \tag{7.1.12}$$

Note that in going from equations for operators Ψ_a and Ψ_a^+ to (7.1.11, 12) only those couplings of operators Ψ_a and Ψ_a^+ were taken into consideration which correspond to one and the same particle, thus neglecting the exchange effects. If the exchange effects are taken into account, the set of equations for operators $\hat{\rho}_a$ and $\hat{\phi}$ has a more complex form [Ref. 7.3, 71].

In some cases it is not practical to employ the operator density matrix in coordinate presentation but rather the corresponding equation for the operator in a mixed presentation. This is defined by the expression

$$\hat{N}_a(x, t) = \frac{N_a}{(2\pi)^3} \int \hat{\rho}_a \left(r + \frac{1}{2} \hbar\gamma, r - \frac{1}{2} \hbar\gamma, t \right) e^{-i\gamma p} \frac{(2\pi\hbar)^3}{V} \, d\gamma \, (x = r, p) . \tag{7.1.13}$$

From (7.1.2, 10, 13)

$$\int \hat{N}_a(x, t) \frac{dx}{(2\pi\hbar)^3} = N_a \int \hat{\rho}_a(r, r, t) \frac{dr}{V} = \int \Psi_a^+(r, t) \Psi_a(r, t) dr = \hat{N}_a. \quad (7.1.14)$$

The relationship between operators $\hat{N}_a$ and $\hat{\rho}_a$ (7.1.13) is similar to that between a Wigner quantum distribution function and the density matrix.

The equations for operators $\hat{N}_a(x, t)$, from (7.1.11, 12), have the form [7.3]

$$\left(\frac{\partial}{\partial t} + v \frac{\partial}{\partial r}\right) \hat{N}_a(x, t) - \frac{ie_a}{\hbar(2\pi)^3} \int \left[\hat{\varphi}\left(r - \frac{1}{2}\hbar\gamma\right) - \hat{\varphi}\left(r + \frac{1}{2}\hbar\gamma\right)\right]$$

$$\times \hat{N}_a(r, \eta, t) e^{i\gamma(\eta-p)} dy \, d\eta = 0, \quad (7.1.15)$$

$$\Delta \hat{\varphi}(r, t) = - 4\pi \sum_a e_a \int \hat{N}_a(r, p, t) \frac{V}{(2\pi\hbar)^3} dp. \quad (7.1.16)$$

In the classical limit this set of equations coincides with the equations for the random functions $N_a(x, t)$ and $E^m(r, t)$ (2.3.1, 5).

The Hamiltonian operator (7.1.7) can also be expressed by the phase density operator N_a. Without allowing for exchange effects,

$$\hat{H} = \sum_a \int \frac{p^2}{2m_a} \hat{N}_a(x, t) \frac{dx}{(2\pi\hbar)^3} + \frac{1}{2} \sum_{ab} \int \Phi_{ab} \hat{N}_a(x, t) \hat{N}_b(x', t) \frac{dx \, dx'}{(2\pi\hbar)^6}. \quad (7.1.17)$$

Unlike (7.1.7), this expression accounts for self-energy.

Still another presentation of the equation for operator $\hat{N}_a(x, t)$ can be of use. It is especially convenient for analyzing a more complicated system when, along with Coulomb interaction, electromagnetic interaction is taken into account. Using a Hamiltonian operator $\hat{H}_a$ of a single particle of kind a in the field of all the surrounding particles

$$\hat{H}_a = \frac{\delta \hat{H}}{\delta \hat{N}_a} = \frac{\hat{p}^2}{2m_a} + e_a \hat{\varphi}_a(r, t), \quad (7.1.18)$$

(7.1.15) can be written in the form

$$\frac{\partial \hat{N}_a}{\partial t} - \frac{i}{\hbar(2\pi)^6} \int \left[\hat{H}_a\left(\rho - \frac{1}{2}\hbar\gamma, \eta + \frac{1}{2}\hbar\tau\right)\right.$$

$$\left. - \hat{H}_a\left(\rho + \frac{1}{2}\hbar\gamma, \eta - \frac{1}{2}\hbar\tau\right)\right] \hat{N}_a(\rho, \eta, t) e^{i\tau(\eta-p)+i\gamma(\rho-r)} d\tau \, dy \, d\rho \, d\eta \quad . \quad (7.1.19)$$

In classical theory the equations for the microscopic functions N_a^m and E^m are taken to be the basic ones for obtaining the equations for moments. The distributions of random functions are not employed in their explicit form. In quantum theory the initial equations for operators $\hat{N}_a$ and $\hat{\varphi}$ are also employed in obtaining the equations for moments. In order to enhance the similarity to the classical approach, we shall retain the classical designations for quantum operators, i.e.,

$$\hat{N}_a,\ \hat{\varphi} \to N_a,\ \varphi^m\ . \tag{7.1.20}$$

Naturally, while $N_a(x, t)$ in classical theory is a random time function which depends on the particle's coordinates and momenta according to (2.1.5), in quantum theory $\hat{N}_a$ is an operator acting upon functions of occupation numbers. We shall next see that this distinction shows up in the construction of the kinetic theory only in the definition fluctuation sources.

7.2 Partially Ionized Plasma

A plasma consisting of free charged particles can be treated as an extreme case of a partially ionized plasma. Another extreme case is represented by a gas of neutral particles, where the degree of ionization is zero and all charged particles are bound in pairs, the total number of pairs being $N_{ab} = N$.

In the description of bound states in classical theory in Chap. 3 we took the equation for the microscopic phase density of pairs of charged particles as the initial equation (see Sects. 3.1, 2). In quantum theory we shall employ a similar approach.

Let us introduce the operator density matrix for a pair of charged particles $\hat{\rho}_{ab}(r'_a, r'_b, r''_a, r''_b, t)$. Subscript a corresponds to electrons, and b to ions ($a = \mathrm{e}$, $b = \mathrm{i}$). Thus, $\hat{\rho}_{ab} \equiv \hat{\rho}_{\mathrm{ei}}$.

The equation for operator $\hat{\rho}_{\mathrm{ei}}$ can be written in analogy to (7.1.11) for a one-particle operator density matrix. It has the following form [7.3, 4]:

$$i\hbar\frac{\partial \hat{\rho}_{\mathrm{ei}}}{\partial t} = \left[-\frac{\hbar^2}{2m_{\mathrm{e}}}(\Delta_{r'_{\mathrm{e}}} - \Delta_{r''_{\mathrm{e}}}) - \frac{\hbar^2}{2m_{\mathrm{i}}}(\Delta_{r'_{\mathrm{i}}} - \Delta_{r''_{\mathrm{i}}}) \right] \hat{\rho}_{\mathrm{ei}}$$
$$+ \{[\Phi_{\mathrm{ei}}(|r'_{\mathrm{e}} - r'_{\mathrm{i}}|) + e_{\mathrm{e}}\,\hat{\varphi}(r'_{\mathrm{e}}, t) + e_{\mathrm{i}}\,\hat{\varphi}(r'_{\mathrm{i}}, t)]$$
$$- [\Phi_{\mathrm{ei}}(|r''_{\mathrm{e}} - r''_{\mathrm{i}}|) + e_{\mathrm{e}}\,\hat{\varphi}(r''_{\mathrm{e}}, t) + e_{\mathrm{i}}\,\hat{\varphi}(r''_{\mathrm{i}}, t)]\} \hat{\rho}_{\mathrm{ei}}\ . \tag{7.2.1}$$

The operator of electric potential is connected to the operator $\hat{\rho}_{\mathrm{ei}}$ through the Poisson equation,

$$\Delta_q\,\hat{\varphi}(q, t) = -4\pi N \int [e_{\mathrm{e}}\delta(q - r_{\mathrm{e}}) + e_{\mathrm{i}}\delta(q - r_{\mathrm{i}})]$$
$$\times \hat{\rho}_{\mathrm{ei}}(r_{\mathrm{e}}, r_{\mathrm{i}}; r_{\mathrm{e}}, r_{\mathrm{i}}, t)\frac{dr_{\mathrm{e}}\,dr_{\mathrm{i}}}{V^2}\ . \tag{7.2.2}$$

In (7.2.1) the interaction of the charged particles in a pair Φ_{ei} is singled out. The electric potential is therefore determined by the distribution of the charged particles singled-out pairs.

From (7.2.1) one can go to a simpler equation for a one-particle density matrix (7.1.11). To do this one should employ the relationship between operators $\hat{\rho}_a$ and $\hat{\rho}_{ab}$, for instance,

$$\hat{\rho}_e(r'_e, r''_e, t) = \int \hat{\rho}_{ei}(r'_e, r'_i; r''_e, r_i, t) \frac{dr_i}{V}. \tag{7.2.3}$$

The operator of the total number of particles is connected with the operator $\hat{\rho}_{ei}$ through formula

$$N \int \hat{\rho}_{ei}(r_e, r_i; r_e, r_i, t) \frac{dr_e \, dr_i}{V^2} = \hat{N}. \tag{7.2.4}$$

For the description of bound states it is less convenient to employ operator $\hat{\rho}_{ei}(r'_e, r'_i, r''_e, r''_i, t)$ than the corresponding operator in the space of variables r and R, which are defined by (3.1.4), i.e., the operator $\rho_{ei}(R', r', R'', r'', t)$. The corresponding equation, from (7.2.1, 2) has the form [7.3, 4]

$$i\hbar \frac{\partial \hat{\rho}_{ei}}{\partial t} = \left[-\frac{\hbar^2}{2M}(\Delta_{R'} - \Delta_{R''}) - \frac{\hbar^2}{2\mu}(\Delta_{r'} - \Delta_{r''}) \right] \hat{\rho}_{ei}$$

$$+ \{[\Phi_{ei}(r') + \hat{U}_{ei}(R', r', t)] - [\Phi_{ei} + \hat{U}_{ei}(R'', r'', t)]\} \hat{\rho}_{ei} \tag{7.2.5}$$

with the equation for $\hat{\varphi}$

$$\Delta \hat{\varphi}(q, t) = -4\pi N \int \left[e_e \delta \left(q - \left(R + \frac{m_i}{M} r \right) \right) + e_i \delta \left(q - \left(R - \frac{m_e}{M} r \right) \right) \right]$$

$$\times \hat{\rho}_{ei}(R, r; R, r, t) \frac{dR \, dr}{V^2}. \tag{7.2.6}$$

In (7.2.5) the designation for the potential energy operator of the chosen pair in the field of surrounding pairs,

$$\hat{U}_{ei}(R, r) = e_e \hat{\varphi} \left(R + \frac{m_i}{M} r, t \right) + e_i \hat{\varphi} \left(R - \frac{m_e}{M} r, t \right), \tag{7.2.7}$$

is used. In the new variables, (7.2.4) becomes

$$N \int \hat{\rho}_{ei}(R, r; R, r, t) \frac{dR \, dr}{V^2} = \hat{N}. \tag{7.2.8}$$

Equations (7.2.5, 6) describe the distribution of pairs of particles in both bound and free states. In order to separate free and bound states, we shall proceed in the following manner. Consider eigenfunctions of the Hamiltonian operator of a pair of particles, given by

$$\left[-\frac{\hbar^2}{2M} \Delta_R - \frac{\hbar^2}{2\mu} \Delta_r + \Phi_{ei}(r) \right] \Psi_{\alpha P}(r, R) = E_{\alpha P} \Psi_{\alpha P}(r, R) . \tag{7.2.9}$$

From this it follows that

$$\Psi_{\alpha P}(r, R) = \Psi_\alpha(r) \Psi_P(R) , \quad E_{\alpha P} = E_\alpha + E_P , \tag{7.2.10}$$

where $\Psi_P(R)$ are the eigenfunctions of the Hamiltonian operator of a free pair

$$\Psi_P(R) = \exp\left(i\frac{PR}{\hbar} \right), \quad \int \Psi_{P''}^*(R)\, \Psi_{P'}(R)\, \frac{dR}{V} = \frac{(2\pi\hbar)^3}{V} \delta(P' - P''), \tag{7.2.11a}$$

$\Psi_\alpha(r)$ are the eigenfunctions of the Hamiltonian operator for the relative motion of particles in a pair, $\alpha = n$ for a discrete spectrum, and $\alpha = p$ for a continuous spectrum. The normalization condition for these functions has the form

$$\int \Psi_\beta^*(r)\, \Psi_\alpha(r)\, \frac{dr}{V}$$

$$= \begin{cases} \delta_{nm} & \text{at} \quad \alpha = n,\ \beta = m, \\ 0 & \text{at} \quad \alpha = p',\ \beta = m;\ \alpha = n,\ \beta = p'', \\ \dfrac{(2\pi\hbar)^3}{V}\delta(p' - p'') & \text{at} \quad \alpha = p',\ \beta = p''. \end{cases} \tag{7.2.11b}$$

Now we expand the operator density matrix of a pair of particles over eigenfunctions,

$$\rho_{ei}(R', r'; R'', r'', t) = \frac{V^2}{(2\pi\hbar)^6} \sum_{\alpha\beta} \int \hat{\rho}_{\alpha\beta}(P', P'', t)$$

$$\times \Psi_\alpha(r')\, \Psi_\beta^*(r'')\, \Psi_{P'}(R')\, \Psi_{P''}^*(R'')\, dP\, dP'' . \tag{7.2.12}$$

Using this expansion together with (7.2.5, 6), we find the set of equations for operator density matrices $[\rho_{\alpha\beta}(P', P'', t)$ [7.3, 4],

$$i\hbar \frac{\partial \hat{\rho}_{\alpha\beta}(P', P'', t)}{\partial t} = (E_\alpha + E_{P'} - E_\beta - E_{P''})\, \hat{\rho}_{\alpha\beta}(P', P'', t)$$

$$+ \frac{V}{(2\pi\hbar)^3} \sum_\gamma \int [U_{\alpha\gamma}(P', P, t)\, \hat{\rho}_{\gamma\beta}(P, P'', t)$$

$$- \hat{\rho}_{\alpha\gamma}(P', P, t)\, \hat{U}_{\gamma\beta}(P, P'', t)]\, dP \tag{7.2.13}$$

where $\alpha = n, p'$, $\beta = m, p''$, and $\gamma = l, p$. The summation over γ means

$$\sum_{\gamma} \ldots = \sum_{l} \ldots + \frac{V}{(2\pi\hbar)^3} \int \ldots dp \; . \tag{7.2.14}$$

The matrix element of the potential energy operator (7.2.7) was designated as

$$\hat{U}_{\alpha\beta}(P',P'',t) = \int \Psi_\alpha^*(r)\, \Psi_{P'}^*(R)\, \hat{u}_{\mathrm{ei}}(R,r,t)\, \Psi_\beta(r)\, \Psi_{P''}(R) \frac{dr\, dR}{V^2} \; . \tag{7.2.15}$$

Thus we have gone from the initial equation (7.2.5) to a set of four equations for operators

$$\hat{\rho}_{nm}(P',P'',t)\,, \quad \hat{\rho}_{p'p''}(P',P'',t)\,, \quad \hat{\rho}_{np''}(P',P'',t)\,, \quad \rho_{p'm}(P',P'',t)\,. \tag{7.2.16}$$

This set of equations must be supplemented by the equation for operator $\hat{U}_{\alpha\beta}(P',P'',t)$, which can be obtained from (7.2.6) in the following way.

Substituting (7.2.7) into (7.2.15) and expanding operator $\hat{\varphi}(q,t)$ into the Fourier integral

$$\hat{\varphi}(q,t) = \frac{1}{(2\pi)^3} \int \hat{\varphi}(k,t) e^{ikq} dk \tag{7.2.17}$$

results in

$$\hat{U}_{\alpha\beta}(P',P'',t) = \frac{1}{V} \int P_{\alpha\beta}(k)\, \delta\left(k - \frac{P'-P''}{\hbar}\right) \hat{\varphi}(k,t)\, dk \; , \tag{7.2.18}$$

with the matrix element designated

$$P_{\alpha\beta}(k) = \int \left[e_{\mathrm{e}} \exp\left(i\frac{m_{\mathrm{i}}}{M} kr \right) + e_{\mathrm{i}} \exp\left(-i\frac{m_{\mathrm{e}}}{M} kr \right) \right] \Psi_\alpha^*(r)\, \Psi_\beta(r) \frac{dr}{V} \; . \tag{7.2.19}$$

Equation (7.2.18) connects operators $\hat{U}_{\alpha\beta}$ and $\hat{\varphi}(k,t)$.

Into the right-hand side of equation (7.2.6) for operator $\hat{\varphi}$ we substitute expansion (7.2.12) and carry out a Fourier transform (7.2.17) over q. Finally, using the designation for the matrix element (7.2.19), we get

$$\hat{\varphi}(k,t) = \sum_{\alpha_1\beta_1} \frac{4\pi N}{k^2} \int P_{\alpha_1\beta_1}^*(k)\, \rho_{\alpha_1\beta_1}(P_1',P_1'',t)$$

$$\times\, \delta(\hbar k - (P_1' - P_1'')) \frac{V}{(2\pi\hbar)^3} dP_1'\, dP_1'' \; , \tag{7.2.20}$$

which connects the operators $\hat{\varphi}(k,t)$ and $\hat{\rho}_{\alpha\beta}(P',P'',t)$.

The set of equations (7.2.13, 20) is closed if (7.2.18) is taken into account. It is a part of the basis for constructing the kinetic theory of partially ionized plasma. Before discussing this theory, we shall consider the basic operator equations for a more general case, when electromagnetic interaction between charged particles in the system is taken into account in addition to Coulomb interaction.

7.3 The Hamiltonian with Electromagnetic Interaction (Extreme Cases)

Our discussion begins once more with a system of free charged particles. In the classical case the Hamiltonian function is given by (2.7.3). We employ it to find the Hamiltonian operator for a single particle in the field of surrounding particles [cf. (7.1.18)],

$$\hat{H}_a = \frac{1}{2m_a}\left[P - \frac{e_a}{c}A^m(r, t)\right]^2 + e_a \varphi^m(r, t). \tag{7.3.1}$$

From here on, the designations given in (7.1.20) will be employed. For the operator N_a (7.1.19) may be used, with the Hamiltonian operator determined by (7.3.1). The equations for the field operators coincide with the form of (2.2.1, 2).

Consider another extreme case of partially ionized plasma, i.e., when all oppositely charged particles are bound in pairs. From (3.5.13), the Hamiltonian operator of a chosen pair of particles in the field of surrounding pairs is (omitting the cap ∧ over N_{ab}),

$$\hat{H}_{ei} = \frac{1}{2m_e}\left[P_e - \frac{e_e}{c}A^m(r_e, t)\right]^2 + \frac{1}{2m_i}\left[P_i - \frac{e_i}{c}A^m(r_i, t)\right]^2$$

$$+ e_e \varphi^m(r_e, t) + e_i \varphi^m(r_i, t). \tag{7.3.2}$$

The equation for the operator N_{ei} can be written in analogy to (7.1.19), but must be supplemented by the equations for field operators. These coincide in form with classical equations, where the operators of charge and current densities are given by equations similar to (3.3.2).

In future descriptions of an atom – field system we shall employ the dipole approximation. In the classical approach the equations of motion have the form of (3.4.5, 6), and the Hamiltonian function is given by (3.5.5). Using the latter we obtain an expression for the Hamiltonian operator of an atom in the field of surrounding atoms,

$$\hat{H}_{ei}(R, \hat{P}, r, \hat{p}) = \frac{1}{2M}\left(\hat{P} + \frac{e}{c}[rB^m(R, t)]\right)^2 + \frac{\hat{p}^2}{2\mu}$$

$$+ \Phi_{ei}(r) - erE^m(R, t), \qquad e \equiv e_i \tag{7.3.3}$$

where $\Phi_{\mathrm{ei}}(r)$ is the potential energy of pair of particles in an atom.

In linear approximation with respect to the magnetic field this expression can be written in the form

$$\hat{H}_{\mathrm{ei}} = \frac{\hat{P}^2}{2M} + \frac{\hat{p}^2}{2\mu} + \Phi_{\mathrm{ei}} - er\left\{E^{\mathrm{m}}(R, t) + \frac{1}{c}[\hat{V}B^{\mathrm{m}}(R, t)]\right\}, \quad V = \frac{P}{M}. \quad (7.3.4)$$

In this approximation the energy of the interaction with surrounding atoms is proportional to the scalar product of displacement and the Lorentz force.

Finally, one more form of presentation can be used,

$$\hat{H}_{\mathrm{ei}} = \frac{\hat{P}^2}{2M} + \frac{\hat{p}^2}{2\mu} + \Phi_{\mathrm{ei}} - erE^{\mathrm{m}}(R, t) - \frac{e}{c}[r \times \hat{V}]B^{\mathrm{m}}(R, t). \quad (7.3.5)$$

Here the last term in the right-hand side is the potential energy of the atom's magnetic moment. The magnetic moment arises from the motion of an atom as a whole (see Sect. 3.3).

The equation for the operator $N_{\mathrm{ei}}(R, P, r, p, t)$ can be written by analogy with (7.1.19). The corresponding equations for field operators coincide in form with the classical field equations (3.3.19, 20). The polarization and magnetization operators P^{m} and M^{m} are connected with operator N_{ei} by (3.3.28).

7.4 The Equations for Operators of Field and Particles

Consider the equations analogous to those discussed in Sect. 7.2 for a more general case, when not only Coulomb interaction, but electromagnetic interaction is also taken into account. Then instead of the Hamiltonian operator for a pair of particles, which (7.2.1) corresponds to, a more general expression (7.3.2) ought to be employed.

Limiting ourselves in (7.3.2) to taking only those terms into account which are linear in A^{m}, and going from variables r_{e} and r_{i} to r and R, like in Sect. 7.2, we can write the equation for the operator matrix of density $\rho_{\mathrm{ei}}(R', r', R'', r'', t)$. This only differs from (7.2.5) in that the operator $\hat{U}_{\mathrm{ei}}$ is now given by a more complex expression,

$$\hat{U}_{\mathrm{ei}}(R, r) = e_{\mathrm{e}}\varphi^{\mathrm{m}}\left(R + \frac{m_{\mathrm{i}}}{M}r, t\right) + e_{\mathrm{i}}\varphi^{\mathrm{m}}\left(R - \frac{m_{\mathrm{e}}}{M}r, t\right)$$

$$- \frac{e_{\mathrm{e}}}{m_{\mathrm{e}}c}A^{\mathrm{m}}\left(R + \frac{m_{\mathrm{i}}}{M}r, t\right)\left(\frac{m_{\mathrm{e}}}{M}\hat{P} + \hat{p}\right)$$

$$- \frac{e_{\mathrm{i}}}{m_{\mathrm{i}}c}A^{\mathrm{m}}\left(R - \frac{m_{\mathrm{e}}}{M}r, t\right)\left(\frac{m_{\mathrm{i}}}{M}\hat{P} - \hat{p}\right). \quad (7.4.1)$$

We have taken advantage of the fact that the operators $\hat{p}_a$ and A commute,

$$\hat{p}_a A(r_\alpha, t) - A \hat{p}_a = -i\hbar \operatorname{div} A = 0.$$

We always employ the Coulomb calibration when $\operatorname{div} A = 0$. Unlike (7.2.7), operator (7.4.1) is complex, so the equation for the operator $\hat{\rho}_{ei}$ now is to be written [cf. (7.2.5)] [7.5]

$$-i\hbar \frac{\partial \rho_{ei}}{\partial t} = \left[-\frac{\hbar^2}{2M}(\Delta_{R'} - \Delta_{R''}) - \frac{\hbar^2}{2\mu}(\Delta_{r'} - \Delta_{r''}) \right] \rho_{ei}$$

$$+ \{ [\Phi_{ei}(r') + \hat{U}_{ei}(R', r', t)] - [\Phi_{ei}(r'') + \hat{U}_{ei}^*(R'', r'', t)] \} \hat{\rho}_{ei}. \qquad (7.4.2)$$

Again using (7.2.12), we go from (7.4.2) to the set of equations for the operator matrices of density $\rho_{\alpha\beta}(P', P'', t)$. This set of equations coincides in form with (7.2.13). The operator $\hat{U}_{\alpha\beta}$ is composed of two parts. The first of them (for the Coulomb field) is determined by (7.2.15), while the second accounts for the contribution from the rotational electromagnetic field and is related to the Fourier component of the vector potential [cf. (7.2.18)]

$$\hat{U}_{\alpha\beta}(P', P'', t) = -\frac{\hbar^3}{V} \int P_{\alpha\beta}(k, P' + P'') \hat{A}(k, t) \delta[\hbar k - (P' - P')] dk. \qquad (7.4.3)$$

The vector matrix element is given by

$$P_{\alpha\beta}(k, P' + P'') = \int \left[\frac{e_e}{e} \exp\left(i\frac{m_i}{M} kr \right) \left(\frac{P' + P''}{2M} + \frac{\hat{p}_\alpha^* + \hat{p}_\beta}{2m_e} \right) \right.$$

$$+ \left. \frac{e_i}{c} \exp\left(-\frac{m_e}{M} kr \right) \left(\frac{P + P''}{2M} - \frac{\hat{p}_\alpha^* + \hat{p}_\beta}{2m_i} \right) \right] \Psi_\alpha^*(r) \Psi_\beta(r) \frac{dr}{V}. \qquad (7.4.4)$$

The operators $\hat{p}_\alpha$ and $\hat{p}_\beta$ act upon functions Ψ_α and Ψ_β, respectively.

In the future we shall need the square modulus of this matrix element. Let us write the corresponding expressions for two extreme cases.

1) Free charged particles approximation:

$$|P_{p'p''}(k, P' + P'')|_{ij}^2 = \left[\frac{e_e^2}{c^2} \frac{(2\pi\hbar)^3}{V} \delta(p_i' - p_i'') \left(\frac{v_e' + v_e''}{2} \right)_{ij}^2 \right.$$

$$+ \left. \frac{e_i^2}{c^2} \frac{(2\pi\hbar)^3}{V} \delta(p_e' - p_e'') \left(\frac{v_i' + v_i''}{2} \right)_{ij}^2 \right]. \qquad (7.4.5)$$

In the right-hand side of this expression the transition is carried out from the variables p and P to variables p_e and p_i [see (8.3.8)].

2) Dipole approximation for a discrete spectrum:

$$|P_{nm}|_{ij}^2 = \frac{e^2}{c^2\mu^2}|p_{nm}|_{ij}^2 .$$ (7.4.6)

In order to close the set of operator equations it is necessary to write one more equation for the Fourier component of the operator of vector potential. It connects the operators $\hat{A}(k, t)$ and $\rho_{\alpha\beta}$ and is similar to the equation for a Coulomb plasma (7.2.20).

7.5 Operator Equations for an Atom – Field System in Dipole Approximation

In considering a system of atoms it is more suitable to employ the operator $N_{nm}(R, P, t)$ instead of operator $N_{ei}(R, P, r, p, t)$. The relation between them is easily found if the connection between the Wigner function and density matrix (7.1.13) and the expansion of the density matrix over the eigenfunctions of the Hamiltonian operator of the immobile atom is used. This expression has the form

$$N_{ei}(R, P, r, p, t) = \frac{1}{(2\pi)^3} \sum_{nm} N_{nm}(R, P, t)$$

$$\times \int \Psi_n \left(r + \frac{1}{2}\hbar\gamma\right) \Psi_m^* \left(r - \frac{1}{2}\hbar\gamma\right) e^{-i\gamma p} \frac{(2\pi\hbar)^3}{V} d\gamma .$$ (7.5.1)

We shall employ the Hamiltonian operator (7.3.4), in which we designate

$$\mathscr{E}^m(R, V, t) = E^m(R, t) + \frac{1}{c}[VB^m(R, t)] .$$ (7.5.2)

The equation for operator $N_{nm}(R, P, t)$ then becomes

$$\left(\frac{\partial}{\partial t} + V\frac{\partial}{\partial R} + i\omega_{nm}\right)N_{nm}(R, P, t) = \frac{i}{\hbar(2\pi)^6}\int\sum_{n_1} dp\, d\gamma\, dP_1\, dR_1$$

$$\times [d_{nn_1}\mathscr{E}^m(R_1 + \tfrac{1}{2}\hbar p, P_1 - \tfrac{1}{2}\hbar\gamma, t)N_{n_1 m}(R_1, P_1, t)$$

$$- N_{nn_1}(R_1, P_1, t)d_{n_1 m}\mathscr{E}^m(R_1 - \tfrac{1}{2}\hbar p, P_1 + \tfrac{1}{2}\hbar\gamma, t)]$$

$$\cdot \exp\left[i\gamma(R_1 - R) + i p(P_1 - P)\right] .$$ (7.5.3)

This is a quantum equation in respect both to the internal variables and to the motion of the atom as a whole.

Naturally, a simpler description is of interest when the atoms' motion is considered from a classical point of view. The corresponding equation follows from (7.5.3) with expansion over $\hbar\rho\,\partial/\partial R$, $\hbar\gamma\,\partial/\partial P$. Although momenta of atoms is then canonical, it is more practical to employ noncanonical momenta. In classical theory, the connection between then is given by (3.5.9). The transition from one distribution to the other in this case presents no difficulties. The equation for phase density in the approximation in question in noncanonical variables has the form of (3.2.7). In quantum theory, (3.5.9) ought to be replaced by the operator relation

$$\hat{P} = MV - \hat{L}_{nm}. \tag{7.5.4}$$

Operator $\hat{L}_{nm}$ is given by

$$\hat{L}_{nm}F_{nm} = \frac{1}{2c}\sum_n ([d_{nn_1}B^m]F_{n_1m} + F_{nn_1}[d_{n_1m}B^m]). \tag{7.5.5}$$

Accordingly,

$$N_{nm}(MV, R, t) = N_{nm}(\hat{P} - \hat{L}_{nm}, R, t), \tag{7.5.6}$$

and the equation for the operator $N_{nm}(V, R, t)$ has the form

$$\left(\frac{\partial}{\partial t} + V\frac{\partial}{\partial R} + i\omega_{nm}\right)N_{nm}$$

$$-\frac{i}{\hbar}\sum_{n_1}[d_{nn_1}\mathscr{E}^m(R, V, t)N_{n_1m} - N_{nn_1}d_{n_1m}\mathscr{E}^m]$$

$$+\frac{1}{2}\sum_{n_1}\left\{\left[\left(d_{nn_1}\frac{\partial}{\partial R}\right)\mathscr{E}^m(R, V, t) + \left[\frac{i\omega_{nn_1}}{c}d_{nn_1}B^m\right]\right]\frac{\partial N_{nm}}{\partial MV} + n \rightleftarrows m\right\}. \tag{7.5.7}$$

Finally, one other form of the equation is useful when a mixture of coordinates and momenta is employed for internal variables. This equation, obtained with (7.5.1), has the form

$$\frac{\partial N}{\partial t} + V\frac{\partial N}{\partial R} + v\frac{\partial N}{\partial r} - \frac{i}{\hbar(2\pi)^3}\int\left[\Phi\left(\left|r + \frac{1}{2}\hbar\tau\right|\right) - \Phi\left(\left|2 - \frac{1}{2}\hbar\tau\right|\right)\right]$$

$$\times N(R, P, r, p', t)e^{i\tau(p'-p)}d\tau\,dp' + e\,\mathscr{E}^m(R, V, t)\frac{\partial N}{\partial p}$$

$$+\left\{\left(r\frac{\partial}{\partial R}\right)\mathscr{E}^m + \frac{e}{c}[v \times B^m(R, t)]\right\}\frac{\partial N}{\partial P} = 0, \qquad P = MV. \tag{7.5.8}$$

This equation is the closest in form to the classical equation (3.2.7) becoming identical at $\hbar \to 0$.

In some cases, e. g., in the theory of solid-state lasers (lasers with uniformly broadened line), the approximation of immobile atoms is good enough. In this case

$$N_{nm}(R, P, t) = N_{nm}(R, t) \frac{(2\pi\hbar)^3}{V} \delta(P),$$

and (7.5.7) becomes

$$\left(\frac{\partial}{\partial t} + i\omega_{nm} \right) N_{nm}(R, t)$$

$$- \frac{i}{\hbar} \sum_{n_1} [d_{nn_1} E^m(R, t) N_{n_1 m} - N_{nn_1} d_{n_1 m} E^m(R, t)] = 0. \qquad (7.5.9)$$

In this approximation the dependence upon coordinates is nonexplicit [only through the function $E(R, t)$].

The operator equations obtained in this chapter serve as the initial ones in the statistical theory of electromagnetic processes in quantum systems.

8. The Kinetic Equations for Partially Ionized Plasma. The Coulomb Approximation

In this chapter we shall determine the expression describing the fluctuations which determine the collision integrals in the kinetic equations for the distribution functions of electrons, ions, and atoms in a three-component plasma. The general properties and the structure of the collision integrals are analyzed. The theory of large-scale (kinetic) fluctuations is developed for partially ionized plasma. In addition, the fluctuations in the concentrations of electrons, ions, and atoms are calculated in an approximation using chemical kinetics.

8.1 The Polarization Approximation

Our next task is to obtain the kinetic equations for the distribution functions of electrons (f_e), ions (f_i), and atoms (f_n). We shall take as our initial equations the set of operators (7.2.13) for four operator density matrices (7.2.16). This set will be considered together with the equation (7.2.20) for the operator of electric potential.

The kinetic equations will be obtained in two steps. First we shall find the kinetic equation for the distribution function of pairs of charged particles, and then we shall introduce the condition that correlations attenuate when the particles pass from bound states into free states. This will enable us to go from an equation for the distribution function of pairs of charged particles to a set of three kinetic equations for the distribution functions of electrons, ions, and atoms.

In order not to complicate our task unnecessarily from the very beginning, we shall first concern ourselves with the calculation of dissipative processes. We may assume the averaged distributions to be spatially uniform. In this case the averaged matrix of density is diagonal, i.e.,

$$\langle \hat{\rho}_{\alpha\beta}(P', P'', t) \rangle = \delta_{\alpha\beta} \frac{(2\pi\hbar)^3}{V} \delta(P' - P'') f_\alpha(P', t) \quad \text{and} \quad \langle U_{\alpha\beta} \rangle = 0 . \tag{8.1.1}$$

Here $f_\alpha(P', t)$ is the distribution function of pairs of charged particles. The normalization condition is

$$\sum_a \int f_\alpha(P, t) \frac{V}{(2\pi\hbar)^3} dP = 1 . \tag{8.1.2}$$

For deviations we use the designations

$$\delta\rho_{\alpha\beta} = \hat{\rho}_{\alpha\beta} - \langle \hat{\rho}_{\alpha\beta} \rangle, \quad \delta U_{\alpha\beta} = U_{\alpha\beta} . \tag{8.1.3}$$

Then from (7.2.13) after averaging over the distribution of the occupation numbers we obtain the following equation for the distribution function of pairs [8.1, 2]:

$$\frac{\partial f_\alpha(P', t)}{\partial t} = -\frac{2V}{(2\pi\hbar)^3\hbar} \sum_\beta \int \mathrm{Im}\{\delta\rho_{\alpha\beta}(P', P'', t)\,\delta U_{\beta\alpha}(P'', P', t)\,dP''\}$$

$$= -\frac{2V}{(2\pi\hbar)^3\hbar} \sum_\beta \int \mathrm{Im}\{\delta\rho_{\alpha\beta}(P', P'')\,\delta U_{\alpha\beta}(P', P'')\}_\omega\,dP''\,\frac{d\omega}{2\pi}$$

$$\equiv I_\alpha(P', t), \tag{8.1.4}$$

introducing the designation for the collision integral for pairs of particles. It is determined by the correlation of fluctuations $\delta\rho_{\alpha\beta}$ and $\delta U_{\alpha\beta}$ or by the relevant spectral density.

To calculate the correlations of fluctuations we again (see Sects. 4.4–6, 6.2) employ the polarization approximation. This supplements the approximation of perturbation theory with respect to interaction (Born's approximation for the interaction of pairs of charged particles) by taking the polarization of the medium into account.

The equation for the function $\delta\rho_{\alpha\beta}$ can be written in analogy to (4.4.10, 6.5). Employing (7.2.13, 20, 8.1.1) we get following equations for fluctuations $\delta\rho_{\alpha\beta}$, $\delta U_{\alpha\beta}$, and $\delta\varphi$ in the polarization approximation:

$$\left\{ i\hbar\frac{\partial}{\partial t} - (E_\alpha + E_{P'} - E_\beta - E_{P''}) \right\} (\delta\rho_{\alpha\beta} - \delta\rho_{\alpha\beta}^{\mathrm{source}})_{P'_1 P'', t}$$

$$= \delta U_{\alpha\beta}(P', P'', t)[f_\beta(P'', t) - f_\alpha(P', t)]. \tag{8.1.5}$$

$$\delta U_{\alpha\beta}(P', P'', t) = \frac{1}{V}\int P_{\alpha\beta}(k)\,\delta\left(k - \frac{P' - P''}{\hbar}\right)\delta\varphi(k, t)\,dk,$$

$$\delta\varphi(k, t) = \sum_{\alpha\beta}\frac{4\pi N}{k^2}\int P_{\alpha\beta}^*(k)\,\delta\rho_{\alpha\beta}(P', P'', t)$$

$$\times \delta(\hbar k - (P' - P''))\frac{V}{(2\pi\hbar)^3}\,dP'\,dP''. \tag{8.1.6}$$

The two-time correlation of the source fluctuations is given by an equation similar to (4.4.11),

$$\left\{ i\hbar\frac{\partial}{\partial t} - (E_\alpha + E_{P'} - E_\beta - E_{P''}) \right\}$$

$$\times \langle \delta\rho_{\alpha\beta}(P', P'', t)\,\delta\rho_{\alpha_1\beta_1}^*(P'_1, P''_1, t') \rangle^{\mathrm{source}} = 0, \tag{8.1.7}$$

to which the initial ($t = t'$) condition should be added, i. e., the expression for the one-time correlation of the source fluctuations. In classical theory the corresponding initial conditions are given by (4.4.8, 6.4).

8.2 The Correlation of the Source Fluctuations

The one-time correlation of source fluctuations for a system of free charged particles is given by (4.4.8). Comparing this equation with that relating the correlation of phase density fluctuations δN_a with the correlation function g_{ab} (see, e. g., [8.1, Eq. (26.10)]), we find

$$\overline{(\delta N_a \delta N_b)}_{x,x';t} = \frac{N_a N_b - \delta_{ab} N_a}{V^2} g_{ab}(x, x', t)$$

$$+ \frac{N_a}{V} \delta_{ab} \left[\delta(x - x') f_a - \frac{1}{V} f_a(x, t) f_b(x', t) \right]. \qquad (8.2.1)$$

This comparison shows that the function $\overline{(\delta N_a \delta N_b)}_{x,x';t}^{\text{source}}$ expresses that part of a one-time correlation of phase density fluctuations which cannot be given with the correlation function g_{ab}, but is determined by one-particle distribution functions f_a. In a similar way we can determine the correlation of the source fluctuations in the quantum system of pairs of charged particles now under consideration.

The quantized wave functions are related to the operator $\hat{\rho}_{ei}$ by an expression similar to (7.1.10) (for the sake of clarity without subscripts),

$$\frac{N_{ei}}{V} \hat{\rho}(r'_e, r'_i; r''_e, r''_i, t) = \Psi^+(r''_e, r''_i, t) \Psi(r'_e, r'_i, t), \quad N_{ei} = N; \qquad (8.2.2)$$

they satisfy the commutative relations similar to (7.1.1).

The one-particle density matrix of pairs of charged particles is given by

$$\frac{N_{ei}}{V} \rho(r'_e, r'_i; r''_e, r''_i, t) = \langle \Psi^+(r''_e, r''_i, t) \Psi(r'_e, r'_i, t) \rangle, \qquad (8.2.3)$$

while the two-particle one is given by

$$\frac{N(N-1)}{V^2} \rho(r'_e, r'_i, r'_{e_1}, r'_{i_1}; r''_e, r''_i, r''_{e_1}, r''_{i_1}, t)$$

$$= \langle \Psi^+(r''_e, r''_i, t) \Psi^+(r''_{e_1}, r''_{i_1}, t) \Psi(r'_{e_1}, r'_{i_1}, t) \Psi(r'_e, r'_i, t) \rangle. \qquad (8.2.4)$$

The expression for the second moment of the operator density matrix of pairs of particles is

$$\left(\frac{N}{V}\right)^2 \langle \hat{\rho}(r'_e, r'_i, r''_e, r''_i, t)\, \hat{\rho}(r'_{e_1}, r'_{i_1}, r''_{e_1}, r''_{i_1}, t)\rangle$$

$$= \langle \Psi^+(r''_e, r''_i, t)\, \Psi(r'_e, r'_i, t)\, \Psi^+(r''_{e_1}, r''_{i_1}, t)\, \Psi(r'_{e_1}, r'_{i_1}, t)\rangle . \tag{8.2.5}$$

Employing the above formulae and the definition of the operator density matrix

$$g_{ei, e_1 i_1} = \rho_{ei, e_1 i_1} - \rho_{ei}\rho_{e_1 i_1} , \tag{8.2.6}$$

we obtain the desired expression for the one-time correlation of the source fluctuations,

$$\langle \delta\rho(r'_e, r'_i ; r''_e, r''_i, t)\, \delta\rho(r'_{e_1}, r'_{i_1} ; r''_{e_1}, r''_{i_1}, t)\rangle^{\text{source}}$$

$$= \frac{1}{N}[V^2\delta(r'_e - r''_{e_1})\,\delta(r'_i - r''_{i_1})\,\rho(r'_{e_1}, r'_{i_1}, r''_e, r''_i, t)$$

$$- \rho(r'_e, r'_i ; r''_e, r''_i, t)\,\rho(r'_{e_1}, r'_{i_1} ; r''_{e_1}, r''_{i_1}, t)] . \tag{8.2.7}$$

This expression's structure is similar to that of (4.4.8). The second term in brackets ensures that

$$\int \delta\rho_{ei}(r_e, r_i; r_e, r_i, t)\,\frac{dr_e\, dr_i}{V^2} = 0 , \tag{8.2.8}$$

which reflects the conservation of the total number of pairs of charged particles.

Here the variables r and R (the coordinates of the relative motion within a pair and the pair's center of mass) may be employed instead of r_e and r_i. This leads to the expression for the correlation of fluctuations $\delta\rho_{ei}^{\text{source}}(r, R, t)$. The one-time correlation of the source fluctuations $\delta\rho_{ei}^{\text{source}}(P', P'', t)$ in (8.1.5) can then be determined, using (7.2.12),

$$\langle \delta\rho_{\alpha\beta}(P', P'', t)\, \delta\rho_{\alpha_1\beta_1}(P'_1, P''_1, t)\rangle^{\text{source}}$$

$$= \frac{1}{N}\left[\delta_{\alpha\beta_1}\frac{(2\pi\hbar)^3}{V}\,\delta(P' - P''_1)\,\rho_{\alpha_1\beta}(P'_1, P'', t) - \rho_{\alpha\beta}(P', P'', t)\,\rho_{\alpha_1\beta_1}(P'_1, P''_1, t)\right] . \tag{8.2.9}$$

The second term in brackets ensures fulfillment of a condition similar to (8.2.8),

$$\sum_\alpha \int \delta\rho_{\alpha\alpha}(P, P, t)\,\frac{V}{(2\pi\hbar)^3}\,dP = 0 . \tag{8.2.10}$$

With the aid of (8.2.7, 9) the correlation of the source fluctuations can be presented in other forms as well. The most interesting of these is that where the motion of a pair as a whole is described in a mixed presentation of coordinates and momenta, while the internal motion is described in the same way as in (8.2.9).

If, in addition, the motion of centers of mass is assumed to be classical, the correlation of the source fluctuations $\delta N_{\alpha\beta}^{\text{source}}(\mathbf{R}, \mathbf{P}, t)$ is

$$\langle \delta N_{\alpha\beta}(\mathbf{R}, \mathbf{P}, t)\, \delta N_{\alpha_1\beta_1}(\mathbf{R}_1, \mathbf{P}_1, t) \rangle^{\text{source}} = N[\delta_{\alpha\beta_1}(2\pi\hbar)^3 \delta(\mathbf{R} - \mathbf{R}_1)$$

$$\times\, \delta(\mathbf{P} - \mathbf{P}_1)\, \rho_{\alpha_1\beta}(\mathbf{R}, \mathbf{P}, t) - \rho_{\alpha\beta}(\mathbf{R}, \mathbf{P}, t)\, \rho_{\alpha_1\beta_1}(\mathbf{R}_1, \mathbf{P}_1, t)]\,. \tag{8.2.11}$$

With summation over the internal variables and designating $f(\mathbf{R}, \mathbf{P}, t) = \sum_\alpha \rho_{\alpha\alpha}(\mathbf{R}, \mathbf{P}, t)$ for the distribution function of pairs' coordinates and momenta $(X = (\mathbf{R}, \mathbf{P}))$, the correlation of the source fluctuations for a single-component gas of pairs of particles (atoms) is defined

$$\langle \delta N(X, t)\, \delta N(X_1, t) \rangle^{\text{source}} = N[(2\pi\hbar)^3 \delta(X - X_1)\, f(X, t)$$

$$- f(X, t)\, f(X_1, t)]\,. \tag{8.2.12}$$

In juxtaposition to the classical formula (8.2.1), one should bear in mind the different normalization of the classical and quantum distribution functions.

The "compensatory" terms in (8.2.7, 9, 11, 12), which ensure fulfillment of (8.2.8, 10), are only significant in the theory of large-scale (kinetic) fluctuations. Our immediate task is obtaining the collision integrals which are determined by small-scale fluctuations (fluctuations with lifetimes much shorter than the relaxation times of the distribution functions). The compensatory terms do not play any role here and can be neglected.

The formulae obtained here for correlations of the source fluctuations are nonsymmetric with respect to variables "1" and "2." The symmetrized expressions are more convenient in obtaining the kinetic equation. Then (neglecting the compensatory terms) we get from (8.2.9)

$$\langle \delta\rho_{\alpha\beta}(\mathbf{P}', \mathbf{P}'', t)\, \delta\rho_{\alpha_1\beta_1}^*(\mathbf{P}_1', \mathbf{P}_1'', t) \rangle^{\text{source}}$$

$$= \frac{1}{2N}\, \frac{(2\pi\hbar)^3}{V}\, [\delta_{\alpha\alpha_1} \delta(\mathbf{P}' - \mathbf{P}_1')\, \rho_{\beta_1\beta}(\mathbf{P}_1'', \mathbf{P}'', t)$$

$$+ \delta_{\beta_1\beta} \delta(\mathbf{P}_1'' - \mathbf{P}'')\, \rho_{\alpha\alpha_1}(\mathbf{P}', \mathbf{P}_1', t)]\,. \tag{8.2.13}$$

In order to facilitate the presentation of the formulae to be obtained, the second multiplier in the left-hand side is replaced by its complex conjugate, and the variables $\alpha_1 \rightleftarrows \beta_1$ and $\mathbf{P}_1' \rightleftarrows \mathbf{P}_1''$ are substituted.

In kinetic equations for nonequilibrium states where off-diagonal elements of the matrix do not play an important role, (8.1.1) may be considered satisfied, and (8.2.13) has a simpler form,

$$\langle \delta\rho_{\alpha\beta}(\mathbf{P}', \mathbf{P}'', t)\, \delta\rho_{\alpha_1\beta_1}^*(\mathbf{P}_1', \mathbf{P}_1'', t) \rangle^{\text{source}} = \frac{1}{2N} \left(\frac{(2\pi\hbar)^3}{V} \right)^2 \delta_{\alpha\alpha_1} \delta_{\beta\beta_1}$$

$$\times\, \delta(\mathbf{P}' - \mathbf{P}_1') \delta(\mathbf{P}'' - \mathbf{P}_1'')[f_\beta(\mathbf{P}'', t) + f_\alpha(\mathbf{P}', t)]\,. \tag{8.2.14}$$

The stationary solution here can be found by returning to (8.1.7). As in classical kinetic theory (see Chaps. 4, 6), the stationary solution applies only to small-scale fluctuations. To separate the small-scale and large-scale regions we substitute in (8.1.7) $\partial/\partial t \rightarrow \partial/\partial t + \Delta$; in the final formula $\Delta \rightarrow 0$. Taking this into account, the spectral density of the fluctuations of the source is

$$[\delta\rho_{\alpha\beta}(\boldsymbol{p}',\boldsymbol{p}'')\delta\rho_{\alpha_1\beta_1}(\boldsymbol{p}_1',\boldsymbol{p}_1'',t)]_\omega^{\text{source}} = 2\pi\delta[\hbar\omega - (E_\alpha + E_{p'} - E_\beta - E_{p''})]$$

$$\times \langle\delta\rho_{\alpha\beta}(\boldsymbol{p}',\boldsymbol{p}'',t)\,\delta\rho^*_{\alpha_1\beta_1}(\boldsymbol{p}_1',\boldsymbol{p}_1'',t)\rangle^{\text{source}}. \tag{8.2.15}$$

This expression relates the spectral density of the source fluctuations to the one-time correlation, which is given by (8.2.14).

8.3 Dielectric Permittivity

Employing (8.1.5, 6) we can obtain the equation for the fluctuation of the potential electric field $\delta E(\boldsymbol{k}, t)$, and with its aid the expression for the dielectric permittivity. In (8.1.5, 6) we carry out the Fourier expansion with respect to time and write the stationary solution in the form

$$\delta\rho_{\alpha\beta}(\boldsymbol{P}',\boldsymbol{P}'',\omega) = \delta\rho_{\alpha\beta}^{\text{source}}(\boldsymbol{P}',\boldsymbol{P}'',\omega)$$

$$+\frac{1}{V}\int\frac{P_{\alpha\beta}(\boldsymbol{k}')\,\delta(\hbar k' - (P' - P''))[\,f_\beta(\boldsymbol{P}'',t) - f_\alpha(\boldsymbol{P}',t)]}{\hbar(\omega - i\Delta) - (E_\alpha + E_{P'} - E_\beta - E_{P''})}\,\delta\varphi(\omega k')d(\hbar k'), \tag{8.3.1}$$

where $\delta\varphi(\boldsymbol{k}, \omega)$ is the Fourier component of the fluctuation of the electric potential. The element of the matrix $P_{\alpha\beta}(\boldsymbol{k})$ is given by (7.2.19).

Substituting this solution into the right-hand side of (8.1.6) for the Fourier component $\delta\varphi(\omega, \boldsymbol{k})$ and integrating over $\boldsymbol{k}'$, the equation for $\delta E(\omega, \boldsymbol{k})$ can be written

$$\varepsilon(\omega, \boldsymbol{k})\,\delta E(\omega, \boldsymbol{k}) = \delta E^{\text{source}}(\omega, \boldsymbol{k}) \tag{8.3.2}$$

with the permittivity designated

$$\varepsilon(\omega, \boldsymbol{k})$$

$$= 1 + \frac{4\pi N}{V}\sum_{\alpha\beta}\int\frac{|P_{\alpha\beta}(\boldsymbol{k})|^2\delta(\hbar k - (P' - P''))[\,f_\alpha(\boldsymbol{P}',t) - f_\beta(\boldsymbol{P}'',t)]}{k^2[\hbar(\omega + i\Delta) - (E_\alpha + E_{P'} - E_\beta - E_{P''})]}$$

$$\times \frac{V}{(2\pi\hbar)^3}dP'\,dP'', \tag{8.3.3}$$

and the source of electric field fluctuations

$$\delta E^{\text{source}}(\omega, k) = \sum_{\alpha\beta} \frac{4\pi N}{k^2} ik \int P^*_{\alpha\beta}(k)\, \delta(\hbar k - (p' - p''))$$

$$\times \delta\rho^{\text{source}}_{\alpha\beta}(p', p'', \omega)\, \frac{V}{(2\pi\hbar)^3}\, dp'\, dp'' . \tag{8.3.4}$$

Let us now discuss the structure of (8.3.3) in greater detail. A double summation is carried out over α and β. According to (7.2.14) each sum can be split into two parts, accounting for the contributions from discrete and continuous parts of the spectrum, respectively. The permittivity (8.3.3) can then be presented as

$$\varepsilon(\omega, k) = 1 + 4\pi(\alpha_{\text{ff}} + \alpha_{\text{fb}} + \alpha_{\text{bf}} + \alpha_{\text{bb}}) , \tag{8.3.5}$$

with subscripts f and b for the free and bound states. Hence the polarizability $\alpha(\omega, k)$ is presented as a sum of four terms. Let us consider each of them separately.

The polarizability $\alpha_{\text{ff}}(\omega, k)$. Summation in (8.3.3) over α and β is replace by integration over p' and p''. With zero-order approximation for wave functions Ψ_p in the continuous spectral region,

$$\Psi_p(r) = \exp\left(\frac{ipr}{\hbar}\right) \tag{8.3.6}$$

and the square modulus of the matrix element (7.2.19) is given by

$$|P_{p'p''}(k)|^2$$

$$= \frac{(2\pi\hbar)^3}{V}\left[e_e^2\delta\left(p' - p'' - \frac{m_i}{M}\hbar k\right) + e_i^2\delta\left(p' - p'' + \frac{m_e}{M}\hbar k\right)\right]. \tag{8.3.7}$$

Returning to the formula for polarizability $\alpha_{\text{ff}}(\omega, k)$ which follows from (8.3.3) at $\alpha = p'$ and $\beta = p''$, we can substitute equation (8.3.7) into this formula and replace the pairs of variables p', P'; p'', P'' by new ones p'_e, p'_i, p''_e, p''_i, which are given by

$$P = p_e + p_i , \qquad p_e = p + \frac{m_e}{M}P ,$$

$$p = \frac{m_e p_i - m_i p_e}{M} \qquad p_i = -p + \frac{m_i}{M}P . \tag{8.3.8}$$

Finally, we take into account that

$$\int f(p_a, p_b, t)\, \frac{V}{(2\pi\hbar)^3}\, dp_b = f(p_a, t) , \qquad a = e, i , \quad b \neq a . \tag{8.3.9}$$

As a result we get

$$\alpha_{\text{ff}} = \sum_a \frac{e_a^2 n_a^2}{k^2} \int \frac{\delta(\hbar k - (p' - p''))[f_a(p', t) - f_a(p'', t)]}{\hbar(\omega + i\Delta) - (p'^2/2m_e - p''^2/2m_a)} \frac{V}{(2\pi\hbar)^3} dp'\, dp''\,.$$

$$(8.3.10)\cdot$$

It is easy to see that this coincides with the corresponding expression for the polarizability of completely ionized quantum plasma [Ref. 8.1, Eq. (74.2)]. In the classical approximation (8.3.10) is

$$\alpha_{\text{ff}} = \sum_a \frac{e_a^2 n_a}{k^2} \frac{k \frac{\partial f_a(p, t)}{\partial p}}{\omega + i\Delta - kv} \frac{V}{(2\pi\hbar)^3} dp\,.$$

$$(8.3.11)$$

A thorough investigation of the polarizability of completely ionized plasma has been performed by many researchers; the results can be found in [8.3 – 8].

The polarizabilities $\alpha_{\text{fb}}(\omega, k)$, $\alpha_{\text{bf}}(\omega, k)$. In general the function $\varepsilon(\omega, k)$ has the property

$$\varepsilon(\omega, k) = \varepsilon^*(-\omega^*, -k)\,.$$

$$(8.3.12)$$

Using this relation it can be easily proven that the polarizabilities α_{fb} and α_{bf} are connected by

$$\alpha_{\text{fb}}(\omega, k) = \alpha_{\text{bf}}^*(-\omega^*, -k)\,.$$

$$(8.3.13)$$

Hence it is sufficient to consider only one of the functions α_{fb} and α_{bf}.

In (8.3.3) we replace summation over α by integration over p' and substitute m for β. We then introduce new variables p_e and p_i for p' and P' according to (8.3.8). This results in the following replacement in (8.3.3):

$$f_\alpha(P', t) \to f_{p'}(P', t) \to f(p_e, p_i, t)\,.$$

$$(8.3.14)$$

Here one must employ the condition of attenuation of correlations of particles in a pair at the transition from a bound state into a free one, which was mentioned in Sect. 8.1.

This condition corresponds to replacing the mean number of pairs of free charged particles by the product of the mean numbers of free charged particles with the same values of momenta,

$$Nf(p_e, p_i, t) \to Nf(p_e, t)Nf(p_i, t)\,.$$

$$(8.3.15)$$

As a result the expression for the polarizability is

$$\alpha_{bf}(\omega, k) = \frac{N}{V} \sum_m \int \frac{V^2}{(2\pi\hbar)^6} \, dp_e \, dp_i \, dP'' \frac{1}{k^2} |P_{p'_i m}(k)|^2$$

$$\times \delta(\hbar k - (p_e - p_i - P'')) \frac{Nf(p_e, t) f(p_i, t) - f_m(P'', t)}{\hbar(\omega + i\Delta) - (p_e^2/2m_e + p_i^2/2m_i - E_m - E_{P''})}. \tag{8.3.16}$$

The momentum p' is connected with p_e and p_i by one of the equations in (8.3.8).

The sequence of subscripts b and f in α_{bf} reflects the direction of the absorption process, which is determined by the imaginary part of polarizability,

$$\mathrm{Im}\{\alpha_{bf}\} = -\pi \frac{N}{V} \frac{V^2}{(2\pi\hbar)^6} \sum_m \int dp_e \, dp_i \, dP'' \frac{1}{k^2} |P_{p'_i m}(k)|^2$$

$$\times \delta(\hbar k - (p_e + p_i - P'')) \delta\left(\hbar\omega - \left(\frac{p_e^2}{2m_e} + \frac{p_i^2}{2m_i} - E_m - E_{P''}\right)\right)$$

$$\times [Nf(p_e, t) f(p_i, t) - f_m(P'', t)]. \tag{8.3.17}$$

This formula describes the process of electron absorption upon atom ionization. For the state of equilibrium its form is

$$\mathrm{Im}\{\alpha_{bf}\} = -\pi \frac{N}{V} \frac{V^2}{(2\pi\hbar)^6} \left[\exp\left(-\frac{\hbar\omega}{\mathcal{X}T}\right) - 1\right] \sum_m \int dp_e \, dp_i \, dP'' \frac{1}{k^2} |P_{p'_i m}(k)|^2$$

$$\times \delta(\hbar k - (p_e + p_i - P''))$$

$$\times \delta\left(\hbar\omega - \left(\frac{p_e^2}{2m_e} - \frac{p_i^2}{2m_i} - E_m - E_{P''}\right)\right) f_m(P'') \geq 0. \tag{8.3.18}$$

When this equals zero, it corresponds to $\omega = 0$.

In carrying out the transition from (8.3.17) to (8.3.18), one should use the expressions for the equilibrium distribution functions of electrons, ions ($a = e, i$), and atoms in the state n:

$$f_a = \frac{1}{N} \exp\left(\frac{\mu_a - p^2/2m_a}{\mathcal{X}T}\right), \quad f_n(P) = \frac{1}{N} \exp\left(\frac{\mu_{ei} - E_n - E_P}{\mathcal{X}T}\right), \tag{8.3.19}$$

where μ_a and μ_{ei} are the chemical potentials given by

$$\mu_a = \mathcal{X}T \ln\left[\frac{N_a}{V}\left(\frac{2\pi\hbar^2}{m_a \mathcal{X}T}\right)^{3/2}\right], \quad \mu_{ei} = \mathcal{X}T \ln\left[\frac{N_{ei}}{V}\left(\frac{2\pi\hbar^2}{M\mathcal{X}T}\right)^{3/2} \frac{1}{Z}\right],$$

$$\tag{8.3.20}$$

and satisfying the requirement of chemical (ionization) equilibrium

$$\mu_e + \mu_i = \mu_{ei} \, . \tag{8.3.21}$$

The distribution functions of electrons, ions, and atoms are normalized with respect to concentration,

$$\frac{V}{(2\pi\hbar)^3} \int f_a dp = \frac{N_a}{N} = c_a \, , \quad \frac{V}{(2\pi\hbar)^3} \sum_n \int f_n dP = \frac{N_{ei}}{N} = c_{ei} \, . \tag{8.3.22}$$

The concentrations are in turn related by

$$c_a + c_{ei} = 1 \, , \quad a = e, i \, . \tag{8.3.23}$$

From (8.3.20) it follows that (8.3.21) can be written in the form (the Saha formula)

$$\frac{c_e c_i}{c_{ei}} \equiv \frac{n_e n_i}{n_{ei} n} = \left(\frac{\mu \mathscr{X}T}{2\pi\hbar^2}\right)^{3/2} \frac{1}{nZ} \, , \quad \mu = \frac{m_e m_i}{M} \, , \quad n = \frac{N}{V} \, . \tag{8.3.24}$$

The distinction between free and bound states is, of course, not strict. In general, a boundary region of width $\Delta E \sim \mathscr{X}T$ exists near $E = 0$ in the energy spectrum, where such a distinction cannot be made. However, the formal choice of $E = 0$ for the dividing point can bring about difficulties which are nonphysical in nature. As is well-known, the statistical sum Z over the discrete spectrum is divergent for hydrogen and hydrogenlike atoms, while the corresponding sum over the whole spectrum (both discrete and continuous) converges. A more detailed discussion of this matter is necessary. For hydrogen it is possible to avoid this difficulty by definiting the statistical sum via the Planck – Larkin formula [8.9 – 11]

$$Z = \sum_{n=1}^{\infty} n^2 \left[\exp\left(-\frac{E_n}{\mathscr{X}T}\right) - 1 + \frac{E_n}{\mathscr{X}T} \right] . \tag{8.3.25}$$

In this formula the two first terms of expansion of the function $\exp(-E_n/\mathscr{X}T)$ are omitted (compensated). Such a "regularization" makes the statistical sum independent of the concentration, and therefore the influence of the medium upon the spectrum of bound states is not accounted for.

Let us now discuss a number of special cases. For immobile atoms in the dipole approximation, (8.3.18) is

$$\text{Im}\{\alpha_{bf}\} = \pi \frac{N_e}{V} \frac{V}{(2\pi\hbar)^3} \left[1 - \exp\left(-\frac{\hbar\omega}{\mathscr{X}T}\right) \right] \sum_m \int dp \frac{|k d_{pm}|^2}{k^2} \frac{1}{Z}$$

$$\times \exp\left(-\frac{E_m}{\mathscr{X}T}\right) \delta\left(\hbar\omega - \left(\frac{p^2}{2\mu} - E_m\right)\right) . \tag{8.3.26}$$

For the hydrogen atom ionized from the ground level ($m = 0$) the square of the matrix element is determined by [8.12]

$$|r_{p0}k|^2 = \frac{2^{10}\pi a_0^3}{Vh^2} \frac{a_0^4(pk)^2}{1+(pa_0/\hbar)^4},$$

(8.3.27)

where a_0 is the Bohr radius. We substitute this expression into (8.3.26) and integrate over p, getting

$$\text{Im}\{\alpha_{bf}\} = \frac{N_{ei}}{V}\frac{2^8 a_0^3}{3\hbar^3}\frac{a_0^4}{\hbar^2}(2\mu)^{5/2}e^2\left[1 - \exp\left(\frac{\hbar\omega}{\mathcal{X}T}\right)\right]\frac{(\hbar\omega - I)^{3/2}}{\left(1 + \dfrac{\hbar\omega - I}{\hbar^2/2\mu a_0^2}\right)^6},$$

$$\hbar\omega > I$$

(8.3.28)

where I is the ionization potential of the hydrogen atom. This takes into account that $\exp(-I/\mathcal{X}T)/Z \approx 1$.

This expression describes the dissipation in the ionization of the hydrogen atom from the ground level, considering, however, only the dissipation due to Coulomb interaction. The processes determined by electromagnetic interaction will be discussed in Chaps. 9, 11.

The polarizability $\alpha_{bb}(\omega, k)$. In (8.3.3) we set $\alpha = n$ and $\beta = m$, and integrate over $P' - P''$, getting as a result

$$\alpha_{bb} = \frac{N}{V}\frac{V}{(2\pi\hbar)^3}\sum_{nm}\int dP\frac{|P_{nm}(k)|^2}{k^2}\frac{f_n(P+\frac{1}{2}\hbar k) - f_m(P-\frac{1}{2}\hbar k)}{\hbar(\omega + i\Delta) - (E_n - E_m) - \hbar k V},$$

$$V = \frac{P}{M}.$$

(8.3.29)

In the dipole approximation we find, from (7.2.19),

$$\frac{|P_{nm}(k)|^2}{k^2} = \frac{|k d_{nm}|^2}{k^2} = \frac{1}{3}|d_{nm}|^2.$$

(8.3.30)

This equation reflects the isotropy of the system.

An interesting case is where the distribution of atoms over momenta is Maxwellian. Then the distribution function $f_n(P)$ can be presented in the form

$$\frac{N}{V}f_n(P) = \frac{N_{ei}}{V}\rho_n f(P),$$

(8.3.31)

where ρ_n is the distribution of atoms in levels, and $f(P)$ is the Maxwell distribution.

In the dipole approximation, taking (8.3.30) into account, we can present (8.3.29) in the form

$$\alpha_{bb} = \frac{N}{3\hbar V}\frac{V}{(2\pi\hbar)^3}\sum_{nm}\int\frac{|d_{nm}|^2[f_n(P+\frac{1}{2}\hbar k)-f_m(P-\frac{1}{2}\hbar k)]}{\omega+i\Delta-\omega_{nm}-kV}\,dP. \quad (8.3.32)$$

This expression will be used often in the following.

8.4 The Spectral Density of the Electric Field Fluctuations

From (8.3.2) it follows that the spectral density of fluctuations is proportional to the spectral density of the source of fluctuations, which is given by (8.2.15, 3.4). As a result,

$$(\delta E\,\delta E)_{\omega,k} = \frac{N}{V}\frac{(4\pi)^2\pi\hbar V}{(2\pi\hbar)^3}\sum_{\alpha\beta}\int dP'\,dP''\,[f_\alpha(P',t)+f_\beta(P'',t)]$$

$$\times\,\delta(\hbar k-(P'-P''))\,\delta(\hbar\omega-(E_\alpha+E_{P'}-E_\beta-E_{P''}))\frac{|P_{\alpha\beta}(k)|^2}{k^2|\varepsilon(\omega,k)|^2}. \quad (8.4.1)$$

The dielectric permittivity and the matrix element entering this equation are given by (7.2.19, 8.3.3).

In the equilibrium state, (8.4.1) can be presented in the form

$$(\delta E\,\delta E)_{\omega,k} = \frac{8\pi}{\omega}\frac{\mathrm{Im}\{\varepsilon(\omega,k)\}}{|\varepsilon(\omega,k)|^2}\left(\frac{1}{2}\hbar\omega+\frac{\hbar\omega}{\exp(\hbar\omega/\mathscr{X}T)-1}\right), \quad (8.4.2)$$

the Callen–Welton formula for the system in question. In order to obtain this expression one may employ the formula for the imaginary part of permittivity, which follows from (8.3.3),

$$\mathrm{Im}\{\varepsilon(\omega,k)\} = -\frac{4\pi^2N}{V}\sum_{\alpha\beta}\int\frac{V}{(2\pi\hbar)^3}dP'\,dP''\,\frac{|P_{\alpha\beta}(k)|^2}{k^2}\,\delta(\hbar k-(P'-P''))$$

$$\times\,\delta(\hbar\omega-(E_\alpha+E_{P'}-E_\beta-E_{P''}))[f_\alpha(P',t)-f_\beta(P'',t)], \quad (8.4.3)$$

and the equation

$$\delta(\hbar\omega-(E_\alpha+E_{P'}-E_\beta-E_{P''}))\frac{f_\alpha(P')+f_\beta(P'')}{f_\alpha(P')-f_\beta(P'')}$$

$$= -\delta(\ldots)2\left(\frac{1}{2}+\frac{1}{\exp(\hbar\omega/\mathscr{X}T)-1}\right). \quad (8.4.4)$$

Equation (8.4.1), like the formula for polarizability, can be presented as a sum of four terms,

$$(\delta E \, \delta E)_{\omega,k} = (\delta E \, \delta E)_{\omega,k}^{ff} + (\delta E \, \delta E)_{\omega,k}^{fb} + (\delta E \, \delta E)_{\omega,k}^{bf} + (\delta E \, \delta E)_{\omega,k}^{bb} . \quad (8.4.5)$$

Let us consider each of the four contributions separately.

The Spectral Density $(\delta E \, \delta E)_{\omega,k}^{ff}$. In (8.4.1) transformations similar to those performed to obtain (8.3.10) from the general expression (8.3.3) produce

$$(\delta E \, \delta E)_{\omega,k}^{ff} = \pi \hbar \frac{V}{(2\pi\hbar)^3} \sum_a \frac{(4\pi)^2 e_a^2 n_a}{k^2 |\varepsilon(\omega,k)|^2} \int dp' \, dp'' \, \delta(\hbar k - (p' - p''))$$

$$\times \delta \left(\hbar\omega - \left(\frac{p'^2}{2m_a} - \frac{p''^2}{2m_a} \right) \right) [f_a(p', t) + f_a(p'', t)] . \quad (8.4.6)$$

This expression differs from the corresponding one for completely ionized plasma [Ref. 8.1, Eq. (74.2)] only in that the dielectric permittivity here includes the polarizability, determined by all the processes in partially ionized plasma.

The Spectral Densities $(\delta E \, \delta E)_{\omega,k}^{fb}$, $(\delta E \, \delta E)_{\omega,k}^{bf}$. The spectral density of the field fluctuations has the property

$$(\delta E \, \delta E)_{\omega,k} = (\delta E \, \delta E)_{-\omega,-k}^* . \quad (8.4.7)$$

It follows in particular that

$$(\delta E \, \delta E)_{\omega,k}^{fb} = (\delta E \, \delta E)_{-\omega,-k}^{bf} . \quad (8.4.8)$$

Transformations similar those performed in obtaining (8.3.16) from the general expression (8.3.3) produce

$$(\delta E \, \delta E)_{\omega,k}^{bf} = \pi \hbar (4\pi)^2 \frac{N}{V} \frac{V^2}{(2\pi\hbar)^6} \sum_m \int dp'_e \, dp'_i \, dP'' \frac{1}{k^2} |P_{p'_i m}(k)|^2$$

$$\times \frac{Nf(p'_e) f(p'_i) + f_m(P'')}{|\varepsilon(\omega,k)|^2} \delta(\hbar k - (p'_e + p'_i - P''))$$

$$\times \delta \left(\hbar\omega - \left(\frac{p'^2_e}{2m_e} + \frac{p'^2_i}{2m_i} - E_m - E_{P''} \right) \right) . \quad (8.4.9)$$

This expression is significantly simplified in the immobile atom approximation.

The Spectral Density $(\delta E \, \delta E)_{\omega,k}^{bb}$. The expression for this spectral density follows from (8.4.1), substituting $\alpha \to n$ and $\beta \to m$. In the dipole approximation, substitution as in (8.3.30) can be carried out.

In many applied matters, e.g., in the theory of gas lasers, special interest is drawn to the case in which the distribution of atoms over momenta is in equilibrium, while nonequilibrium over energy levels (8.3.31). Then in the dipole approximation from (8.4.1) we find

$$
(\delta E \, \delta E)_{\omega, k}^{bb} = \frac{16\pi^3}{3} \frac{N}{V} \sum_{n_1 m_1} \int \frac{|d_{n_1 m_1}|^2}{|\varepsilon(\omega, k)|^2} \delta(\omega - \omega_{n_1 m_1} - k\,V)
$$

$$
\times \left[f_{n_1}\left(P + \frac{1}{2}\hbar k\right) + f_{m_1}\left(P - \frac{1}{2}\hbar k\right) \right] \frac{V}{(2\pi\hbar)^3} dP \,. \tag{8.4.10}
$$

In the immobile atom approximation the distribution function can be presented in the form

$$
f_n(P, t) = \rho_n(t) \frac{(2\pi\hbar)^3}{V} \delta(P) \,, \tag{8.4.11}
$$

and integration over momenta is carried out.

8.5 The Collision Integral

Using (8.3.1, 2) and the equation for the spectral density of the source fluctuations (8.2.15), we proceed to find the spectral density of fluctuations $\delta\rho_{\alpha\beta}$ and δU, which determines the collision integral in (8.1.4).

From the structure of (8.3.1, 2) it is clear that the collision integral can be presented as a sum of two parts: the "induced" one, which is proportional to the spectral density of the field fluctuations, and the "spontaneous" one, proportional to the imaginary part of permittivity. Skipping simple transformations, the final result is [8.1, 2]:

$$
I_\alpha(P', t) = \frac{1}{(2\pi)^3 \hbar} \sum_\beta \int d\omega \, dk \, dP'' \frac{|P_{\alpha\beta}(k)|^2}{k^2} \delta(\hbar k - (P' - P''))
$$

$$
\times \delta(\hbar\omega - (E_\alpha + E_{P'} - E_\beta - E_{P''})) \left\{ (\delta E \, \delta E)_{\omega, k} [f_\beta(P'', t) - f_\alpha(P', t)] \right.
$$

$$
\left. - \frac{4\pi\hbar \, \mathrm{Im}\{\varepsilon(\omega, k)\}}{|\varepsilon(\omega, k)|^2} [f_\beta(P'', t) + f_\alpha(P', t)] \right\} \,. \tag{8.5.1}
$$

This presentation of the collision integral corresponds to the expression for the Balescu – Lenard collision integral for completely ionized plasma (4.4.20). Another presentation is also of use; it follows from (8.5.1) after substituting (8.4.1, 3) into it. It corresponds to the classical expression (4.4.21) and has the form [8.1, 2]

$$I_\alpha(P',t) = 4\frac{N}{V}\frac{V}{(2\pi\hbar)^3}\sum_{\alpha_1\beta_1\beta}\int dP_1' \, dP_1'' \, dP'' \, d\omega \, dk \, \frac{|P_{\alpha\beta}(k)|^2|P_{\alpha_1\beta_1}(k)|^2}{k^4|\varepsilon(\omega,k)|^2}$$

$$\times \delta(\hbar k - (P' - P''))\,\delta(\hbar\omega - (E_\alpha + E_{P'} - E_\beta - E_{P''}))\,\delta(P' + P_1''$$

$$- (P_1' + P''))\,\delta(E_\alpha + E_{P'} + E_{\beta_1} + E_{P_1''} - (E_\beta + E_{P''} + E_{\alpha_1} + E_{P_1'}))$$

$$\times [f_{\alpha_1}(P_1')f_\beta(P'') - f_{\beta_1}(P_1')f_\alpha(P')]\,. \tag{8.5.2}$$

Thus we have obtained two equivalent expressions for the collision integral in the kinetic equation for the distribution function of pairs of charged particles $f_\alpha(P,t)$. In order to derive from this equation the set of kinetic equations for the distribution functions of electrons, ions, and atoms in a partially ionized plasma, we next consider separately the distributions over the discrete and continuous spectra. Then, in accordance with (7.2.14), the collision integrals (8.5.1, 2) are split into eight parts each – eight collision integrals, each accounting for a separate process. In the collision integrals which are determined by transitions from the discrete spectrum into the continuous one, one must require that the correlations of charged particles bound in pairs be damped [condition (8.3.15)]. Carrying out this program, we first obtain the kinetic equations for free charged particles. The more convenient method is to employ the collision integral in the form of (8.5.1).

In (8.1.4, 5.1) we set $\alpha = p'$ and substitute $p', P' \to p_e, p_i$ according to (8.3.8). This results in the substitution

$$I_{p'}(P',t) \to I(p_e, p_i, t)\,. \tag{8.5.3}$$

Integrating successively over p_e and p_i and using the definitions of the distribution functions (8.3.9) for electrons and ions produces the following kinetic equations:

$$\frac{\partial f_a(p_a, t)}{\partial t} = \frac{V}{(2\pi\hbar)^3}\int I(p_a, p_b, t)\,dp_b \equiv I_a(p_a, t)\,,$$

$$a \neq b\,, \qquad a = e, i\,. \tag{8.5.4}$$

Now setting $\alpha = n$ in (8.1.4, 5.1), we get the kinetic equation for the distribution function of atoms,

$$\frac{\partial f_n(P', t)}{\partial t} = I_n(P', t)\,. \tag{8.5.5}$$

The normalization conditions for the distribution functions of electrons, ions, and atoms are given by (8.3.22).

In the next section we shall discuss in greater detail the structure of the collision integrals in the kinetic equations (8.5.4, 5). Here we are concerned with the general properties of the collision integrals. We multiply (8.5.2) by an arbitrary function $\varphi_\alpha(P')$ and by N/V, sum over α, and integrate over P'. Then, as in the investigation of the properties of the Boltzmann collision integral, we symmetrize the integrand in two steps:

1) $\alpha \rightleftarrows \beta_1$, $P' \rightleftarrows P_1''$; $\alpha_1 \rightleftarrows \beta$, $P_1' \rightleftarrows P''$, and

2) $\alpha \rightleftarrows \beta$, $P' \rightleftarrows P''$; $\alpha_1 \rightleftarrows \beta_1$, $P_1' \rightleftarrows P_1''$, $\omega \rightarrow -\omega$, $k \rightarrow -k$.

The result is

$$I(t) = \frac{N}{4V} \frac{V^2}{(2\pi\hbar)^6} \int dP' \, dP'' \, dP_1' \, dP_1'' \, d\omega \, dk \, [\varphi_\alpha(p') + \varphi_{\beta_1}(P_1'')$$

$$- \varphi_{\alpha_1}(P_1') - \varphi_\beta(P'')](\dots) . \tag{8.5.6}$$

The integrand from (8.5.2) should be substituted into the parentheses $(\dots)$.
From (8.5.6) it follows that

$$I(t) = 0 \quad \text{at} \quad \varphi_\alpha(P') = 1, P', E_\alpha + E_{P'}, \tag{8.5.7}$$

which can be proven by employing (8.5.2). The properties given here ensure fulfillment of all the conservation laws (for both free and bound states) of the number of particles, momentum, and energy. Finally,

$$I(t) \geqslant 0 \quad \text{at} \quad \varphi_\alpha = - \mathcal{K} \ln f_\alpha(P', t) . \tag{8.5.8}$$

The equal sign corresponds to the equilibrium state.
This last equation permits defining the entropy of partially ionized plasma as

$$S(t) = - \frac{N}{V} \mathcal{K} \sum_\alpha \int \frac{V}{(2\pi\hbar)^3} dP' f_\alpha(P', t) \ln f_\alpha(P', t) . \tag{8.5.9}$$

In the kinetic equations obtained here the interaction of particles only determines the dissipative processes. The nondissipative characteristics, e.g., the thermodynamic functions, coincide with those of an ideal plasma. The interaction of particles can be accounted for more consistently by obtaining the kinetic equations with the aid of the method presented in [8.1]. As a result the equations for nonideal partially ionized plasma can be obtained. The fulfillment of this program will be facilitated by expanding over the eigenfunctions of atoms in the environment instead of over the eigenfunctions of isolated atoms (Sect. 7.2). This requires the definition of the corresponding effective potential.

8.6 The Structure of Collision Integrals

It is convenient to present the collision integrals in the kinetic equations (8.5.4) for electrons and ions as sums of two parts each

$$I_a(p_a, t) = [I_a(p_a, t)]_1 + [I_a(p_a, t)]_2, \quad a = e, i. \tag{8.6.1}$$

They both follow from (8.5.1) upon substituting $\alpha \to p'$ and p', $P' \to p'_e$, p'_i. In order to single out the first term we set $\beta = p''$ in the sum over β, replace variables p'', $P'' \to p''_e$, P''_i, and integrate over the ions' (or electrons') variables p'_i, p''_i, getting

$$[I_a(p'_a, t)]_1 = \frac{e_a^2}{(2\pi)^3 \hbar} \int dp''_a \, d\omega \, dk \frac{1}{k^2} \delta(\hbar k - (p'_a - p''_a))$$

$$\times \delta \left(\hbar \omega - \left(\frac{p'^2}{2m_a} - \frac{p''^2_a}{2m_a} \right) \right) \left\{ (\delta E \, \delta E)_{\omega, k} \, [f_a(p''_a, t) - f_a(p'_a, t)] \right.$$

$$\left. - \frac{4\pi \hbar \, \text{Im}\{\varepsilon(\omega, k)\}}{|\varepsilon(\omega, k)|^2} \, [f_a(p''_a, t) + f_a(p'_a, t)] \right\}. \tag{8.6.2}$$

This result for completely ionized plasma coincides with the quantum Balescu – Lenard collision integral for completely ionized plasma [Ref. 8.1, Eq. (74.7)].

We know that the spectral density of the field fluctuations and the imaginary part of permittivity can be presented as a sum of four parts each. Correspondingly, the collision integral (8.6.2) can be presented as a sum of four collision integrals, determined by the following processes[1]:

1) $\quad p'_a + p''_{1b} \leftrightarrow p''_a + p'_{1b}$

Scattering of free charged particles. Further on we indicate this process in the direction from left to right.

2) $\quad p'_a + m_1 P''_1 \leftrightarrow p''_a + p'_{1a} + p'_{1b}$

Ionization of atoms from the state $m_1 P''_1$ by the impact of an electron ($a = e$) or an ion ($a = i$).

3) $\quad p'_a + p''_{1a} + p''_{1b} \leftrightarrow p''_a + n_1 P'_1$

Recombination in the presence of an electron ($a = e$) or an ion ($a = i$).

4) $\quad p'_a + m_1 P''_1 \leftrightarrow p''_a + n_1 P'_1$

Inelastic collision of an electron or an ion with an atom.

We see that in each of these four processes the number of particles with subscript a does not change. Integral (8.6.2) therefore has the property

$$\frac{V}{(2\pi\hbar)^3} \int [I_a(p'_a, t)]_1 \, dp'_a = 0. \tag{8.6.3}$$

---

[1] In the presentation of chemical reactions adopted here the states of interacting particles before and after the reaction are indicated.

To prove this one must replace the variables in the integrand $p'_a \leftrightarrow p''_a$; $\omega, k \to -\omega, -k$, and take advantage of the properties of functions $(\delta E \, \delta E)_{\omega, k}$, $\text{Im}\{\varepsilon(\omega, k)\}$. Thanks to (8.6.3) the collision integral $[I_a]_1$ does not contribute to the equations for the concentration of electrons and ions (see Sect. 8.7).

In order to obtain the expression for the second term in the right-hand side of (8.6.1), we replace β with m in the sum $\sum_{\beta}$ entering the collision integral I_a, getting

$$[I_a(p'_a, t)]_2 = \frac{1}{(2\pi)^3 \hbar} \frac{V}{(2\pi\hbar)^3} \sum_m \int dp'_b \, dP'' \, d\omega \, dk \, \frac{1}{k^2}$$

$$\times |P_{\frac{m_b p'_a - m_a p'_b}{M} m}(k)|^2 \delta(\hbar k - (p'_a + p'_b - P'')) \delta\left(\hbar\omega - \left(\frac{p'^2_a}{2m_a} + \frac{p'^2_b}{2m_b}\right.\right.$$

$$\left.\left. - E_m - E_{P''}\right)\right)\left\{(\delta E \, \delta E)_{\omega, k} [f_m(P'', t) - N f_a(p'_a, t) f_b(p'_b, t)]\right.$$

$$\left. - \frac{4\pi\hbar \, \text{Im}\{\varepsilon(\omega, k)\}}{|\varepsilon(\omega, k)|^2} [f_m(P'', t) + N f_a(p'_a, t) f_b(p'_b, t)]\right\}. \tag{8.6.4}$$

This collision integral also describes four processes, to which we assign numbers 5 to 8.

5) $p'_a + p'_b + p''_{1b} \leftrightarrow mP'' + p'_{1a}$

Recombination of particles a, b $(a \neq b)$.

6) $p'_a + p'_b + m_1 P''_1 \leftrightarrow mP'' + p'_{1a} + p'_{1b}$

The process of inelastic scattering accompanied by the exchange of particles.

7) $p'_a + p'_b + p''_{1a} + p''_{1b} \leftrightarrow mP'' + n_1 P'_1$

The process of double recombination.

8) $p'_a + p'_b + m_1 P''_1 \leftrightarrow mP'' + n_1 P'_1$

The process of recombination in the presence of an atom.

In all these processes the initially free particles with momenta p'_a, p'_b enter bound states, and the initial number of particles is not conserved. Therefore,

$$\frac{V}{(2\pi\hbar)^3} \int [I_a(p'_a, t)]_2 \, dp'_a \neq 0. \tag{8.6.5}$$

Consider now the kinetic equation for the distribution function of atoms. The collision integral in this equation can also be presented in the form

$$I_n(P', t) = [I_n(P', t)]_1 + [I_n(P', t)]_2. \tag{8.6.6}$$

The first term in the right-hand side follows from (8.5.1) after substituting $\alpha \to n$ and $\beta \to m$. It describes the following four processes:

1) $nP' + p''_{1b} \leftrightarrow mP'' + p'_{1b}$

Inelastic scattering of atoms on free charged particles.

2) $nP' + m_1 P''_1 \leftrightarrow mP'' + p'_{1a} + p'_{1b}$

Ionization on impact of an atom.

3) $nP' + p''_{1a} + p''_{1b} \leftrightarrow mP'' + n_1 P'_1$

Recombination with participation of an atom.

4) $nP' + m_1 P''_1 \leftrightarrow mP'' + n_1 P'_1$

Inelastic scattering of atoms.

Since in these processes the initial number of atoms does not change,

$$\sum \int [I_n(P', t)]_1 \frac{V}{(2\pi\hbar)^3} dP' = 0. \tag{8.6.7}$$

In order to obtain the expression for the second part of the collision integral (8.5.6), one should set $\alpha = n$ and $\beta = p''$ in (8.5.1) and replace the variables p'', $P'' \to p''_e$, p''_i according to (8.3.8). The result is

$$[I_n(P', t)]_2 = \frac{1}{(2\pi)^3 h} \frac{V}{(2\pi\hbar)^3} \int dp''_e \, dp''_i \frac{d\omega \, dk}{k^2} |P_{n_1(m_i p''_e - m_e p'_i)/M}(k)|^2$$

$$\times \delta(\hbar k - (P' - p''_e - p''_i)) \delta \left(\hbar\omega - \left(E_n + E_{P'} - \frac{p''^2_e}{2m_e} - \frac{p''^2_i}{2m_i} \right) \right)$$

$$\times \left\{ (\delta E \, \delta E)_{\omega, k} [N f_e(p''_e) f_i(p''_i) - f_n(P')] \right.$$

$$\left. - \frac{4\pi\hbar \, \text{Im}\{\varepsilon(\omega, k)\}}{|\varepsilon(\omega, k)|^2} [N f_e(p''_e) f_i(p''_i) + f_n(P')] \right\}. \tag{8.6.8}$$

This collision integral is determined by four processes, which are reciprocal to processes 5 – 8 above

5) $nP + p''_{1a} \leftrightarrow p''_a + p''_b + p'_{1b};$

6) $nP' + p''_{1a} \leftrightarrow p''_a + p''_b + p'_{1a} + p'_{1b};$

7) $nP' + m_1 P''_1 \leftrightarrow p''_a + p''_b + p'_{1a} + p'_{1b};$

8) $nP' + m_1 P''_1 \leftrightarrow p''_a + p''_b + n_1 P'_1 .$

Because the number of atoms «nP'» is changed in these processes,

$$\sum_n \int [I_n(P', t)]_2 \frac{V}{(2\pi\hbar)^3} dP' \neq 0 . \tag{8.6.9}$$

In the equilibrium state the collision integrals $I_a I_n$ become zero if the distribution functions f_a and f_n are determined by (8.3.19, 22) and the concentrations n_e, n_i, and n_{ei} satisfy the required chemical equilibrium (8.3.24). Other common properties of the collision integrals were discussed in Sect. 8.5.

The kinetic equations obtained here can be employed to construct the set of gasdynamic equations for partially ionized plasma. Considering only an exemplary set of these equations, let us assume that the equilibrium distribution has been established over translatory and inner degress of freedom, i.e., the distribution functions are given by (8.3.19, 22), but the ionization equilibrium has not yet been established, i.e., the concentrations are not bound by (8.3.24). This state can be singled out due to the fact that the equilibrium state in a system of chemically reacting gases is established roughly in three stages — first the equilibrium over translatory degrees of freedom, then over the inner degrees of freedom, and finally the ionization (chemical) equilibrium. The process of establishing equilibrium in a plasma is in general, of course, much more complicated; this is due, in particular, to the great difference in mass between electrons and ions.

8.7 The Equations for Concentrations

We have seen that the collision integral can be presented in either of two equivalent forms (see Sect. 8.5). Correspondingly, there are two equivalent presentations of equations for the concentrations of free and bound (in atoms) charged particles. Since the concentrations of charged particles are interrelated via the two conditions

$$n_e = n_i , \quad n_e + n_{ei} = N/V \equiv n , \tag{8.7.1}$$

it suffices to consider only one equation, say, for the concentration of electrons n_e.

The collision integrals in (8.5.4) are determined by (8.6.2, 4). In order to obtain the equation for concentration n_a, we multiply (8.5.4) by $(N/V)V/(2\pi\hbar)^3$ and integrate over p'_a. The resulting equation can be written in the form

$$\frac{dn_a}{dt} = \frac{1}{(2\pi)^3 \pi\hbar} \int d\omega \, dk \, [\mathrm{Im}\{\alpha_{\mathrm{bf}}(\omega, k)\}(\delta E \, \delta E)_{\omega, k}$$
$$- \mathrm{Im}\{\alpha(\omega, k)\}(\delta E \, \delta E)_{\omega, k}^{\mathrm{bf}}] . \tag{8.7.2}$$

Here we have employed (8.3.17, 4.9). It can be easily proven that in the equilibrium state the right-hand side of this equation is zero.

If the distribution over translatory and internal degrees of freedom is in equilibrium, the formulae for the spectral densities and polarizabilities are significantly simplified.

An interesting case is when the deviation from the chemical equilibrium is determined by a significant nonequilibration of the spectral distribution of field fluctuations. Such an example will be discussed in Chap. 9, in the description of the photoionization process.

To describe processes of ionization and recombination in the collisions of charged particles and atoms we employ the collision integral in the form (8.5.2), resulting in the following equation for the concentration of electrons:

$$\frac{dn_e}{dt} = (\alpha n_e n_{ei} - \beta n_e^2 n_i) + (\alpha_1 n_{ei}^2 - \beta_1 n_e n_i n_{ei})$$
$$+ (\alpha_2 n_{ei}^2 - \beta_2 n_e^2 n_i^2) + (\alpha_3 n_e n_i n_{ei} - \beta_3 n_e n_i n_{ei}) . \tag{8.7.3}$$

This equation should be supplemented by (8.6.1).

In this equation α is the coefficient of shock ionization, and β is the coefficient of triple recombination. They are connected by the relation

$$\beta = \left(\frac{2\pi\hbar^2}{\mu \mathcal{X} T}\right)^{3/2} Z\alpha , \tag{8.7.4}$$

which reflects the condition of detailed balance for this pair of processes.

The same relation also connects the coefficients α_1 and β_1, where α_1 is the ionization coefficient for the collision of two atoms, and β_1 is the corresponding recombination coefficient. The coefficients of double ionization in two-atom collisions (α_2) and of the corresponding recombination process (β_2) are linked by the relation

$$\beta_2 = (\beta/\alpha)^2 \alpha_2 . \tag{8.7.5}$$

Since the coefficients of the exchange reaction, α_3 and β_3, are equal, the last term in (8.7.3) drops out by virtue of (8.7.1).

The general expression for the coefficient of shock ionization is [8.2]

$$\alpha = \frac{4V}{(2\pi\hbar)^3} \sum_{c=e,i} e_c^2 \sum_m \int dp' \, dP' \, dP'' \, dp_c' \, dp_c'' \, d\omega \, dk \, \frac{|P_{p'm}(k)|^2}{k^4 |\varepsilon(\omega,k)|^2}$$
$$\times \delta(P' + p_c'' - (P'' - p_c')) \delta\left(\frac{p'^2}{2\mu} + \frac{P'^2}{2M} + \frac{p_c''^2}{2m_c} - \left(E_m + \frac{P''^2}{2M} + \frac{p_c'^2}{2m_c}\right)\right)$$
$$\times \delta(\hbar k - (P' + P'')) \delta\left(\hbar\omega - \left(\frac{p'^2}{2\mu} + \frac{P'^2}{2M} - E_m - \frac{P''^2}{2M}\right)\right)$$
$$\times \frac{1}{[2\pi(m_c M)^{1/2} \mathcal{X} T]^3 Z} \exp\left(-\frac{E_m + P''^2/2M + p_c'^2/2m_c}{\mathcal{X} T}\right) . \tag{8.7.6}$$

Consider the special case when the effects of polarization can be neglected [$\varepsilon(\omega, k) = 1$], atoms are immobile ($M = \infty$), ionization starts from the hydrogen atom ground level, and the matrix elements are taken in the dipole approximation (8.3.27). The result is the following expression:

$$\alpha = \frac{2^{10}}{3^7} 35 e^{-4/3} \frac{a_0^3 \mu e^4}{\hbar^3} \frac{\varkappa T}{I} e^{-I/\varkappa T}. \tag{8.7.7}$$

This formula differs from the classical result [Ref. 8.13, Eq. (83.4)] by the pre-exponential factor ($\varkappa T/I$ instead of ($\varkappa T/I)^{1/2}$ in [8.13]). The dependence of the effective cross section near the threshold upon transgressing the generation threshold is nonlinear. This is the consequence of employing the Born approximation.

8.8 The Kinetic Theory of Fluctuations in Partially Ionized Plasma

In obtaining the kinetic equations for the distribution functions of electrons, ions, and atoms in partially ionized plasma we have actually employed the condition that the initial correlations be completely damped. Therefore, as was also the case for the kinetic theory of completely ionized plasma (Chap. 4), the fluctuations of the distribution functions are not taken into account. In other words, employing the condition of total damping of initial correlations means neglecting large-scale fluctuations.

The kinetic theory of fluctuations in a partially ionized plasma can be developed along the same lines as the kinetic theory of fluctuations in gases and completely ionized plasma (see Sect. 4.8 and [Ref. 8.1, Chaps. 4, 11]). As our initial equations we shall employ those for smoothed (with respect to small-scale) operator density matrices. They are similar to (4.8.1, 4), and account for the dissipative processes which are determined by small-scale fluctuations.

If large-scale fluctuations are taken into account, the structure of the equations for the distribution functions $f_a(P', t)$ is changed, becoming [cf. (4.8.7)] [8.1, 14]

$$\frac{\partial f_a(P', t)}{\partial t} = I_a(P', t) + \tilde{I}_a(P', t). \tag{8.8.1}$$

The collision integral $I_a(P', t)$ is given by (8.5.1, 2) and

$$\tilde{I}_a(P', t) = -\frac{2}{\hbar} \sum_\beta \int \text{Im}\{\langle \delta \tilde{\rho}_{\alpha\beta}(P', P'', t) \delta \tilde{U}^*_{\alpha\beta}(P', P'', t)\rangle\} \frac{V}{(2\pi\hbar)^3} dP'' \tag{8.8.2}$$

is the additional contribution to the collision integral and is determined by the large-scale fluctuations of the operator density matrix $\delta\tilde{\rho}_{\alpha\beta}$.

Recall that in the kinetic theory of fluctuations in gases [Ref. 8.1, Chap. 4] as well as in completely ionized plasma (Sect. 4.8) [Ref. 8.1, Chap. 11] the intensity of the source of large-scale fluctuations is given by the sum of two contributions (4.8.20). One of these is determined by the additional collision integral $\tilde{I}_\alpha$ (4.8.21) and vanishes when $\tilde{I}_\alpha = 0$. The second contribution is due to the atomic structure, and is therefore present even in the equilibrium state, when both collision integrals are zero ($\tilde{I}_\alpha = 0$, $I_\alpha = 0$).

The contribution from the collision integral $\tilde{I}_\alpha$ is important at large deviations from equilibrium. Here we consider only the states where it may be assumed that $\tilde{I}_\alpha = 0$. Since under this condition the principal contribution is determined by the diagonal elements of the density matrix $\delta \tilde{\rho}_{\alpha\beta}$, we set [cf. (8.1.1)]

$$\delta \tilde{\rho}_{\alpha\beta}(P', P'', t) = \delta_{\alpha\beta} \frac{(2\pi\hbar)^3}{V} \delta(P' - P'') \delta f_\alpha(P', t) , \tag{8.8.3}$$

where δf_α is the fluctuation of the distribution function f_α, i.e., $\delta f_\alpha = f_\alpha - \langle f_\alpha \rangle$.

In this approximation the equation for the distribution function can be written in the form of a Langevin equation [cf. (7.15.10) at $\tilde{I} = 0$]. Thus [8.1, 14, 15],

$$\frac{\partial f_\alpha(P', t)}{\partial t} = I_\alpha(P', t) + y_\alpha(P', t) . \tag{8.8.4}$$

The moments of the Langevin source y_α have the form

$$\langle y_\alpha(P', t) \rangle = 0 , \quad \langle y_\alpha(P', t) y_\beta(P'', t) \rangle = A_{\alpha\beta}(P', P'', t) \delta(t - t') . \tag{8.8.5}$$

The intensity of the random source can given as a formula similar to those cited in [Ref. 8.1, Sects. 22, 63]; this formula can be found in [8.15]. Here we shall only discuss the common properties of the correlator of the Langevin source in (8.8.5). They play an important role in the transition from the kinetic equations to the gasdynamic description, in particular, in obtaining equations for the concentrations of charged particles and atoms in partially ionized plasma.

If we multiply the correlator (8.8.5) by arbitrary functions $\varphi_\alpha(P')$, $\Psi_\beta(P'')$, integrate over P' and P'', and sum over α and β, we can obtain {by also employing [Ref. 8.15, Eq. (10)}

$$\sum_{\alpha\beta} \int \varphi_\alpha(P') \Psi_\beta(P'') \langle y_\alpha(P', t) y_\beta(P'', t) \rangle \frac{V^2}{(2\pi\hbar)^6} dP' dP''$$

$$= \frac{2}{V} \delta(t - t') \sum_{\alpha\beta\gamma\eta} \int \frac{|P_{\alpha\eta}(k)|^2 |P_{\beta\gamma}(k)|^2}{k^4 |\varepsilon(, k)|^2} \delta(\hbar k - (P' - P_1'))$$

$$\times \delta(P' + P'' - P_1' - P_1'') \delta(\hbar\omega - (E_\alpha + E_{P'} - E_\eta - E_{P_1'}))$$

$$\times \delta(E_\alpha + E_{P'} + E_\beta + E_{P''} - E_\gamma - E_{P_1} - E_\eta - E_{P_1'})$$

$$\times [\varphi_\alpha(P') + \varphi_\beta(P'') - \varphi_\gamma(P_1') - \varphi_\eta(P_1'')] [\Psi_\alpha(P') + \Psi_\beta(P'')$$

$$- \Psi_\gamma(P_1') - \Psi_\eta(P_1'')] f_\alpha(P', t) f_\beta(P'', t) d\omega dk$$

$$\times \frac{V^2}{(2\pi\hbar)^6} dP' dP'' dP_1' dP_1'' . \tag{8.8.6}$$

Hence it follows that

$$\sum_{\alpha\beta} \int \varphi_\alpha(\boldsymbol{P}') \, \Psi_\beta(\boldsymbol{P}'') \langle y_\alpha(\boldsymbol{P}', t) y_\beta(\boldsymbol{P}'', t) \rangle \frac{V^2}{(2\pi\hbar)^6} \, d\boldsymbol{P}' \, d\boldsymbol{P}'' = 0 \qquad (8.8.7)$$

at

$$\varphi_\alpha(\boldsymbol{P}) \,, \quad \Psi_\alpha(\boldsymbol{P}) = 1, \boldsymbol{P} \,, \quad E_\alpha + E_P \,. \qquad (8.8.8)$$

These properties are similar to those in (8.5.7) for the collision integral I_α. Naturally, the result (8.8.6) is also true for the extreme cases of completely ionized plasma as well as a gas (zero degree of ionization). For the latter case, (8.8.7) corresponds to [Ref. 8.1, Eq. (23.17)].

Using (8.8.4) we proceed to find the equations for the concentrations of charged particles in a plasma. Instead of (8.7.1, 3) we get the corresponding Langevin equations,

$$\frac{dn_e}{dt} - [(\alpha n_e n_{ei} - \beta n_e^2 n_i) + (\alpha_1 n_{ei}^2 - \beta_1 n_e n_i n_{ei}) + (\alpha_2 n_{ei}^2 - \beta_2 n_e^2 n_i^2)] = y_e \,,$$
$$(8.8.9)$$

$$n_e = n_i \,, \quad n_e + n_{ei} = N/V \equiv n \,. \qquad (8.8.10)$$

Since it has been taken into account that $\alpha_3 = \beta_3$, the last term in (8.7.3) drops out.

It is important that the second equation of (8.7.1) remains unchanged and coincides with (8.8.10). This corresponds to the condition that the total concentration $n_e + n_{ei} = n$ and is considered set in advance. Thus only those fluctuations of concentration are taken into account which are produced by chemical reactions.

The moments of the Langevin source in (8.8.9) are given by

$$\langle y_e(t) \rangle = 0 \,, \quad \langle y_e(t) y_e(t') \rangle = A_{ee}(t) \, \delta(t - t') \,. \qquad (8.8.11)$$

The intensity $A_{ee}(t)$ can be determined from [Ref. 8.15, Eq. (10)].

For the state in which equilibrium is already established over the translatory and internal degrees of freedom (Maxwell–Boltzmann distribution), but the chemical equilibrium has not yet been attained (the Saha condition is not satisfied), the expression for intensity has the form

$$A_{ee}(t) = \frac{1}{V} [(\alpha n_e n_{ei} + \beta n_e^2 n_i) + (\alpha_1 n_{ei}^2 + \beta_1 n_{ei} n_e e_i) + 2(\alpha_2 n_{ei}^2 + \beta_2 n_e^2 n_i^2)] \,.$$
$$(8.8.12)$$

The functions n_e, n_i, and n_{ei} are determined here by (8.7.1, 3).

Thus, the intensity of the Langevin source combines three contributions, each of which corresponds to one of three pairs (direct and reciprocal) of processes responsible for changes in concentration. In the state of complete equilibrium, i.e., when the Saha formula is also true, the first and the second terms in each pair in (8.8.12) are equal.

For fluctuations in concentrations in the state of equilibrium we find from (8.8.10, 11)

$$\frac{d\delta n_a}{dt} + \lambda_a \delta n_a = y_a(t), \quad a = e, i,$$

$$\delta n_e = \delta n_i, \quad \delta n_a = - \delta n_{ei}. \tag{8.8.13}$$

This dissipative constant is given by

$$\lambda_a = \left(1 + 2\frac{n_{ei}}{n_a}\right)(\alpha n_a + \alpha_1 n_{ei} + 2\alpha_2 n_{ei}). \tag{8.8.14}$$

From (8.8.13) it follows that the dispersion of fluctuations δn_a is linked to the intensity of the random source by

$$\langle(\delta n_a)^2\rangle = A_{aa}/2\lambda_a. \tag{8.8.15}$$

For the state of equilibrium, (8.8.12) takes the form

$$A_{aa} = \frac{2n_{ei}}{V}(\alpha n_a + \alpha_1 n_{ei} + 2\alpha_2 n_{ei}). \tag{8.8.16}$$

From (8.8.14, 16) the dispersion of fluctuations in concentrations of charged particles and atom is

$$\langle(\delta n_a)^2\rangle = \langle(\delta n_{ei})^2\rangle = \frac{1}{V}\frac{n_{ei}n_a}{2n_{ei} + n_a} = \frac{1}{V}n_a\frac{n - n_a}{2n - n_a}. \tag{8.8.17}$$

The last part employs the second expression in (8.8.13).

The expression for fluctuations in the total number of charged particles $\delta N_a = V\delta n_a$ and atoms $\delta N_{ei} = V\delta n_{ei}$ is

$$\langle(\delta N_a)^2\rangle = \langle(\delta N_{ei})^2\rangle = N_a\frac{N - N_a}{2N - N_a}. \tag{8.8.18}$$

Equations (8.8.17, 18), naturally, do not depend upon the coefficients of ionization and recombination.

The second moments of fluctuations in concentrations, which correspond to different components, are determined by the same formula, but with opposite signs. Indeed, from (8.8.13),

$$\langle(\delta N_a)^2\rangle = - \langle\delta N_a \delta N_{ei}\rangle. \tag{8.8.19}$$

In the extreme cases of partially ionized plasma – completely ionized plasma and gas (zero degree of ionization) – the dispersions of fluctuations (8.8.17, 18) become zero. This is natural since, according to (8.8.10), the quantity $N = Vn$ (total number of pairs of charged particles in both free and bound states) is considered set in advance. Equations (8.8.17, 18), therefore, determine the fluctuations in the number of charged particles in partially ionized plasma where processes of ionization and recombination can take place.

We have discussed fluctuations in spatially uniform plasma. If the distribution is nonuniform, the diffusion of particles must be taken into account. Of course, this gives rise to additional terms in the Langevin sources.

Finally, let us point out that the results obtained describe fluctuations produced by the atomic structure of the system under consideration. For states far from equilibrium, an additional term $\tilde{I}_\alpha$ appears in the kinetic equation, determined by the large-scale fluctuations [Ref. 8.1, Sects. 22, 62]. A corresponding additional term also appears in the right-hand side of the Langevin equation (8.8.4).

The Langevin equation (8.8.9) is dissipatively nonlinear (Sect. 5.9). Therefore, as in Sect. 5.9, the random source can be introduced here whose intensity depends on random variables n_e, n_i, and n_{ei} [cf. (5.9.13, 14)]. In this connection it should be noted once more that the functions n_e, n_i, and n_{ei} in (8.8.12) are determined by dynamic equations (8.7.1, 3). It was already mentioned at the end of Sect. 5.9 that this approach, if consistently pursued, leads to kinetic equations with higher derivatives (see the references in Chap. 14).

Some authors employ other techniques to be calculate fluctuations in concentration. For instance, in [8.16, 17] the authors employed the kinetic equation for the distribution function of concentrations ("master equation") from the very beginning. This approach is efficient when the state of the system is entirely determined by the values of concentrations. In more general cases, when (8.8.4) are to be used instead of the kinetic equations for concentrations (8.8.9), such an approach is of course not impossible, but leads to immense difficulties.

9. Kinetic Equations for Partially Ionized Plasma. The Processes Conditioned by a Transverse Electromagnetic Field

This chapter deals with kinetic processes in partially ionized plasma which result from the interaction of charged particles with a transverse electromagnetic field. The consequences of cooling the atoms by a resonant electromagnetic field are considered as one of the theory's applications.

9.1 Dielectric Permittivity

Here, as in Chap. 8, fluctuations will be calculated according to the polarization approximation. In this approximation, the fluctuations of transverse and longitudinal fields are mutually independent in the absence of strong external fields. The initial equations for calculating the fluctuations follow from the equations in Sect. 7.4. Since the method of calculation is similar to that employed in Chap. 8, we shall present the main results immediately, and discuss their physical content.

The principal characteristics here are also the dielectric permittivity $\varepsilon^\perp(\omega, k)$ and the spectral density of field fluctuations $(\delta E^\perp \delta E^\perp)_{\omega, k}$. The sign $\perp$ indicates that we are dealing with fluctuations of a transverse electromagnetic field.

The dielectric permittivity is now given by [cf. (8.3.3)]

$$\varepsilon^\perp(\omega, k) = 1 - \frac{\omega_L^2}{\omega^2} + \frac{2\pi N}{V} \frac{c^2}{\omega^2} \sum_{\alpha\beta} \int \frac{V}{(2\pi\hbar)^3} \, dP' \, dP'' \, |P_{\alpha\beta}^\perp(k, P' + P'')|^2$$

$$\times \frac{\delta(\hbar k - (P' - P''))[f_\alpha(P', t) - f_\beta(P'', t)]}{\hbar(\omega + i\Delta) - (E_\alpha + E_\alpha + E_{P'} - E_\beta - E_{P''})}, \tag{9.1.1}$$

where $P_{\alpha\beta}^\perp = P_{\alpha\beta} - k(k P_{\alpha\beta})/k^2$ is the transverse component of the vector matrix element, and ω_L^2 is the square of the Langmuir frequency. The expression for ω_L^2 can conveniently be written in the form

$$\omega_L^2 = \omega_L^2 (c_a + c_{ei}),$$

where c_a and c_{ei} are the concentrations of free charged particles and atoms.

Like in case of a Coulomb plasma, the polarizability can be presented as a sum of four terms [cf. (8.3.5)]. The polarizability $\alpha_{ff}(\omega, k)$ is now given by

$$4\pi\alpha_{ff}^{\perp} = -\frac{\omega_L^2}{\omega^2}c_a$$

$$+\frac{V}{(2\pi\hbar)^3}\sum_a\frac{2\pi e_a^2 n_a}{\omega^2 k^2\hbar}\int\frac{[k\times V]^2[f_a(P+\frac{1}{2}\hbar k)-f_a(P-\frac{1}{2}\hbar k)]}{\omega+i\varDelta-kV}\,dP. \tag{9.1.2}$$

In the case of completely ionized plasma, $c_a = 1$. In the classical case it can be presented in the form

$$4\pi\alpha^{\perp}(\omega,k) = \sum_a\frac{2\pi e_a^2 n_a}{k^2\omega}\int\frac{[k\times[V\times k]]}{\omega+i\varDelta-kV}\frac{\partial f_a}{\partial P}\frac{V}{(2\pi\hbar)^3}\,dP, \tag{9.1.3}$$

which corresponds to the classical expression (4.5.3).

The polarizability $\alpha_{bb}^{\perp}(\omega,k)$ in the dipole approximation from the general formula (9.1.1), taking (7.4.6) into account, is

$$4\pi\alpha_{bb} = -\frac{\omega_L^2}{\omega^2}c_{ei} + 2\pi\frac{N}{V}\frac{V}{(2\pi\hbar)^3}\frac{e^2}{\mu\omega^2}\sum_{nm}\int dP'\,dP''\frac{[k\times p_{nm}]^2}{k^2}$$

$$\times\frac{\delta(\hbar k-(P'-P''))(f_n(P',t)-f_m(P'',t))}{\hbar(\omega+i\varDelta)-(E_n+E_{P'}-E_m-E_{P''})}. \tag{9.1.4}$$

Considering the connection of the matrix elements

$$\frac{p_{nm}}{\mu} = \frac{i}{\hbar}(E_n+E_{P'}-E_m-E_{P''})r_{nm}, \tag{9.1.5}$$

the rule of summation

$$\frac{2\mu}{\hbar}\sum_m\omega_{mn}|r_{nm}|_{ij}^2 = \delta_{ij}, \tag{9.1.6}$$

and the normalization condition of the distribution function $f_n(P,t)$ (8.3.22), replacing the variables $P'\to P+\frac{1}{2}\hbar k'$ and $P''\to P-\frac{1}{2}\hbar k'$, and integrating over k', leads to the following expression, from (9.1.4),

$$4\pi a_{bb}(\omega,k)$$

$$= 2\pi\frac{N}{V}\sum_{nm}\int\frac{V}{(2\pi\hbar)^3}\,dP\frac{[k\times d_{nm}]^2}{k^2}\frac{f_n(P+\frac{1}{2}\hbar k)-f_m(P-\frac{1}{2}\hbar k)}{\hbar[(\omega+i\varDelta)-kV]-(E_n-E_m)}. \tag{9.1.7}$$

This form of presentation is especially convenient for various applications (see Sects. 10.7, 8).

Let us finally present the expressions for polarizabilities, determined by transitions from bound states to free ones and vice versa. Taking into account

(8.3.13) and the condition of the attenuation of fluctuations at the transition from bound states into free ones (8.3.15), we obtain from the general formula (9.1.1) the following expression:

$$4\pi\alpha_{bf}(\omega, k) = 4\pi\alpha_{fb}^*(-\omega^*, -k)$$

$$= 2\pi\frac{N}{V}\frac{V^2}{(2\pi\hbar)^6}\frac{c^2}{\omega^2}\sum_m \int dp_e\, dp_i\, dP\, |P^\perp_{\frac{m_i p_e - m_e p_i}{M}, m}(k, p_e + p_i + P)|^2$$

$$\times \frac{\delta(\hbar k - (p_e + p_i - P))[Nf(p_e)\,f(p_i) - f_m(P)]}{\hbar(\omega + i\Delta) - (p_e^2/2m_e + p_i^2/2m_i - E_m - E_P)}\,,\tag{9.1.8}$$

which corresponds to (8.3.16).

9.2 The Spectral Density of Transverse Field Fluctuations

The general expression for the spectral density of transverse field fluctuations follows from (8.4.1) after the substitution

$$\frac{|P_{\alpha\beta}(k)|^2}{k^2|\varepsilon(\omega, k)|^2} \to \frac{c^2\omega^2|P^\perp_{\alpha\beta}(k, P' + P'')|^2}{|\omega_\varepsilon^{2\perp}(\omega, k) - c^2 k^2|^2}\,.\tag{9.2.1}$$

In the equilibrium state the expression for spectral density is [cf. (8.4.2)]

$$(\delta E^\perp \delta E^\perp)_{\omega, k} = \frac{16\pi\omega^3 \mathrm{Im}\{\varepsilon^\perp(\omega, k)\}}{|\omega_\varepsilon^{2\perp}(\omega, k) - c^2 k^2|^2}\left(\frac{1}{2}\hbar\omega + \frac{\hbar\omega}{\exp(\hbar\omega/\mathscr{K}T) - 1}\right).\tag{9.2.2}$$

This spectral density can also be presented as a sum of four terms. The equation for the function $(\delta E^\perp \delta E^\perp)^{\mathrm{ff}}_{\omega, k}$ follows from (8.4.6) after the subsitution

$$\frac{1}{|\varepsilon(\omega, k)|^2} \to \frac{\omega^2[k \times V]^2}{|\omega^2\varepsilon^\perp(\omega, k) - c^2 k^2|^2}\,,\tag{9.2.3}$$

which is similar to the relationship observed in the classical case of completely ionized plasma (4.4.14, 18, 5.1, 2).

The expression for the spectral density $(\delta E^\perp \delta E^\perp)^{\mathrm{bb}}_{\omega, k}$ follows from the general expression (8.4.1, 9.2.1) after substituting $\alpha = n$ and $\beta = m$. In the dipole approximation it is, taking (9.1.5) into account, substituting $P' \to P + \frac{1}{2}\hbar k'$ and $P'' \to P - \frac{1}{2}\hbar k'$, and integrating over k',

$$(\delta E^\perp \delta E^\perp)^{bb}_{\omega,k} = \frac{16\pi^3 N}{V} \sum_{nm} \int \frac{V}{(2\pi\hbar)^3} dP \delta(\omega - kV - \omega_{nm})$$

$$\times \frac{\omega^4 [k \times d_{nm}]^2}{k^2 |\omega^2 \varepsilon^\perp - c^2 k^2|^2} [f_n(P + \tfrac{1}{2}\hbar k) + f_m(P - \tfrac{1}{2}\hbar k)]. \quad (9.2.4)$$

The expressions for spectral densities $(\delta E^\perp \delta E^\perp)^{bf}_{\omega,k}$ and $(\delta E^\perp \delta E^\perp)^{fb}_{\omega,k}$, follow from (8.4.8, 9) after substituting in (9.2.1)

$$\alpha = (m_i p'_e - m_e p'_i)/M, \quad \beta = m.$$

The four terms which determine the total spectral density $(\delta E^\perp \delta E^\perp)_{\omega,k}$ are not independent since each of them includes the total dielectric permittivity.

In nonrelativistic plasma the processes of collisions resulting from interaction via a transverse electromagnetic field are much less probable than corresponding processes in a Coulomb plasma. Nevertheless, in some cases they can play a decisive role. For example, interaction via the transverse field determines the effect of radiation capture (Sect. 10.8), which together with other mechanisms can significantly influence the broadening of the spectral lines of radiation emitted by atoms in plasma.

9.3 The Collision Integral

As in the case of a Coulomb plasma, the collision integral for pairs of particles can be written in two ways. First we consider the form which corresponds to (8.5.1). Taking (9.2.1) into account, we get

$$I^\perp_\alpha(P', t) = \frac{1}{16\pi^3 \hbar} \sum_\beta \int d\omega\, dk\, dP'' \frac{c^2}{\omega^2} |P^\perp_{\alpha\beta}(k, P' + P'')|^2$$

$$\times \delta(\hbar k - (P' - P'')) \delta(\hbar\omega - (E_\alpha + E_{P'} - E_\beta - E_{P''}))$$

$$\times \{(\delta E^\perp \delta E^\perp)_{\omega,k} [f_\beta(P'', t) - f_\alpha(P', t)]$$

$$- A(\omega, k) [f_\beta(P'', t) + f_\alpha(P', t)]\}. \quad (9.3.1)$$

Hereafter the sign $\perp$ will be omitted for I_α if this does not lead to a possible misunderstanding. In (9.3.1) the following abbreviation was introduced:

$$A(\omega, k) = \frac{8\pi\hbar\omega^4 \text{Im}\{\varepsilon^\perp(\omega, k)\}}{|\omega^2 \varepsilon^\perp(\omega, k) - c^2 k^2|^2}. \quad (9.3.2)$$

With the aid of (9.3.1) and following the scheme of Sect. 8.5, we obtain the expressions for the collision integrals in the kinetic equations for distribution functions of electrons, ions, and atoms. For the analysis of induced (forced) and spontaneous kinetic processes it is more convenient, however, to employ the collision integral in the form of (9.3.1). The alternative form is necessary, e. g., for the analysis of radiation capture (imprisonment of radiation).

We shall see that along with "normal" kinetic processes there are also "abnormal" ones, e. g., induced and spontaneous ionization with emission of photons. Some of the anomalous effects described below have been investigated earlier [9.2, 3].

As in Sect. 8.6, the collision integral $I_a(p_a, t)$ is presented as a sum of two parts (8.6.1). The expression for $[I_a]_1$ follows from (8.6.2) after substituting

$$(\delta E\,\delta E)_{\omega,k} \to (\delta E^\perp \delta E^\perp)_{\omega,k}\,; \quad \frac{4\pi\hbar\,\mathrm{Im}\{\varepsilon(\omega,k)\}}{|\varepsilon(\omega,k)|^2} \to A(\omega,k)\,;$$

$$\frac{1}{k^2} \to \frac{1}{2k^2}\frac{[k\times V]^2}{\omega^2}\,. \tag{9.3.3}$$

This integral describes processes 1 to 4 on page 210 except that they are now conditioned by the interaction via the transverse field. The integral $[I_a(p_a, t)]_1$ here also has the property given in (8.6.3).

The expression for the collision integral $[I_a]_2$ follows from (8.6.4) after the first and the second substitutions in (9.3.3) and the substitution

$$\frac{1}{k^2}|p_{P;m}(k)|^2 \to \frac{c^2}{2\omega^2}\,|p_{P;m}(k, p' + p'')|^2\,. \tag{9.3.4}$$

This collision integral describes the processes 5 to 8 on page 211 and has the property of (8.6.5).

The collisions integral in the kinetic equation for atoms $I_n(P', t)$ is also presented as a sum of two parts (8.6.6). The expression for the first part follows from (9.3.1) at $\alpha = n$ and $\beta = m$. This collision integral describes processes 1 to 4 on page 212 and has the property of (8.6.7). The expression for the collision integral $[I_n]_2$ follows from (8.6.8) after the first and the second substitutions given in (9.3.3) and the substitution

$$\frac{1}{k^2}|P_{n,p''}(k)|^2 \to \frac{c^2}{2\omega^2}\,|P_{n,p''}^\perp(k, P' + P'')|^2\,. \tag{9.3.5}$$

The collision integral $[I_n]_2$ describes the four processes 5 to 6 on page 212 and has the property of (8.6.9).

All the processes referred to in this section are determined by the integral contribution at all permissible values of ω and k. In the transparency region, when the quantities ω and k are connected by the dispersional equation

$$\omega^2\mathrm{Re}\{\varepsilon^\perp(\omega,k)\} - c^2k^2 = 0\,, \tag{9.3.6}$$

the spectral density of fluctuations generally cannot be expressed by one-particle distribution functions. A similar situation is also encountered in the case of a

Coulomb plasma when the influence of plasma oscillations (plasmons) upon the kinetic processes is significant ([Ref. 9.1, Sect. 57] and the references cited there).

It is important that in the transition to the turbulent state the level of fluctuations for the values of ω and k which correspond to the transparency region can substantially exceed the level of equilibrium fluctuations. For that reason the classification and analysis of the kinetic processes in the transparency region are matters of special interest.

9.4 The Structure of the Collision Integrals for the Transparency Region

In the transparency region a limiting procedure can be performed, with the extinction (sign$\{\omega\}$Im$\{\varepsilon\}$) tending to zero (Sect. 6.1). Then the function $A(\omega, k)$, given by (9.3.2), takes the form

$$A(\omega, k) = 8\pi^2 \hbar\omega|\omega|\delta(\omega^2 \mathrm{Re}\{\varepsilon^\perp\} - c^2 k^2). \tag{9.4.1}$$

Therefore only the functions at those values of ω and k which are related via the dispersion equation contribute to the collision integral. The kinetic processes are thus determined by the interaction of particles with photons in the medium.

Consider the processes determined by the collision integral $[I_a(p_a, t)]_1$ [see (8.6.2, 9.3.3)]. It describes the processes which comply with the conservation laws,

$$\pm \hbar\omega = \frac{p_a'^2}{2m_a} - \frac{p_a''^2}{2m_a}, \quad \pm \hbar k = p_a' - p_a'', \quad \omega > 0. \tag{9.4.2}$$

In the transparency region, the phase velocity of photons is $\omega/k = c/\sqrt{\mathrm{Re}\{\varepsilon\}}$ if spatial dispersion is neglected. Hence at $\sqrt{\mathrm{Re}\{\varepsilon\}} > 1$ the distribution functions of free charged particles may change because of Čerenkov emission and absorption. The conservation laws with a plus sign permit normal processes, i.e., a particle p_a' is accelerated after absorption of a quantum of energy, and retarded after emission. With a minus sign the conservation laws allow anomalous processes, i.e., upon emission of a quantum or energy the particle p_a' is accelerated, and at absorption retarded.

In order also to include the processes of bremsstrahlung in the collision integral $[I_a]_1$, more exact expressions for matrix elements $|P_{p'p''}|^2$ should be employed than (7.4.6) to obtain the expression for $[I_a]_1$.

The collision integral $[I_a(p_a', t)]_2$ (8.6.4, 9.3.4) describes processes for which the following conservation laws are true:

$$\pm \hbar\omega = \frac{p_a'^2}{2m_a} + \frac{p_b'^2}{2m_b} - E_m - E_{p''}, \quad \pm \hbar k = p_a' + p_b' - P'', \quad \omega > 0. \tag{9.4.3}$$

With a plus sign these conservation laws allow normal processes — photoionization occurs upon the absorption of a quantum, and photorecombination is accompanied by the emission of a quantum of energy. With a minus sign the conservation laws are satisfied provided

$$V'' \cos \theta > c \sqrt{\text{Re}\{\varepsilon\}} \tag{9.4.4}$$

and allow the occurrence of anomalous processes — the dissociation of an atom
is accompanied by emission of a quantum, and recombination by absorption.

Turning to processes described by the collision integrals in the kinetic equation for atoms, the collisions integral $[I_n(P', t)]_1$ (9.3.1) is determined by processes allowed by the conservation laws,

$$\pm \hbar\omega = E_n + E_{P'} - E_m - E_{P''}, \quad \pm \hbar k = P' - P'', \quad \omega > 0 . \tag{9.4.5}$$

The plus sign signifies that excitation processes are allowed upon absorption of
quanta, and the transition to lower states upon emission. These are normal processes. The minus sign signifies processes in which the atom is excited upon
emission of a photon (anomalous Doppler effect), and the atom passes to a lower
state upon absorption of a quantum. These anomalous processes are also only
possible if condition (9.4.4) is satisfied.

Finally, the processes which determine the collision integral $[I_n]_2$ (8.6.8,
9.3.5) obey the conservation laws

$$\pm \hbar\omega = E_n + E_{P'} - \frac{p_a''^2}{2m_a} - \frac{p_b''^2}{2m_b}, \quad \pm \hbar k = P' - p_a'' - p_b'' . \tag{9.4.6}$$

These processes are reciprocal to those allowed by the conservation laws given in
(9.4.3).

Processes which are determined by the interaction of particles with plasma
waves (plasmons) can also be singled out in the kinetic equations for a Coulomb
plasma [Ref. 9.1, Sect. 57].

The kinetic processes determined by the interactions of atoms with radiation
play an important role in many physical phenomena. In particular, they determine the broadening of atoms' emission lines. In the following sections we shall
discuss these processes in greater detail.

9.5 The Evolution of the Distribution Function of Atoms

Recall that the collision integral in the kinetic equation for atoms is presented as
a sum of two parts. The first part is given by (9.3.1) at $\alpha = n$ and $\beta = m$. In the
dipole approximation we employ (7.4.6) for the matrix element $|P_{nm}^\perp|^2$. Also
taking into account (9.1.5) and the identity

$$\frac{1}{k^2} [k \times d_{nm}]^2 = |d_{nm}^\perp|^2 = \frac{2}{3} |d_{nm}|^2, \quad d_{nm} = er_{nm}, \tag{9.5.1}$$

we get the following expression for the collision integral:

$$[I_n(P', t)]_1 = \frac{1}{3(2\pi)^3 \hbar^2} \sum_m \int d\omega\, dk\, dP'' \, |d_{nm}|^2 \delta(\omega - kV - \omega_{nm})$$

$$\times \delta(\hbar k - (P' - P'')) \{ (\delta E^\perp \delta E^\perp)_{\omega,k} [f_m(P'', t) - f_n(P', t)]$$

$$- A(\omega, k)[f_m(P'', t) + f_n(P', t)] \} . \qquad (9.5.2)$$

The function $A(\omega, k)$ is given by (9.3.2).

The integrand (the collision operator) does not explicitly depend on the velocities of atoms provided two conditions are satisfied:

$$\hbar k \ll M v_\mathrm{T}, \quad kV \ll \omega_{nm} \quad (V = (P' + P'')/2M) . \qquad (9.5.3)$$

The first condition implies that the principal contribution in integration over k comes from the region of wavelengths greater than $\hbar/Mv_\mathrm{T}$, or the de Broglie wavelength for $P = Mv_\mathrm{T}$. At $v_\mathrm{T} \sim 10^5$ cm/s and $M \sim 10^{-24}$ g, the wavelength $\hbar/Mv \sim 10^{-8}$ cm. Hence it follows that this condition is well satisfied for the optical region. The second inequality is also well satisfied for the optical region.

Thus, if the conditions in (9.5.3) are satisfied, a simpler expression for the collision integral can be employed than (9.5.2),

$$[I_n(P, t)]_1 = \frac{1}{3\hbar^2(2\pi)^3} \sum_m \int dk\, |d_{nm}|^2 \{ (\delta E^\perp \delta E^\perp)_{\omega_{nm}} [f_m(P, t) - f_n(P, t)]$$

$$- A(\omega_{nm}, k)[f_m(P, t) + f_n(P, t)] \} . \qquad (9.5.4)$$

Of course, switching from (9.5.2) to (9.5.4) leaves many interesting phenomena beyond our scope, e. g., the problem of heating and cooling gas by laser radiation (see Sects. 9.7, 8).

Singling out the contribution in (9.5.4) coming from the transparency region, we employ (9.4.1) and take into account the value of the integral

$$\frac{\pi\omega|\omega|}{(2\pi)^3} \int \delta(\omega^2 \mathrm{Re}\{\varepsilon(\omega)\} - c^2 k^2)\, dk = \frac{\omega^3 \sqrt{\mathrm{Re}\{\varepsilon\}}}{4\pi c^3} . \qquad (9.5.5)$$

As a result we get

$$[I_n(P, t)]_1 = \frac{1}{3\hbar^2} \sum_m |d_{nm}|^2 \left\{ (\delta E^\perp \delta E^\perp)_{\omega_{nm}} [f_m(P, t) - f_n(P, t)] \right.$$

$$\left. - \frac{2\hbar\omega_{nm}^2}{c^3} [f_m(P, t) + f_n(P, t)] \right\} . \qquad (9.5.6)$$

Another convenient form of this expression is where the terms determined by zero oscillations are excluded. To reach to this form we employ the expression for the spectral density of field fluctuations in the state of equilibrium:

$$(\delta E^\perp \delta E^\perp)_\omega = \frac{4\hbar\omega^3\sqrt{\mathrm{Re}\{\varepsilon\}}}{c^3}\left(\frac{1}{2} + \frac{1}{\exp(\hbar\omega/\mathscr{K}T) - 1}\right)$$

$$\equiv 4\pi^2\rho_\omega^\mathrm{T} + \frac{2\hbar\omega^3}{c^3}\sqrt{\mathrm{Re}\{\varepsilon\}}. \tag{9.5.7}$$

In the classical approximation this formula coincides with (6.1.30).

In the right-hand side of (9.5.7) we have singled out the temperature part of the Planck formula ρ_ω. The second term describes the contribution of zero oscillations. The corresonding separation can be carried out for the nonequilibrium state as well,

$$\rho_\omega = \left[\frac{(\delta E^\perp \delta E^\perp)_\omega}{4\pi^2} - \frac{\hbar\omega^3}{2\pi^2 c^3}\sqrt{\mathrm{Re}\{\varepsilon\}}\right]. \tag{9.5.8}$$

Since the second term in brackets is determined by zero oscillations, the function ρ_ω coincides in the equilibrium state with the temperature part of the Planck formula ρ_ω^T.

Substituting the function $(\delta E^\perp \delta E^\perp)_\omega$ in (9.5.6) for ρ_ω according to (9.5.8), and employing the designations for the Einstein coefficients

$$B_m^n = \frac{4\pi^2|d_{nm}|^2}{3\hbar^2}, \quad A_m^n = \frac{4|d_{nm}|^2\omega_{nm}^3}{3\hbar c^3}\sqrt{\mathrm{Re}\{\varepsilon\}}, \tag{9.5.9}$$

(9.5.6) takes the form

$$[I_n(P, t)]_1 = \sum_{m<n} [B_m^n \rho_{\omega_{nm}}(f_m - f_n) - A_m^n f_n] + \sum_{m>n} [B_n^m \rho_{\omega_{nm}}(f_m - f_n) + A_n^m f_m]. \tag{9.5.10}$$

In this expression the contributions determined by induced and spontaneous radiation are clearly indicated. The spontaneous processes may both reduce the level population n (in "downward" transitions) as well as increase the level population n at the cost of transitions from upper $(m > n)$ levels.

In the state of equilibrium, when the distribution over levels is the Boltzmann distribution, and if the function ρ_ω coincides with the temperature part of the Planck formula, (9.5.10) turns into zero.

Equations (9.5.6, 10) do not explicitly depend on the momenta of atoms. This means that the distribution over momenta, and in particular the equilibrium Maxwell distribution, is established by dint of other processes, for instance, elastic collisions of atoms (see Sect. 10.7).

Considering the same approximation, the second part of the collisions integral in the kinetic equation for atoms is generally determined by (8.6.8, 9.3.3, 5).

In the expression for $[I_n(P', t)]_2$ we turn to dipole approximation (7.4.6), replace the variables p_a'', $p_b'' \to p''$, P'' according to (8.3.8), and employ (9.1.5). The result is

$$[I_n(P',t)]_2 = \frac{1}{2(2\pi)^3\hbar} \frac{V}{(2\pi\hbar)^3} \int d\omega \, dk \, dp'' \, dP'' \frac{[k \times d_{np''}]^2}{2k^2}$$

$$\times \delta(\hbar\omega - (E_n + E_{P'} - E_{P''} - E_{P'})) \, \delta(\hbar k - (P' - P''))$$

$$\times \{ (\delta E^\perp \delta E^\perp)_{\omega,k} [Nf(p_e'',t)f(p_i'',t) - f_n(P',t)]$$

$$- A(\omega,k)[Nf(p_e'',t)f(p_i'',t) + f_n(P',t)]\} . \tag{9.5.11}$$

Here it is necessary to substitute the values of p_e'' and p_i'' from (8.3.8). The function $A(\omega,k)$ is given by (9.3.2).

We impose restrictions on the values of the wave vector similar to the relations in (9.5.3),

$$\hbar k \ll M v_T, \quad kV \ll \omega_{np''} \text{ (or } \hbar kV \ll \hbar\omega_{np''}) . \tag{9.5.12}$$

These inequalities are consistent when $Mv_T^2 \sim \hbar\omega_{np''}$. Hence in the optical region the inequalities (9.5.12) are satisfied for discrete levels with

$$\hbar\omega_{np''} \geqslant \mathscr{K}T . \tag{9.5.13}$$

For smaller values of $\hbar\omega_{np''}$ the very concept of a discrete level makes no sense.

Under the conditions in (9.5.12) it is possible to carry out transformations in (9.5.11) similar to those which enabled going from (9.5.2) to (9.5.4). The result is

$$[I_n(P,t)]_2 = \frac{1}{3\hbar^2(2\pi)^3} \frac{V}{(2\pi\hbar)^3} \int dk \, dp \, |d_{np}|^2$$

$$\times \{ (\delta E^\perp \delta E^\perp)_{\omega_{np},k} [Nf(p_e,t)f(p_i,t) - f_n(P,t)]$$

$$- A(\omega,k)[Nf(p_e,t)f(p_i,t) + f_n(P,t)]\}, \quad \text{where} \tag{9.5.14}$$

$$\hbar\omega_{np} = E_n - p^2/2\mu . \tag{9.5.15}$$

The momenta p_e and p_i are linked to p and P by (8.3.8).

Singling out the contribution in (9.5.14) coming from the transparency region we employ again (9.4.1) and the value of the integral (9.5.5). As a result we get

$$[I_n(P,t)]_2 = \frac{1}{3\hbar^2} \frac{V}{(2\pi\hbar)^3} \int dp \, |d_{np}|^2$$

$$\times \left\{ (\delta E^\perp \delta E^\perp)_{\omega_{pn}} [Nf(p_e,t)f(p_i,t) - f_n(P,t)] \right.$$

$$\left. - \frac{2\hbar\omega_{np}^3}{c^3} \sqrt{\mathrm{Re}\{\varepsilon\}} [Nf(p_e,t)f(p_i,t) + f_n(P,t)] \right\}. \tag{9.5.16}$$

The momenta p_e, p_i are linked again to p and P by (8.3.8).

As in case of (9.5.6), another presentation is also possible here so that the contributions of zero oscillations can be excluded. Introducing the function ρ_ω in place of spectral density $(\delta E^\perp \delta E^\perp)_\omega$ [see (9.5.8)], the result is

$$[I_n(P, t)]_2 = \frac{V}{(2\pi\hbar)^3} \int dp \{B_n^p \rho_{\omega_{pn}} [Nf(p_e, t) f(p_i, t) - f_n(P, t)]$$

$$+ A_n^p N f(p_e, t) f(p_i, t)\}. \tag{9.5.17}$$

Naturally, only transitions "upwards" (into the continuous spectrum) are possible from level n; the spontaneous term (9.5.17) thus enters this equation with a plus sign. The Einstein coefficients, which determine the transitions from the discrete spectrum into the continuous one, are given by (9.5.9) with the substitution $n \to p$.

In the equilibrium state (9.5.17) turns into zero. This can be proven by substituting into (9.5.17) the Maxwell distribution over momenta p_e, p_i, and P, the Boltzmann distribution over energies E_n, and the temperature part of the Planck formula ρ_ω^T, and by employing the condition of ionization equilibrium (8.3.24).

9.6 The Equations for Concentrations of Free Charged Particles and Atoms. The Contribution from the Interaction of Particles and Waves

Closed equations for concentrations of free charged particles and atoms can be obtained in regard to the approximation – as was done in Sect. 8.7 – concerning the state of local equilibrium, i.e., the equilibrium with respect to translatory and inner degrees of freedom, but without chemical (ionization) equilibrium. By virtue of the relations in (8.7.1) it suffices to obtain a single equation, e.g., for the concentration of atoms n_{ei}. Let us analyse the contribution to this equation from the interaction of particles with radiation:

$$n_{ei} = \frac{N}{V} \frac{V}{(2\pi\hbar)^3} \sum_n \int dP f_n(P, t), \quad \text{and therefore} \tag{9.6.1}$$

$$\frac{dn_{ei}}{dt} = \frac{N}{V} \frac{V}{(2\pi\hbar)^3} \sum_n \int \{[I_n(P, t)]_1 + [I_n(P, t)]_2\} dP. \tag{9.6.2}$$

The collision integrals $[I_n]_1$ and $[I_n]_2$ are determined by (9.5.6, 16) [or (9.5.10, 17)].

It is easy to see that (9.5.6) does not contribute to (9.6.2). Of the two expressions (9.5.16, 17), the latter is more suitable. Having carried out trivial transformations, we obtain the desired equation.

Two forms of presentation are convenient. One of them is

$$\frac{dn_{ei}}{dt} = V \sum_n \int dp \, B_n^p f(p) [C \exp(\hbar\omega_{pn}/\mathcal{K}T) - 1]$$

$$\times \left(\frac{\omega_{pn}\sqrt{\operatorname{Re}\{\varepsilon\}} \hbar\omega_{pn}}{\pi^2 c^3 [C \exp(\hbar\omega_{pn}/\mathcal{K}T) - 1]} - \rho_{\omega_{pn}}^T \right) n_e n_i, \tag{9.6.3}$$

where ρ_ω^T is the temperature part of the Planck formula (9.5.7), and

$$\hbar\omega_{pn} = \left(\frac{p^2}{2\mu} - E_n\right), \quad C(t) = \frac{n_{ei}}{n_e n_i} \frac{(2\pi\mu\mathcal{H}T)^{3/2}}{(2\pi\hbar)^3} \frac{1}{Z}. \tag{9.6.4}$$

The function $C(t)$ characterizes the degree of deviation from the ionization equilibrium; at the state of equilibrium $C = 1$.

A more convenient form of (9.6.2) is

$$\frac{dn_{ei}}{dt} = (b^{\,ind} + b^{\,sp})n_e n_i - a n_{ei}, \tag{9.6.5}$$

where the coefficients of induced ($b^{\,ind}$) and spontaneous ($b^{\,sp}$) photorecombination and the coefficient of induced photoionization a are introduced. The spontaneous photoionization coefficient equals zero, of course, since only "downward" spontaneous processes are allowed. These coefficients are given by

$$b^{\,ind} = V\sum_n \int dp\, B_n^p \rho_{\omega_{pn}} f(p), \quad b^{\,sp} = V\sum_n \int A_n^p f(p)\,dp,$$

$$a = \frac{V}{(2\pi\hbar)^3}\sum_n \int dp\, B_n^p \exp\left(-\frac{\hbar\omega_{pn}}{\mathcal{H}T}\right)\rho_{\omega_{pn}} f(p), \tag{9.6.6}$$

where $f(p)$ is the Maxwell distribution, $\int f(p)\,dp = 1$.

To calculate these coefficients the matrix elements $|r_{pn}|^2$ of the transition from the continuous spectrum into the discrete, and vice versa, must be known. Consider for example the transitions from the ground level in the hydrogen atom. In this case the square of matrix element is given by [9.4]

$$|r_{p0}|^2 = \frac{2\pi^2 2^8 a_0^5}{Vp^*}\exp\left(-\frac{4}{p^*}\arctan p^*\right)(1 + p^*)^{-5}\left[1 - \exp\left(-\frac{2\pi}{p^*}\right)\right]^{-1},$$

$$p^* = \frac{p a_0}{\hbar}, \tag{9.6.7}$$

where a_0 is the Bohr radius.

The main contribution to the expressions for $b^{\,ind}$ and a comes from the region of small values of p^* (if $I \gg \mathcal{H}T$). In this approximation we find from (9.6.7)

$$|r_{p0}|^2 = \frac{2\pi^2 2^8 e^{-4}}{Vp}\hbar a_0^4, \quad p \ll \frac{\hbar}{a_0}. \tag{9.6.8}$$

Using this formula and integration over p in (9.6.6), we get

$$b^{\,ind} = \frac{4\pi^2}{3\hbar}\frac{(2\pi\hbar)^3}{(2\pi\mu\mathcal{H}T)^{3/2}}2^8 e^{-4}a_0^3\int_{I/\hbar}^\infty \rho_\omega^T \exp\left(-\frac{\hbar\omega-I}{\mathcal{H}T}\right)d\omega, \tag{9.6.9}$$

$$a = \frac{4\pi^2}{3\hbar} 2^8 e^{-4} a_0^3 \int\limits_{I/\hbar}^{\infty} \rho_\omega^T d\omega .$$ (9.6.10)

In the latter expression, integration over ω can be carried out (at $I \gg \varkappa T$); the result is

$$a = \frac{2^{10} e^{-4}}{3} a_0^3 \sqrt{\mathrm{Re}\left\{\varepsilon\left(\frac{I}{\hbar}\right)\right\}}; \frac{I^3}{\hbar^3 c^3} \frac{\varkappa T}{\hbar} \exp\left(-\frac{I}{\varkappa T}\right).$$ (9.6.11)

This expression only differs by the numerical factor from the one obtained using the quasi-classical method [9.5, 6].

9.7 Cooling and Heating of Atoms by a Resonant Field. Classical Theory

The kinetic equations obtained here for the distribution function of atoms $f_n(P, t)$ will now be used to analyze gas cooling under laser radiation.

Consider the motion of a system of atom oscillators with eigenfrequency ω_0 under the effect a monochromatic field with frequency ω. We shall show that — depending upon the sign of mistuning $\Delta = \omega - \omega_0$ — the action of the field may cause either the cooling or heating of atoms [9.7–11]. Solving this problem is facilitated by employing the kinetic equation for the distribution function of atoms as the initial equation.

In Sects. 6.1, 2, (3.2.8) was employed as the microscopic phase density of atoms in obtaining the kinetic equation. If the effect of the field on the atoms is taken into account, a more general equation for the phase density of atoms N_{ab} (3.2.7) should be used in place of (3.2.8). The corresponding kinetic equation has the form (the terms determining the frequency shift omitted)

$$\left(\frac{\partial}{\partial t} + V\frac{\partial}{\partial R} + v\frac{\partial}{\partial r} - m\omega_0^2 r\frac{\partial}{\partial p} + e\mathscr{E}\frac{\partial}{\partial p} + F\frac{\partial}{\partial P}\right)$$

$$\times f(R, P, r, p, t) = I(R, P, r, p, t) ,$$ (9.7.1)

where $\mathscr{E}(R, V, t)$ is the mean value of the Lorentz force (7.5.2), F is the force acting upon the atom

$$F = e(r\,\mathrm{grad}_R)\,\mathscr{E} + \frac{e}{c}[v \times B] ,$$ (9.7.2)

and I is the collision integral, accounting for the contribution from all the dissipative processes. We shall assume $\mathscr{E} = eE$ (zero-order approximation with respect to $\mu_{\mathrm{at}} = v_T/c$).

With the aid of (9.7.1), the equation for the mean density of the kinetic energy of atoms for a spatially uniform gas has the form

$$\frac{\partial}{\partial t} \left\langle n \frac{MV^2}{2} \right\rangle = \langle nFV \rangle, \quad n = \frac{N}{V}.$$

(9.7.3)

The angular brackets designate averaging over the velocity distribution of atoms. The definitions of the vectors of polarization and current are

$$P(R, V, t) = en \int rf \, dr \, dp, \quad J = en \int vf \, dr \, dp.$$

(9.7.4)

Then, also employing (9.7.2), the right-hand side of (9.7.3) can be presented in the form

$$n \langle FV \rangle = \langle P \, \text{grad}_R \, (F \cdot E) \rangle + \frac{1}{c} \langle V [J \times B] \rangle.$$

(9.7.5)

Thus, in order to obtain a closed equation for the mean kinetic energy of atoms it is necessary to calculate the polarization. This calculation is performed under the assumption that the local equilibrium distribution over of velocities is established sooner than polarization. Then the Maxwell distribution can be used for the distribution function of velocities $f(V)$.

In this approximation the collision integral in the kinetic equation (9.7.1) loses the contribution from elastic collisions. The polarization can then be calculated on the basis of the kinetic equation (6.2.17). After trivial calculations we get

$$n \frac{d}{dt} \left\langle \frac{MV^2}{2} \right\rangle = \frac{1}{(2\pi)^4} \int \langle KV \, \text{Im}\{\alpha(\Omega, K, V)\} \rangle (E \cdot E)_{\Omega, k} \, d\Omega \, dK.$$

(9.7.6a)

Here we designate the imaginary part of the polarizability of atoms having velocity V

$$\text{Im}\{\alpha(\omega, k, V)\} = \frac{e^2 n(V) (\gamma/2)}{2 \mu \omega_0 [(\omega - kV - \omega_0)^2 + (\gamma/2)^2]}.$$

(9.7.7)

In (8.7.6a) $(E \cdot E)_{\Omega, K}$ is the spectral density of the field acting upon atoms. For the sake of convenience we have substituted variables ω, k for Ω, K. The change of the mean kinetic energy is thus determined by two functions: by the imaginary part of the permittivity of a group of atoms with velocity V, and by the spectral density of the strength of the electric field acting upon the atoms.

Take the example of a field set in the form of a planar wave. The spectral density is

$$(E \cdot E)_{\Omega, K} = \frac{(2\pi)^4}{2} |E_{\Omega, K}|^2 \delta(\Omega - \omega) \, \delta(K - k),$$

(9.7.8)

and (9.7.6a) takes the form

$$n \frac{d}{dt} \left\langle \frac{MV^2}{2} \right\rangle = \frac{1}{2} \langle k \, V \operatorname{Im}\{\alpha(\omega, k, V)\}\rangle |E_{\omega, k}|^2 . \tag{9.7.6b}$$

In averaging over velocities one can integrate over two components of velocity at once, since the function being averaged only depends on the scalar product $(k \cdot V)$. Assuming $k \parallel x$ we reduce the operation of averaging to a one-dimensional integral with the distribution function

$$f(V_x) = \frac{1}{\sqrt{\pi} u} \exp \left(-\frac{V_x^2}{u^2} \right), \quad u^2 = \frac{2 \mathscr{K} T}{M} . \tag{9.7.9}$$

In (9.7.6b) we substitute $V_x^2/2 \to u^2/4$.

The character of the temperature dependence on time is determined by relative values of γ and ku (the Doppler width). Consider two extreme cases.

1) $\gamma \ll ku$. Equation (9.7.7) can then be presented in the form

$$\operatorname{Im}\{\alpha(k, \omega, V)\} = \frac{\pi e^2 n}{2 \mu \omega_0} \delta(\omega - kV - \omega_0) . \tag{9.7.10}$$

Averaging over velocities with the function (9.7.9) we find

$$\operatorname{Im}\{\alpha(\omega, k)\} = \frac{\sqrt{\pi} e^2 n}{2 \mu \omega_0 ku} \exp \left(-\frac{(\omega - \omega_0)^2}{(ku)^2} \right) . \tag{9.7.11}$$

As a result the equation for u^2 (9.7.6b) takes the form

$$nM \frac{d}{dt} u^2 = 2(\omega - \omega_0) \operatorname{Im}\{\alpha(\omega, k)\} |E_{\omega, k}|^2 . \tag{9.7.12}$$

Thus at mistunings $\Delta = \omega - \omega_0 < 0$, the temperature of atoms decreases with time.

Notice that the interaction mechanism between atoms and field in this approximation is similar to the Landau extinction mechanism in plasma theory. Indeed, (9.7.12) can be written in the form

$$\frac{du^2}{dt} = -\frac{\pi e^2}{\mu \omega_0} u \left(\frac{\partial F}{\partial V} \right)_{V = (\omega - \omega_0)/k} |E_{\omega, k}|^2 , \quad F = f(V_x) . \tag{9.7.13}$$

Hence, as in the case of plasma, the character of the exchange between atoms and field is determined by the sign of the derivative of the distribution function at resonant interaction. If the derivative is negative (positive mistuning), the atoms are heated. In the opposite case (negative mistuning), the atoms are cooled.

From (9.7.11, 12) it follows that the cooling rate is maximal at the condition

$$|\omega - \omega_0| = ku/\sqrt{2} .\tag{9.7.14}$$

2) $\gamma \gg ku$ and $F(V_x) = \delta(V_x)$. In this extreme the rate of cooling becomes equal to zero.

We see that within the framework of classical theory there exists no limit for the cooling of atoms by the field.

9.8 Cooling and Heating of Atoms by a Resonant Field. Quantum Theory

Equation (7.5.3) [or (7.5.8)] can be chosen as the basic equation. It only holds, however, in the first-order approximation in $\hbar k \, \partial/\partial P$. Since this approximation is, of course, not accurate enough to determine the cooling limit, we shall take another approach.

Turning to the kinetic equation for the distribution function of atoms $f_n(P', t)$, obtained in Sect. 9.5, we recall that the collision integral, determined by processes where the number of atoms does not change, is given by (9.5.2). Here we are interested in that part of this expression which is determined by the spectral density of field fluctuations (induced processes).

To write the balance equation for the density of kinetic energy we multiply (9.5.2) by $nP'^2/2M$ and $V/(2n\hbar)^3$, integrate over P', and sum over n. As a result and after symmetrization over P', P'' and n, m, we get an equation which can be presented

$$\frac{\partial}{\partial t} n \left\langle \frac{P^2}{2M} \right\rangle = \frac{n}{(2\pi)^3 2\hbar} \sum_{nm} \int d\omega \, dk \, \frac{V dP}{(2\pi\hbar)^3} \frac{|d_{nm}|^2}{3} kV$$
$$\times \delta(\omega - \omega_{nm} - kV)(\delta E \, \delta E)_{\omega,k} [f_m(P - \tfrac{1}{2}\hbar k) - f_n(P + \tfrac{1}{2}\hbar k)] .\tag{9.8.1}$$

For the spectral density of the field we employ (9.7.8), and single out the resonance transition ($n = a$, $m = b$). The equation obtained in this way is similar to (9.7.6b),

$$\frac{\partial}{\partial t} n \left\langle \frac{P^2}{2M} \right\rangle = \frac{1}{2} \int \frac{V}{(2\pi\hbar)^3} dP (kV \operatorname{Im}\{\alpha(\omega, k, V)\}) |E_{\omega,k}|^2 .\tag{9.8.2}$$

Here we have introduced the designation for the imaginary part of the dielectric permittivity of atoms with velocity V,

$$\operatorname{Im}\{\alpha(\omega, k, V)\} = \frac{\pi n}{3\hbar} |d_{ab}|^2 \delta(\omega - \omega_{ab} - kV)$$
$$\times [f_b(P - \tfrac{1}{2}\hbar k) - f_a(P + \tfrac{1}{2}\hbar k)] .\tag{9.8.3}$$

In order to find the cooling limit we take the finiteness of the resonance width into account in this expression by substituting

$$\pi \delta(\omega - \omega_{ab} - kV) \to \frac{\gamma_{ab}}{(\omega - \omega_{ab} - kV)^2 + \gamma_{ab}^2}. \tag{9.8.4}$$

Now it becomes clear that at negative mistuning and $D < 0$ the temperature of atoms first decreases [the right-hand side of (9.8.2) is negative], the right-hand side of the equation then becomes zero, and the cooling stops. The minimal temperature for any given mistuning is found by setting the right-hand side of (9.8.2) to zero.

Let us consider the example where $\rho_a = 0$ and $\rho_b = 1$ (atoms in the ground state). In the integrand we replace $P - \frac{1}{2}\hbar k \to p$, assume k to be parallel to the x axis, integrate over p_y and p_z, and use (9.7.9) for the one-dimensional Maxwell distribution. The result is the following equation for determining the minimal temperature:

$$\int_{-\infty}^{\infty} \left(kV + \frac{\hbar k^2}{2M}\right) \frac{\gamma_{ab} f(V)}{\left(\omega - \omega_{ab} - kV - \dfrac{\hbar k^2}{2M}\right)^2 + \gamma_{ab}^2}\, dV = 0. \tag{9.8.5}$$

At $\gamma_{ab} = 0$ (infinitesimally narrow resonance) then, the left-hand side is negative at $\omega - \omega_{ab} < 0$, while at $\gamma_{ab} \neq 0$ and $f(V) = \delta(V)$ (infinitesimally narrow distribution over momenta), the left-hand side is positive for any mistuning.

Suppose that the minimal temperature $T_{\min}$ is such that

$$k u_{\min} \ll \gamma_{ab}. \tag{9.8.6}$$

Equation (9.8.5) can then be written in the form

$$2 \frac{|\omega - \omega_{ab}| + \dfrac{\hbar k^2}{2M}}{\left(\omega - \omega_{ab} - \dfrac{\hbar k^2}{2M}\right)^2 + \gamma_{ab}^2} \int_{-\infty}^{\infty} (k \cdot V)^2 f(V)\, dV = \frac{\hbar k^2}{2M} \tag{9.8.7}$$

hence

$$\varkappa T_{\min} \sim \hbar \gamma_{ab} \quad \text{at} \quad |\omega - \omega_{ab}| \sim \gamma_{ab}, \tag{9.8.8}$$

and therefore the minimal temperature is determined by the width of the line γ_{ab}. To prove that the assumption in (9.8.6) is then realized, we substitute $u_{\min}$ from (9.8.8) into the left-hand side of (9.8.6). Thus,

$$k u_{\min} \sim \sqrt{\frac{\hbar k^2}{M \gamma_{ab}} \gamma_{ab}} \ll \gamma_{ab},$$

since for the optical region the value under the radical is of the order $1/(10A)$, where A is the atomic weight.

To estimate the time of cooling we turn to (9.7.12), in which the imaginary part of permittivity is determined from the quantum expression (9.8.3). Since the cooling rate is maximal at the condition given in (9.7.14),

$$|\omega - \omega_{ab}| \, \text{Im} \, \alpha \sim |d_{ab}|^2 nD/\hbar.$$

Because final temperature is much less than the initial one, the minimal cooling time can be estimated from (9.7.12), also taking (9.8.3, 4) into account,

$$(\Delta t)_{\min} \sim \frac{\hbar \mathcal{H} T}{|d_{ab}|^2 E^2 |D|}. \tag{9.8.9}$$

This expression can be written in a more convenient form if the saturation parameter is used (Sect. 12.2),

$$a = \frac{|d_{ab}|^2}{3\hbar^2 \gamma^2}, \quad \gamma^2 = 2\frac{\gamma_a \gamma_b}{\gamma_a + \gamma_b}\gamma_{ab}. \tag{9.8.10}$$

Here γ_a, γ_b, and γ_{ab} are dissipative constants of two-level system. Using them the estimate for $(\Delta t)_{\min}$ can be written in the form

$$(\Delta t)_{\min} \sim \frac{1}{\gamma_{ab}} \frac{\mathcal{H} T}{\hbar \gamma_{ab}} \frac{1}{|D| a E^2}. \tag{9.8.11}$$

Since the value $|D| \sim 1$,

$$(\Delta t)_{\min} \sim \frac{1}{\gamma_{ab}} \frac{T}{T_{\min}} \quad \text{at} \quad aE^2 \sim 1. \tag{9.8.12}$$

The minimal cooling time is therefore $T/T_{\min}$ times greater than the relaxation time $1/\gamma_{ab}$ − the time of spontaneous transition. At $\gamma_{ab} = 10^8$ and $T = 10^2$ from (9.8.9), $T_{\min} \sim 10^{-3}\,\text{K}$, and therefore $(\Delta t)_{\min} \sim 10^{-3}\,\text{s}$. Notice that the dependence of $(\Delta t)_{\min}$ on the field magnitude is determined by the factor $1/(|D| a E^2)$ in this approximation. For a more complete description of the process of cooling in a strong field a more general expression for the collision integral should be employed than (9.5.2).

We have analyzed the cooling of a gas assuming a given field intensity. If extinction is taken into account, the condition in (9.7.14) is no longer optimal; greater values of mistuning are then more advantageous, since the field extinction coefficient is then reduced [9.11].

The possibility of simultaneous self-focusing of atomic and light beams is dealt with in an article by *Klimontovich* and *Luzgin* [9.12]. It gives one more example of an application of the kinetic equations for atoms and a field, discussed in Chaps. 6, 9.

10. Spectral Emission Line Broadening of Atoms in Partially Ionized Plasma

On the basis of the kinetic equations, the theory of an atom's emission line broadening is developed. The shifts and widths of spectral lines are related to the nonequilibrium field fluctuations and the polarization characteristics of plasma. Calculations are carried out for spectral line broadening resulting from electronic and ionic effects (Holtzmark broadening), transverse field fluctuations (induced and spontaneous processes), and the resonant interaction of atoms. The chapter ends with a concise review of the problems facing the kinetic theory of emission line broadening in a nonequilibrium plasma.

10.1 The Foundation of the Kinetic Theory of Spectral Line Broadening

The theory of spectral line broadening has long been a subject of intense investigation [10.1 – 5]. Of course, we do not aspire to give an up-to-date account of the modern theory of line broadening. The aim of this chapter is more specific. We shall show that the theory of line broadening can be considered as part of the general kinetic theory of inequilibrium processes.

The kinetic approach to the theory of broadening seems indispensable today. Indeed, the main effort until recently was in investigating the emission spectra of atoms in an equilibrium plasma. Now, however, the situation has radically changed; the study of spectral line broadening under inequilibrium conditions is becoming increasingly important. For instance, theories of broadening in a turbulent plasma and in strong external fields are being developed. A new branch has emerged which is connected with the study of the emission spectra of quantum optical generators (see Chap. 12).

Naturally, here we shall not be able to dwell upon many important features of the theory of broadening, such as those connected with the fine structure of atomic spectra.

We shall be primarily concerned with those processes of broadening which are due to the long-range nature of the interaction of charged particles in a plasma, as well as those due to the electromagnetic field. This will enable us to employ the kinetic equations in the polarization approximation, obtained in Chaps. 8, 9. At the end of this chapter we shall discuss the possibility of taking strong pair interactions at small distances and weak collective interactions at great distances into account simultaneously.

The processes of broadening connected with small-scale fluctuations of an electromagnetic field in a plasma will be the main object of our concern. We shall widely employ the results of calculations carried out in preceding chapters. In particular, the expressions obtained earlier for the spectral densities of fluctuations will play an important role,

$$(\delta E \, \delta E)_{\omega, k} = (\delta E^{\parallel} \delta E^{\parallel})_{\omega, k} + (\delta E^{\perp} \delta E^{\perp})_{\omega, k} \, . \tag{10.1.1}$$

The first term in the right-hand side is determined by fluctuations of the longitudinal, and the second by fluctuations of the transverse field.

We shall see that there is a connection between the spectral linewidth γ_{nm} and the spectral density (10.1.1). It will be possible to separate two contributions due, respectively, to collisions of particles in a plasma and to radiation.

1) *The Collisions Region.* In this case the spectral density of fluctuations is determined by the distribution functions of particles according to (8.4.1, 9.2.1). Each of these equations can be presented as a sum of four parts, see (8.4.5). Accordingly, it is possible to single out four contributions to the width of spectral lines.

2) *The Transparency Region.* Significant in this case are the spectral densities of field fluctuations for ω, k, which are linked through dispersional equations,

$$\mathrm{Re}\{\varepsilon^{\parallel}(\omega, k)\} = 0 \, , \quad \omega^2 \mathrm{Re}\{\varepsilon^{\perp}(\omega, k)\} - c^2 k^2 = 0 \, . \tag{10.1.2}$$

Recall the $\varepsilon^{\parallel}$ and $\varepsilon^{\perp}$ are dielectric permittivities for longitudinal and transverse waves, respectively. Naturally, the fluctuations of the transverse field play an important role in a gas in optical range.

The parameters of the various processes will be introduced in due course.

The polarization vector of a system of atoms can be expressed through the density matrix $f_{nm}(R, P, t)$,

$$P(R, t) \equiv \sum_{nm} P_{nm} = n_{\mathrm{ei}} \sum_{nm} \int d_{mn} f_{nm}(R, P', t) \frac{V}{(2 \pi \hbar)^3} dP' \, , \tag{10.1.3}$$

where $P_{nm}(R, t)$ is the polarization vector for the $n - m$ transition, and n_{ei} is the average concentration of atoms.

In order to obtain the kinetic equation for the function f_{nm} we take the equation for the operator density matrix (7.5.7) as our starting point. By definition,

$$N f_{nm}(R, P, t) = \langle N_{nm}(R, P, t) \rangle, \quad N \sum_{n} \int f_{nn}(R, P, t) \frac{dR \, dP}{(2 \pi \hbar)^3} = N_{\mathrm{ei}} \, , \tag{10.1.4}$$

where N_{ei} is the total number of atoms. Using the operator equations (7.5.7) we get, after averaging, the following equations for the density matrix:

$$\left(\frac{\partial}{\partial t} + V\frac{\partial}{\partial R} + i\omega_{nm}\right)f_{nm}(R, P, t) - \frac{i}{\hbar}\sum_{n_1}(d_{nn_1}f_{n_1m} - f_{nn_1}d_{n_1m})\,\mathscr{E}$$

$$+ \frac{1}{2}\sum_{n_1}\left\{\left[\left(d_{nn_1}\frac{\partial}{\partial R}\right)\mathscr{E}(R, V, t) + \frac{i}{c}[\omega_{nn_1}d_{nn_1}B]\right]\right.$$

$$\left.\times\frac{\partial f_{n_1m}}{\partial P'} + n \rightleftarrows m\right\} = I_{nm}\,. \qquad (10.1.5)$$

Here we employ the designation $\mathscr{E}(R, V, t)$ for the mean field (7.5.2) and introduce the designation $I_{nm}(R, P, t)$ for the collision integral. The latter is determined by the correlation of fluctuations δN_{nm} and $\delta\mathscr{E}$ [cf. (8.1.4)].

Our next task is to find the collision integral I_{nm}. We shall see that at a preset intensity of field fluctuations and a preset polarizability of the medium it is a non-Markovian linear integral operator (with time lagging), acting upon the function $f_{nm}(R, P, t)$. In the resonance approximation ($\omega \approx \omega_{nm}$) the Fourier components of functions f_{nm} and I_{nm} are linked through the relation

$$I_{nm}(R, P, \omega) = [-\gamma_{nm}(\omega - \omega_{nm}) + i\Delta\omega_{nm}(\omega - \omega_{nm})]f_{nm}(R, P, \omega)\,, \quad (10.1.6)$$

which serves as a definition of the linewidth γ_{nm} and the frequency shift of the transition $\Delta\omega_{nm}$. Generally both of these characteristics are functions of $\omega - \omega_{nm}$, that is, of mistuning with respect to the frequency of the transition ω_{nm}.

Calculation of the collision integral will be carried out with the following simplifying assumptions:

1) In calculating the correlations of the small-scale fluctuations which determine the collision integral, the influence of the mean field can be neglected.

2) The calculation is carried out in the polarization approximation.

3) We employ zero approximation with respect to the parameters μ_{at} and μ_T [see (3.1.9, 12)], so that in (7.5.7) the terms with $d\,\partial/\partial R, B$ can be neglected.

As a result, we get the following expression for the collisions integral I_{nm}:

$$I_{nm}(R, P, t) = \frac{i}{N\hbar}\sum_{n_1}(d_{nn_1}\overline{\delta N_{n_1m}(R, P, t)\delta E(R, t)} - \overline{\delta N_{nn_1}\delta E}d_{n_1m})\,, \quad (10.1.7)$$

and the equation for the fluctuation δN_{nm} can be written in the form [10.6] [cf. (8.1.5)]

$$\left(\frac{\partial}{\partial t} + V\frac{\partial}{\partial R} + i\omega_{nm} + \Delta\right)[\delta N_{nm}(R, P, t) - \delta N_{nm}^{\text{source}}]$$

$$= \frac{iN}{\hbar}\sum_{n_1}[(d_{nn_1}f_{n_1m}(R, P, t) - f_{nn_1}d_{n_1m})\,\delta E(R, t)]\,. \qquad (10.1.8)$$

As before, a dissipative term $\Delta \delta N_{nm}$ is introduced here in order to separate the regions of small-scale and large-scale fluctuations. In the final results $\Delta \to 0$.

The correlation of source fluctuations is determined by [cf. (8.1.7)]

$$\left(\frac{\partial}{\partial t} + V\frac{\partial}{\partial R} + i\,\omega_{nm} + \Delta\right)\langle\delta N_{nm}(R, P, t)\,\delta N^*_{n_1 m_1}(R', P', t)\rangle^{\text{source}} = 0. \tag{10.1.9}$$

This equation must be supplemented by the initial condition – the expression for the one-time correlator of the source fluctuations.

Calculation of these fluctuations was carried out in Sect. 8.2. To take advantage of the results obtained there we turn to (8.2.11), set $\alpha = n$, $\beta = m$, $\alpha_1 = n_1$, and $\beta_1 = m_1$, drop the "compensatory" term, and symmetrize, getting as a result

$$\overline{(\delta N_{nm}\delta N^*_{n_1 m_1})}^{\text{source}}_{R, P, R', P', t} = \frac{(2\pi\hbar)^3}{2}N\delta(P - P')\,\delta(R - R')$$

$$\times\,[\delta_{nn_1}\,f_{m_1 m}(R, P, t) + f_{nn_1}(R, P, t)\,\delta_{m_1 m}]. \tag{10.1.10}$$

This expression corresponds to (8.2.13).

Later we shall employ the lowest order approximation with respect to off-diagonal elements of the density matrix f_{nm}. Then in (10.1.10) we may substitute

$$f_{nm} \to \delta_{nm}\,f_n. \tag{10.1.11}$$

As a result we get

$$\overline{(\delta N_{nm}\delta N^*_{n_1 m})}^{\text{source}} = \frac{(2\pi\hbar)^3}{2}N\delta(P - P')\,\delta(R - R')\,\delta_{nn_1}\,\delta_{mm_1}\,(f_n + f_m). \tag{10.1.12}$$

Let us now write the solution of (10.1.9) at the initial conditions (10.1.10, 12). Taking into account that f_n is a slowly changing function of time, and that f_{nm} at $n \neq m$ describes both slow and fast processes,

$$f_n = f_n(\varepsilon t), \quad f_{nm} = f_{nm}(t, \varepsilon t) = f_{nm}(\varepsilon t)e^{-i\omega_{nm}t}, \tag{10.1.13}$$

where $\varepsilon = \gamma_{nm}/|\omega_{nm}|$ is the small parameter.

In calculating the spectral functions one may assume that the function f_{nm} does not depend on coordinates, which corresponds to zero approximation in $l_{\text{ph}}\,\partial/\partial R$.

In zero approximation with respect to slow variations the solution of (10.1.9) at the initial condition (10.1.10) has the form

$$\overline{(\delta N_{nm}\delta N_{n_1 m_1})}^{\text{source}}_{k, P, P', \tau, t - \tau} = \frac{(2\pi\hbar)^3}{2}N\delta(P - P')\exp\left[-i(\omega_{nm} + kV)\tau\right]$$

$$\times\,[\delta_{nn_1}f_{m_1 m}(t - \tau, \varepsilon t, P) + f_{nn_1}(t - \tau, \varepsilon t, P)\,\delta_{m_1 m}]. \tag{10.1.14a}$$

The corresponding solution at the initial condition (10.1.12) for the spectral density over both ω and k is

$$(\delta N_{nm} \delta N_{n_1 m_1})_{\omega, k, P, P'}^{\text{source}} = \frac{(2\pi\hbar)^3}{2} N 2\pi \delta(\omega - \omega_{nm} - kV)$$

$$\times \delta(P - P') \delta_{nn_1} \delta_{mm_1} (f_n - f_m)_{R, P, t} . \tag{10.1.14b}$$

Now it is no problem to find the expression for the collision integral I_{nm}. The equation for the function δN_{nm} follows from (10.1.8), given attenuation of initial correlations, (for the sake of convenience $n_1 \to m_1$)

$$\delta N_{nm}(k, P, t) = \delta N_{nm}^{\text{source}} + \frac{iN}{\hbar} \int_0^\infty \exp\left[-\Delta\tau - i(\omega_{nm} + kV)\tau\right]$$

$$\times \sum_{m_1} [d_{nm_1} f_{m_1 m}(R, P, t - \tau) - f_{nm_1}(R, P, t - \tau) d_{m_1 m}]$$

$$\times \delta E(k, t - \tau) d\tau . \tag{10.1.15}$$

This expression is the sum of two parts: the induced part, proportional to δE, and $\delta N_{nm}^{\text{source}}$. Correspondingly, the collision integral I_{nm} in (10.1.7) can also be presented as a sum of two parts,

$$I_{nm}(R, P, t) = I_{nm}^{\text{ind}} + I_{nm}^{\text{source}} . \tag{10.1.16}$$

From (10.1.7) it follows that the collision integrals I_{nm} and I_n can be expressed with the spatial spectral densities of fluctuations δN_{nm} and δE,

$$I_{nm} = \frac{i}{\hbar N} \sum_{n_1} \int [d_{nn_1} (\delta N_{n_1 m} \delta E)_k - (\delta N_{nn_1} \delta E)_k d_{n_1 m}] \frac{dk}{(2\pi)^3} , \tag{10.1.17}$$

$$I_n = \frac{2}{\hbar N} \sum_m \int \text{Im}\{(\delta N_{nm} \delta E)_k\} d_{mn} \frac{dk}{(2\pi)^3} . \tag{10.1.18}$$

With the aid of (10.1.15, 17) we can find the expression for the induced part of the collision integral I_{nm},

$$I_{nm}^{\text{ind}} = -\frac{1}{\hbar^2} \sum_{n_1} \sum_{m_1} \int_0^\infty d\tau \int \frac{dk}{(2\pi)^3} \exp\left[-\Delta\tau - i(\omega_{n_1 m} + kV)\tau\right] (d_{nn_1})_i$$

$$\times [(d_{n_1 m_1})_j f_{m_1 m}(P, t - \tau) - f_{n_1 m_1}(P, t - \tau)(d_{m_1 m})_j]$$

$$\times (\delta E_j \delta E_i)_{k, t - \tau, t} + n \rightleftarrows m , \quad * , \tag{10.1.19}$$

which can be simplified in the following way.

1) Quick changes of the collision integral are determined by both quick variations of the distribution functions f_{nm} and "quick" components of the spectral

density of field fluctuations. The latter have no direct connection with spectral line broadening and are therefore not taken into account here. The investigation of processes connected with fast changes is a separate problem. Accordingly, for the spectral densities of field fluctuations we employ the approximation

$$(\delta E_j \delta E_i)_{t-\tau, t, k} = (\delta E_j \delta E_i)_{-\tau, k} .$$ (10.1.20)

2) In the right-hand side of (10.1.19) only the resonant terms are retained, that is, those terms which are proportional to $\exp(-i\omega_{nm}t)$. Taking (10.1.13) into account, we notice that the terms

$$f_{m_1 m} \to f_{nm} \delta_{nm_1} , \quad f_{n_1 m_1} \to f_{nm} \delta_{n_1 n} \delta_{m_1 m}$$ (10.1.21)

must be singled out in (10.1.19) in summations over n_1 and m_1 in order to separate the resonant terms (under assumption 1). This results in the following expression for the Fourier component of $I_{nm}(\omega)$:

$$I_{nm}^{\text{ind}}(\omega, \mathbf{R}, \mathbf{P}) = -\frac{1}{\hbar^2} \int_0^\infty d\tau \int \frac{d\mathbf{k}}{(2\pi)^3} \exp[-\Delta\tau + i((\omega - \omega_{nm}) - \mathbf{k}\mathbf{V})\tau]$$

$$\times \left[\sum_{n_1} |d_{nn_1}|_{ij}^2 \exp[i\omega_{nn_1}\tau] - (d_{nn_1})_i (d_{mm})_i \right] (\delta E_j \delta E_i)_{-\tau, k} f_{nm}(\mathbf{R}, \mathbf{P}, \omega)$$

$$+ n \rightleftarrows m, \quad *, \quad \omega \to -\omega .$$ (10.1.22)

In the middle of the line ($\omega = \omega_{nm}$) this expression is somewhat simplified.

10.2 The Dissipative Matrix. The Frequency Shift

Consider first the expression for the dissipative matrix. Combining (10.1.6, 22) and taking into account that

$$|d_{nm}|_{ij}^2 = \tfrac{1}{3}\delta_{ij}|d_{nm}|^2 , \quad (d_{nn})_i (d_{mm})_j = \tfrac{1}{3}\delta_{ij}(d_{nm}d_{mm})$$ (10.2.1)

in the isotropic medium, after integration over $\mathbf{k}$ we get the following expression for the induced part of the dissipative matrix:

$$\gamma_{nm}^{\text{ind}}(\omega - \omega_{nm})$$

$$= \frac{1}{6\hbar^2} \left[\sum_{n \neq n_1} |d_{nn_1}|^2 (\delta E \delta E)_{\omega_{nn_1}} + \sum_{n_1 \neq m} |d_{n_1 m}|^2 (\delta E \delta E)_{\omega_{n_1 m}} \right]$$

$$+ \frac{1}{6\hbar^2} (d_{nm} - d_{mm})^2 \int (\delta E \delta E)_{\omega - \omega_{nm} - \mathbf{k}\mathbf{V}, k} \frac{d\mathbf{k}}{(2\pi)^3} .$$ (10.2.2)

We have simplified the terms which include off-diagonal matrix elements, based on the inequalities

$$\omega - \omega_{nm}, \quad \gamma_{nm}, \quad k v_T \ll \omega_{nm}. \tag{10.2.3}$$

Thus the quantity γ_{nm} is determined by two contributions, one of which is determined by the spectral density at high frequencies (frequencies of transitions) and the other by that at low frequencies of the order $k v_T$.

If in (10.2.2) we neglect spatial dispersion and assume mistuning equal to zero ($\omega - \omega_{nm} = 0$), (10.2.2) has the form

$$\gamma_{nm}^{ind} = \frac{1}{6 \hbar^2} \left[\sum_{n_1 \neq n} |d_{nn_1}|^2 (\delta E \, \delta E)_{\omega_{nn_1}} + \sum_{n_1 \neq m} |d_{n_1 m}|^2 (\delta E \, \delta E)_{\omega_{n_1 m}} \right]$$

$$+ \frac{1}{6 \hbar^2} (d_{nn} - d_{mm})^2 (\delta E \, \delta E)_{\omega = 0}. \tag{10.2.4}$$

The distinction between the two contributions is even more pronounced here. In the first term one can single out the resonant contribution, which is only determined by the transition between the two levels n, m. Then

$$\gamma_{nm}^{ind} = \frac{1}{3 \hbar^2} |d_{nm}|^2 (\delta E \, \delta E)_{\omega_{nm}}. \tag{10.2.5}$$

Consider now the corresponding expression for the frequency shift. From (10.1.6, 22) with the condition $\omega - \omega_{nm} \ll k v_T$ we find

$$(\Delta \omega_{nm})^{ind} = \frac{1}{3 \hbar^2} \int \frac{dk}{(2\pi)^3} \oint \frac{d\omega}{2\pi} \left\{ \left(\sum_{n_1 \neq n} |d_{nn_1}|^2 \frac{(\delta E \, \delta E)_{\omega, k}}{\omega_{nn_1} + k V - \omega} \right. \right.$$

$$+ \sum_{n_1 \neq m} |d_{n_1 m}|^2 \frac{(\delta E \, \delta E)_{\omega, k}}{\omega_{n_1 m} + k V - \omega} \Bigg)$$

$$+ \left. [|d_{nn}|^2 + |d_{mm}|^2 - (2 d_{nn} d_{mm})] \frac{(\delta E \, \delta E)_{\omega, k}}{k V - \omega} \right\}. \tag{10.2.6}$$

Hence it follows that (the Doppler effect not taken into account)

$$(\Delta \omega_{nm})^{ind} = \frac{1}{3 \hbar^2} \oint \frac{d\omega}{2\pi} \left(\sum_{n_1 \neq n} |d_{nn_1}|^2 \frac{(\delta E \, \delta E)_{\omega}}{\omega_{nn_1} - \omega} + \sum_{n_1 \neq n} |d_{n_1 m}|^2 \frac{(\delta E \, \delta E)_{\omega}}{\omega_{n_1 m} - \omega} \right). \tag{10.2.7}$$

The magnitude of the frequency shift can be presented as a sum of changes of the energy levels themselves. Then, for instance, from (10.2.7) we get the following expression for ΔE_n,

$$\Delta E_n = \frac{1}{3 \hbar} \sum_{n = n_1} \oint \frac{d\omega}{2\pi} |d_{nn_1}|^2 \frac{(\delta E \, \delta E)_{\omega}}{\omega_{nn_1} - \omega}. \tag{10.2.8}$$

This expression describes the quadratic Stark effect due to the fluctuationals field. From here one can also obtain the expression for the Stark effect in a monochromatic field. In this case

$$(\delta E \, \delta E)_\omega = \tfrac{1}{2} |E_{\omega_0}|^2 2\pi [\delta(\omega - \omega_0) + \delta(\omega + \omega_0)] \, , \qquad (10.2.9)$$

with (10.2.8) becoming

$$\Delta E_n = \frac{1}{6\hbar} \sum_{n \neq n_1} |d_{nn_1}|^2 \left(\frac{1}{\omega_{nn_1} - \omega_0} + \frac{1}{\omega_{nn_1} + \omega_0} \right) |E_{\omega_0}|^2 \, . \qquad (10.2.10)$$

This expression coincides with [Ref. 10.2, Eq. (2.8, 70)].

The spectral density of field fluctuations in the above equations corresponds to fluctuations of both the longitudinal and transverse fields. In order to separate them one can present the spectral density of field fluctuations in the form of (10.1.1). The expressions for the spectral densities which enter this equation were obtained in Sect. 8.4 for a Coulomb partially ionized plasma and in Sect. 9.2 for transverse field fluctuations.

10.3 The Influence of the Source Fluctuations on Linewidth and Frequency Shift

The second term in the collision integral (10.1.16), I_{nm}^{source}, is determined by the correlation $\langle \delta N_{nm}^{\text{source}} \delta E \rangle$ (10.1.15, 17). In order to calculate this correlation, one must express the field fluctuations through the function $\delta N_{n_1 m_1}^{\text{source}}$ with the aid of the field equations, and then employ(2.1.14). Since the equations for longitudinal and transverse field fluctuations are different, the corresponding contributions to the collision integral I_{nm}^{source} will also be different.

Let us discuss the scheme for calculating the contribution of the transverse field fluctuations. The equation for $\delta E^{\perp}(\omega, k)$ has the form

$$\delta E^{\perp}(\omega, k) = \frac{\omega^2 4\pi [\delta P^{\perp}(\omega, k)]^{\text{source}}}{\omega^2 \varepsilon^{\perp}(\omega, k) - c^2 k^2} \, . \qquad (10.3.1)$$

The function $\delta P^{\perp}(\omega, k)$ entering this equation is connected with $\delta N_{n_1 m_1}^{\text{source}}$,

$$[\delta P^{\perp}(\omega, k)]^{\text{source}} = \sum_{n_1 m_1} \int d_{m_1 n_1}^{\perp} \delta N_{n_1 m_1}^{\text{source}}(\omega, k, P) \frac{V}{(2\pi\hbar)^3} dP \, . \qquad (10.3.2)$$

With the aid of (10.3.1, 2) and the expression for the fluctuations' density (10.1.14), we find the spectral density, which determines the collision integral I_{nm}^{source}. Using the same approximations as in Sect. 10.2, we get

$$I_{nm}^{\text{source}}(R, P, t) = - i \frac{4\pi}{3\hbar} \sum_{n_1} \int \frac{dk}{(2\pi)^3} [\,|d_{nn_1}|^2 B(\omega_{nn_1} + kV, k)$$

$$- |d_{n_1 m}|^2 B(\omega_{n_1 m} + kV, k)] f_{nm}(R, P, t), \tag{10.3.3}$$

with

$$B(\omega, k) = \frac{1}{2\varepsilon^{\|}(\omega, k)} + \frac{\omega^2}{\omega^2 \varepsilon^{\perp}(\omega, k) - c^2 k^2}. \tag{10.3.4}$$

The first term corresponds to the contribution of the longitudinal field, and the second to that of the transverse field.

In (10.3.3, 4) the contributions determined by the diagonal elements of the matrix, d_{nn} and d_{mm}, were omitted. They depend upon the values of the function $B(\omega, k)$ at low frequencies [cf. (10.2.2, 4)].

From juxtaposing (10.1.6, 3.3) we get the expressions which determine the contributions of the source fluctuations in the formulae for γ_{nm} and $\Delta\omega_{nm}$. In zero approximation with respect to kV (the Doppler effect not taken into account) the expression for $\gamma_{nm}^{\text{source}}$ has the form

$$\gamma_{nm}^{\text{source}} = \frac{1}{6\hbar^2} \sum_{n_1} \int \frac{dk}{(2\pi)^3} \left[|d_{nn_1}|^2 \left(\frac{4\pi\hbar \, \text{Im}\{\varepsilon^{\|}(\omega_{nn_1}, k)\}}{|\varepsilon^{\|}(\omega_{nn_1}, k)|^2} \right. \right.$$

$$\left. \left. + \frac{8\pi\hbar\omega_{nn_1}^2 \, \text{Im}\{\varepsilon^{\perp}(\omega_{nn_1}, k)\}}{|\omega_{nn_1}^2 \varepsilon^{\perp}(\omega_{nn_1}, k) - c^2 k^2|^2} \right) - (nn_1 \to n_1 m) \right]. \tag{10.3.5}$$

The corresponding expression for the frequency shift is as easily obtained.

Now we have the expressions for the collision integrals in the kinetic equations for the distribution functions. From the structure of (10.1.22, 3.3) it is evident that the collision integral I_{nm} for partially ionized plasma consists of four parts. At the same time, the collision integral I_n consists of eight parts (Sect. 8.6 and Chap. 9). This difference is due to the fact that the integral I_n account both for the processes where the number of atoms is conserved (processes 1 to 4 on p. 212 in Sect. 8.6), and for processes where the number of atoms is altered (processes 5 to 8 on p. 212). The collision integral I_{nm}, however, does not change the number of atoms since it only determines the transitions from one state to another in the discrete spectrum. The transitions from the discrete spectrum levels and back are described by the collision integrals $I_{np''}$ and $I_{p'm}$. The expressions for these can be obtained following the same scheme, with the condition (8.3.15).

Let us analyze the correspondence between the first part of the collision integral I_n (the expression $[I_n]_1$) and the dissipative part of the collision integral I_{nm}. Finding their relationship will enable us to obtain the expressions for the collision integrals $I_{np''}$ and $I_{p'm}$ with the aid of the formula for the integral $[I_n]_2$, skipping time-consuming calculations.

10.4 The Probabilities of the Transition. The Broadening at Spontaneous and Induced Processes

For the case of a Coulomb plasma the expression for the collision integral $[I_n]_1$ follows from (8.5.1) after the substitutions $\alpha \to n$, $\beta \to m$. The corresponding expression for $[I_n]_1$, determined by the transverse field fluctuations, follows from (9.3.1) after the same substitution. In the dipole approximation it can be presented in the form of (9.5.2). Recall that at the conditions given in (9.5.3) the collision operator does not depend on momenta, so a simpler expression (9.5.4) can be used in place of (9.5.2).

Carrying out similar transformations in the collision integral $[I_n]_1$ for a Coulomb plasma results in

$$[I_n(P,t)]_1 = \frac{1}{3\hbar^2} \sum_m \int \frac{dk}{(2\pi)^3} |d_{nm}|^2 \left\{ (\delta E^{\parallel} \delta E^{\parallel})_{\omega_{nm},k} [f_m(P,t) - f_n(P,t)] \right.$$

$$\left. - \frac{4\pi\hbar \, \text{Im}\{\varepsilon^{\parallel}(\omega,k)\}}{|\varepsilon^{\parallel}(\omega,k)|^2} [f_m(P,t) - f_n(P,t)] \right\}. \tag{10.4.1}$$

The superscript $\parallel$ again denotes longitudinal field.

Let us present the sum of integrals (9.5.4, 10.4.1) in the form

$$[I_n]_1 = \sum_m [W_{mn} f_m(P,t) - W_{nm} f_n(P,t)] . \tag{10.4.2}$$

The probabilities of the transition introduced here can be split into two parts,

$$W_{nm} = W_{nm}^{\parallel} + W_{nm}^{\perp} . \tag{10.4.3}$$

By comparing (10.4.2) and (9.5.4, 10.4.1) we find the expressions for the probabilities of the transition,

$$W_{nm}^{\parallel} = \frac{1}{3\hbar^2} \int \frac{dk}{(2\pi)^3} |d_{nm}|^2 \left[(\delta E^{\parallel} \delta E^{\parallel})_{\omega_{nm},k} + \frac{4\pi\hbar \, \text{Im}\{\varepsilon^{\parallel}(\omega_{nm},k)\}}{|\varepsilon^{\parallel}(\omega_{nm},k)|^2} \right], \tag{10.4.4}$$

$$W_{nm}^{\perp} = \frac{1}{3\hbar^2} \int \frac{dk}{(2\pi)^3} |d_{nm}|^2 \left[(\delta E^{\perp} \delta E^{\perp})_{\omega_{nm},k} \right.$$

$$\left. + \frac{8\pi\hbar\omega_{nm}^4 \, \text{Im}\{\varepsilon^{\perp}(\omega,k)\}}{|\omega_{nm}^2 \varepsilon^{\perp}(\omega_{nm},k) - c^2 k^2|^2} \right]. \tag{10.4.5}$$

From these formulae and (10.2.4, 3.5) [dropping the contribution determined by the diagonal elements d_{nn} and d_{mm} in (10.2.4)] one can see the connection between the dissipative matrix γ_{nm} and the probabilities of the transition,

$$\gamma_{nm} = \frac{1}{2} \sum_{n_1} (W_{nn_1} + W_{mn_1}) .$$

(10.4.6)

Here we take into account that $\text{Im}\{\varepsilon(\omega, k)\} = - \text{Im}\{\varepsilon(-\omega, -k)\}$.

For partially ionized plasma the probability of the transition can be presented as a sum of four terms. Distinguishing these terms is possible by virtue of the presentation of the spectral density of field fluctuations and the polarizability in the form of (8.3.5, 4.5). The contribution which falls into the transparency region calls for special attention. Here we present the corresponding expression for the probabilities of the transition W_{nm}. It follows from (9.5.6, 10.4.2) and has the form [10.7]

$$W_{nm} = \frac{|d_{nm}|^2}{3 \hbar^2} \left[(\delta E^{\perp} \delta E^{\perp})_{\omega_{n,m}} + \frac{2 \hbar \omega_{n,m}^3}{c^3} \sqrt{\text{Re}\{\varepsilon(\omega_{nm})\}} \right] .$$

(10.4.7a)

For the state of equilibrium the right-hand side of this equation is conveniently expressed via Einstein's coefficients (9.5.9). Taking (9.5.7) into account we get

$$W_{nm} = B_m^n \rho_{\omega_{nm}}^{T} + A_m^n .$$

(10.4.7b)

When using this expression one must be aware of the properties of Einstein coefficients,

$$B_m^n = B_n^m , \quad A_m^n = - A_n^m .$$

(10.4.8)

At zero temperature (10.4.7b) becomes

$$W_{nm} = \begin{cases} A_m^n , & n > m , \\ 0 , & n < m . \end{cases}$$

(10.4.9)

Using this result, we find from (10.4.6),

$$\gamma_{nm} = \frac{1}{2} \left(\sum_{n > n_1} A_{n_1}^n + \sum_{m > n_1} A_{n_1}^m \right) .$$

(10.4.10)

Hence follows the expression for the resonant transition,

$$\gamma_{nm} = \tfrac{1}{2} W_{nm} = \tfrac{1}{2} A_m^n .$$

(10.4.11)

Equations (10.4.10, 11) are well known [Ref. 10.2, Eq. (35.5)].

The ad hoc expressions for probabilities of the transition employed here correspond to the collision integrals (9.5.4, 10.4.1). They describe the transitions between different states in a discrete spectrum.

The most general expressions for the probabilities of a transition in partially ionized plasma correspond to the collision integrals (8.5.1, 9.3.1). Let us present the sum of these integrals,

$$I_\alpha(P', t) = \sum_\beta \int \frac{V}{(2\pi\hbar)^3} dP'' [W_{\beta\alpha}(P'', P', t) f_\beta(P'', t)$$

$$- W_{\alpha\beta}(P', P'', t) f_\alpha(P', t)] . \tag{10.4.12}$$

The probabilities of the transition $W_{\alpha\beta}$ here can also be presented in the form

$$W_{\alpha\beta}(P', P'', t) = W_{\alpha\beta}^{\parallel} + W_{\alpha\beta}^{\perp} . \tag{10.4.13}$$

The probabilities of the transition for the Coulomb plasma can be found from a juxtaposition of (8.5.1, 10.4.12),

$$W_{\alpha\beta}(P', P'', t) = \frac{1}{(2\pi)^3\hbar} \int d\omega \, dk \, \frac{|P_{\alpha\beta}(k)|^2}{k^2} \delta(\hbar k - (P' - P''))$$

$$\times \delta(\hbar\omega - (E_\alpha + E_{P'} - E_\beta - E_{P''})) \left[(\delta E^{\parallel} \delta E^{\parallel})_{\omega, k} + \frac{4\pi\hbar \, \text{Im}\{\varepsilon^{\parallel}(\omega, k)\}}{|\varepsilon(\omega, k)|^2} \right] . \tag{10.4.14}$$

The expression for the probabilities of the transition $W_{\alpha\beta}^{\perp}$ follows from a comparison of (9.3.1, 10.4.12). In (10.4.14) the spectral density of fluctuations and dielectric permittivity are given by (8.3.3, 4.1), and the matrix element $P_{\alpha\beta}(k)$ by (7.2.19).

Since each of the indices α and β may correspond either to continuous or to discrete spectrum, (10.4.14) determines four probabilities of transition,

$$W_{nm}(P', P''), \qquad W_{np''}(P', P''),$$
$$W_{p'm}(P', P''), \qquad W_{p'p''}(P', P''), \tag{10.4.15}$$

each of which consists of four parts (8.3.5, 4.5).

The dissipative density matrix for partially ionized plasma is given by a formula similar to (10.4.6),

$$\gamma_{\alpha\beta}(P', P'') = \frac{1}{2} \sum_\eta \int \frac{V}{(2\pi\hbar)^3} dP [W_{\alpha\eta}(P', P) + W_{\beta\eta}(P'', P)] . \tag{10.4.16}$$

In turn, this expression determines four dissipative matrices: $\gamma_{nm}, \gamma_{np''}, \gamma_{p'm}, \gamma_{p'p''}$. Naturally, the relationship between these matrices and the corresponding collision integrals is more complex than that given by (10.1.6).

Notice that all the above formulae for probabilities of transition and dissipative matrices also hold for nonequilibrium states, except (10.4.7b) which is true only for the state of equilibrium. The time dependence in the transparency region is determined by the temporal trend of the spectral density of field fluctuations.

In the collision region, however, where ω and k are not connected via dispersional equations (10.1.2), the temporal dependence enters the calculations through the distribution functions of electrons, ions, and atoms. These functions satisfy the kinetic equations discussed in Chaps. 8, 9. Thus it is possible to employ the solutions of the kinetic equations to calculate atoms' emission line broadening for inequilibrium plasma, as well. Naturally, this task is very complicated in general; a way to succed is to employ model collision integrals.

10.5 Spectral Line Broadening by a Plasma's Electrons

Let us now consider examples of spectral line broadening due to collisions of atoms with charged particles in a plasma. The broadening here will depend upon the ratio of the concentrations of charged particles and atoms.

We shall discuss some examples of applications of the above equations to spectral linewidth, starting with emission line broadening due to electrons in a plasma [10.7].

We can examine the role of fast field fluctuations by employing the simplest equation (10.2.5), which only accounts for the transition between two chosen levels n and m. The calculation will be carried out under the following assumptions.

1) The degree of ionization is high, so when calculating the spectral density we shall take only the contribution given by free charged particles into account in (10.2.5). For the same reason the permittivity in (8.4.6) is determined only by free charged particles.

2) The distribution of free charged particles is in equilibrium. Then the expression (8.4.6) for the spectral density of field fluctuations can be taken in the form of (8.4.2); for the imaginary part of permittivity one may employ the expression

$$\text{Im}\{\varepsilon(\omega, k)\} = -\sum_a \frac{4\pi^2 e_a^2 n_a}{\hbar k^2} \int \frac{V}{(2\pi\hbar)^3} \, dp \, [f_a(p + \tfrac{1}{2}\hbar k) - f_a(p - \tfrac{1}{2}\hbar k)] \,.$$

$$(10.5.1)$$

3) The function $(|\varepsilon(\omega, k)|)^{-2}$, which accounts for the dynamic polarization of the medium in (8.4.2, 6), is replaced by its average over ω. This corresponds to employing the interaction between charged particles in calculating the correlations of the effective potential [Ref. 10.8, Sect. 47, 55]. Thus the potential of interaction is replaced by the effective potential, which corresponds to the replacement

$$\frac{1}{|\varepsilon(\omega, k)|^2} \to \frac{(\delta E \, \delta E)_k}{(\delta E \, \delta E)_k^0},$$

$$(10.5.2a)$$

where $(\delta E \, \delta E)_k$ is the spatial spectral density of field fluctuations with collective interactions taken into account, and $(\delta E \, \delta E)_k^0$ is the same function but not accounting for the polarization. In particular, it follows from (10.5.2a) for the equilibrium state for free charged particles, when the degree of ionization is close to one, that

$$\frac{1}{|\varepsilon(\omega,k)|^2} \rightarrow \frac{r_D^2 k^2}{1+r_D^2 k^2} . \tag{10.5.2b}$$

Thus, the interaction which determines correlations of charged particles is screened at distances of the order of the Debye radius.

With the above assumptions we get the following expression for the line-width:

$$\gamma_{nm} = \frac{|d_{nm}|^2}{3\hbar^2} \frac{4\pi e^2 n}{(2\pi)^{3/2}} \left(\frac{m_e}{\varkappa T}\right)^{1/2} \left[\exp\left(\frac{\hbar\omega_{nm}}{2\varkappa T}\right) + \exp\left(-\frac{\hbar\omega_{nm}}{2\varkappa T}\right)\right]$$

$$\times \int_0^\infty \exp\left(-\frac{m_e \omega_{nm}^2}{2\varkappa T k^2} - \frac{\hbar^2 k^2}{8 m_e \varkappa T}\right) \frac{4\pi dk}{k[1+(r_D^2 k^2)^{-1}]} . \tag{10.5.3}$$

Here only the contribution of electrons is left ($m = m_e$). In case of ions this approximation holds only for small concentrations of charged particles (see Sect. 10.11). Provided this condition is satisfied, the contribution of ions is given by the same equation with $m_e \rightarrow m_i$.

Notice that the integral in (10.5.3) converges both at large and at small values of k. The actual cutoff of the integration range must be determined from the conditions governing the applicability of the theory. This is either the condition defining where perturbation theory is applicable or the condition governing the pairedness of interactions in collisions between atoms and electrons. The first limit is determined by the Landau length $l_L \sim e^2/\varkappa T$ [see (4.1.4)], and the second by the Weisskopf length for electron–atom interaction $\rho_{W_e} \sim ed/\hbar v_{T_e}$ [see (4.2.4)]. Finally, in (10.5.3) there is also a cutoff at the de Broglie length.

Consequently, the lowest limit (in length) of the integration range in (10.5.3) is determined by the largest of the three values,

$$r_{min} = \max\left(\frac{\hbar}{m_e v_{T_e}}; \ l_L; \ \rho_{W_e}\right). \tag{10.5.4}$$

As a rule, it is the Weisskopf radius.

At large distances (small values of k) the integration limit is determined by the smaller of the two parameters v_{T_e}/ω_{nm} and r_D,

$$r_{max} = \min\left(r_D; \ v_{T_e}/\omega_{nm}\right). \tag{10.5.5}$$

As a result, the integrand in (10.5.3) can be approximated by $4\pi dk/k$, and the integration range restricted by the conditions

$$1/r_{min} > k > 1/r_{max} . \tag{10.5.6}$$

If at the same time we assume that $\hbar\omega_{nm}/\mathscr{K}T \ll 1$ (high temperatures), then

$$\gamma_{nm} = \frac{8\sqrt{2\pi}}{3} \frac{|d_{nm}|^2}{\hbar^2} \frac{e^2 n}{v_{T_e}} \ln\left(\frac{r_{max}}{r_{min}}\right). \tag{10.5.7}$$

This expression describes the broadening due to the high-frequency fluctuations (with $\omega \sim \omega_{nm}$) of the electric field created by the electrons.

The contribution of low-frequency field fluctuations is determined by the second term in (10.2.2). Under the same assumptions we get the following result instead of (10.5.3),

$$\gamma_{nm} = \frac{(d_{nn} - d_{mm})^2}{\sqrt{2\pi}\,\hbar^2} \frac{e^2 n}{v_{T_e}} \int \left\{ \left[\exp\left(\frac{\hbar\omega}{2\mathscr{K}T}\right) + \exp\left(-\frac{\hbar\omega}{2\mathscr{K}T}\right) \right] \right.$$

$$\left. \times \exp\left(-\frac{m\omega^2}{2\mathscr{K}Tk^2} - \frac{\hbar^2 k^2}{8m_e\mathscr{K}T}\right) \frac{4\pi\,dk}{k[1 + (r_D^2 k^2)^{-1}]} \right\}_{\omega = \omega - \omega_{nm}}. \tag{10.5.8}$$

Here we again have to decide upon the position of the limit of the integration range at small distances (large values of k). This limit is determined by the same condition (10.5.4).

However, the condition that determines the effective cutoff at large distances (small values of k) now differs from (10.5.5). Indeed, in place of ω_{nm} we have low frequency $\omega - \omega_{nm}$. This is demonstrated by the conditions under which Doppler broadening can be neglected (low temperatures, $\omega - \omega_{nm} \gg k v_T$). The integrand in (10.5.8) can again be approximatively replaced by $4\pi\,dk/k$. As a result we get, instead of (10.5.7), the following expression:

$$\gamma_{nm} = \frac{4\sqrt{2\pi}}{3} \frac{(d_{nn} - d_{mm})^2}{\hbar^2} \frac{e^2 n}{v_{T_e}} \ln\left(\frac{r_{max}}{r_{min}}\right). \tag{10.5.9}$$

The value of r_{max} is determined by

$$r_{max} = \min\left(\frac{v_{T_e}}{|\omega - \omega_{nm}|}, r_D\right). \tag{10.5.10}$$

It differs from (10.5.5) by the substitution

$$\omega_{nm} \to |\omega - \omega_{nm}|. \tag{10.5.11}$$

The quantity $v_{T_e}/|\omega - \omega_{nm}|$ is called the Lewis parameter. We see that the value of γ_{nm} now depends on the mistuning $|\omega - \omega_{nm}|$. However, this dependence is not strong.

10.6 Resonant Broadening of Spectral Lines Due to Atoms' Collisions

Consider now the spectral line broadening of atoms in another extreme case, i.e., when the degree of ionization is close to zero and the collisional broadening is determined by the collisions of atoms. As in Sect. 10.5, we shall consider the influence of high-frequency ($\omega \sim \omega_{nm}$) and low-frequency field oscillations on the linewidth. The influence of high-frequency oscillations will be evaluated with the aid of (10.2.5), in order to demonstrate the principal tendencies.

To determine the spectral density of field fluctuations (8.4.10) must now be used, instead of (8.4.6). For the state of equilibrium this formula, thanks to a low degree of ionization, can be simplified to (8.4.2).

In calculating γ_{nm}, the polarization effects will be neglected since they are insignificant at a low degree of ionization. Thus, $\varepsilon(\omega, k) = 1$ in (8.4.10). To single out the resonant interaction of atoms we set $\omega = \omega_{nm}$, $n_1 = n$, $m_1 = m$ in (8.4.10). Then from (8.4.2) we get the following expression for γ_{nm}:

$$\gamma_{nm} = \frac{|d_{nm}|^4}{3\hbar^2} \frac{16\pi^3 n}{3\sqrt{\pi} v_T k} (\rho_m + \rho_n) \int \frac{dk}{(2\pi)^3} \exp\left(-\frac{\hbar^2 k^2}{4M^2 v_T^2}\right), \quad v_T = \sqrt{\frac{2 \mathscr{K} T}{M}}$$

$$(10.6.1)$$

where ρ_n and ρ_m are the populations of the corresponding levels.

The formula obtained for γ_{nm} differs markedly from (10.5.3) in the structure of the integral. Indeed, in (10.5.3) the behavior of the integrand closely resembles the behavior of the function $1/k$. The limits of the integration range therefore enter (10.5.7) under the logarithm sign. In (10.6.1) the situation is entirely different.

The value of the integral in (10.6.1) is determined by the behavior of the spectral density of field fluctuations with large values of k (small distances). The integral is cut off in this case at distances of the order of the de Broglie wavelength for atoms $\hbar/Mv_T$, which at $v_T = 10^5$ cm/s is of the order of an atom's size. However, even at much greater distances the condition governing the applicability of perturbation theory, which was actually employed in obtaining (10.6.1), is violated.

Taking this circumstance into account, we restrict the integration range by the condition $k > 1/r_{min}$, where the value of r_{min} is determined by the condition of the applicability of perturbation theory [Ref. 10.9, Sect. 126]. In particular, for dipole – dipole interaction we find

$$r_{min} = \frac{1}{k_{max}} = \left(\frac{2}{3} \frac{|d_{nm}|^2}{\hbar v_T}\right)^{1/2}.$$

$$(10.6.2)$$

This value is of the order of the Weisskopf radius (4.2.4). As a result we get from (10.6.1) the following expression for γ_{nm}:

$$\gamma_{nm} = \frac{2\sqrt{\pi}}{3} \frac{|d_{nm}|^2}{\hbar} n(\rho_n + \rho_m).$$

$$(10.6.3)$$

If level m is the ground level, we get for the linewidth (γ_{nm} is the line halfwidth)

$$2\gamma_{nm} = \frac{4\sqrt{\pi}}{3\hbar}|d_{nm}|^2 n = 2\sqrt{\pi}\,\frac{e^2 n}{m\,\omega_{nm}}f_{nm}, \tag{10.6.4}$$

where f_{nm} is the oscillator strength. This only differs from the well-known result obtained by *Vlasov* and *Fursov* [10.10] in numeric factor ($\sqrt{\pi}$ instead of $\pi/3$). The numerical factor here, in contrast to the equations in the preceding section, displays a strong dependence upon the value of $r_{\min}$. Therefore the contribution of interactions at small distances $r < r_{\min}$ must be accounted for. This contribution can be estimated with the aid of basic kinetic theory, giving

$$2\gamma_{nm} = \frac{1}{\tau_{\text{col}}} = \pi r_{\min}^2 n v_{\text{T}} = \frac{2\pi}{3}\,\frac{|d_{nm}|^2}{\hbar}\,n, \tag{10.6.5}$$

where τ_{col} is the average time between two successive collisions of spheres with cross section $\pi r_{\min}^2$.

We see that both contributions (10.6.4, 5) are of the same order. Adding together the two results, we get the final expression,

$$2\gamma_{nm} = (\pi + 2\sqrt{\pi})\,\frac{2}{3}\,\frac{|d_{nm}|^2}{\hbar}\,n = (\pi + 2\sqrt{\pi})\,\frac{e^2 n}{m\,\omega_{nm}}f_{nm}. \tag{10.6.6}$$

Various methods have been employed to calculate resonant broadening [10.11 – 13], and the results obtained differ only in the numeric coefficient, which varies from 5 to 7.5.

We have thus investigated three processes of atoms' spectral line broadening in partially ionized plasma. Let us compare their contributions. The ratio of the magnitude of broadening from free charged particles and atoms depends strongly on the degree of ionization. If the concentrations of atoms and electrons are about the same order, then the ratio

$$\frac{\gamma_{\text{free}}}{\gamma_{\text{bound}}} \sim \frac{e^2}{\hbar}\sqrt{\frac{m}{\mathscr{K}T}} = \zeta_{\text{B}}, \tag{10.6.7}$$

i.e., is determined by the Born parameter which includes the thermal velocity of electrons. At $T = 4 \cdot 10^4\,$K this ratio is approximately equals to five.

For a plasma with a low degree of ionization, the ratio of broadening due to radiational interaction to broadening by atoms is (10.4.11, 6.6)

$$\frac{\gamma_{\text{rad}}}{\gamma_{\text{bound}}} \sim \frac{\omega_{nm}^2}{nc^3} \sim \frac{1}{\lambda^3 n} = \frac{1}{\varepsilon_{\text{em}}}, \tag{10.6.8}$$

i.e., this ratio is determined by the "optical density parameter" (4.2.18).

Finally, for a plasma with a high degree of ionization, when broadening by electrons is significant, (10.4.11, 5.7)

$$\frac{\gamma_{\text{rad}}}{\gamma_{\text{free}}} \sim \frac{\hbar \omega_{nm}^3}{c^3 e^2 n} \sqrt{\frac{\varkappa T}{m}} = \frac{1}{\varepsilon_{\text{em}} \zeta_{\text{B}}} . \tag{10.6.9}$$

This ratio is thus determined by two dimensionless parameters, the Born parameter ζ_{B} and the optical density parameter (4.2.18).

Finally, let us point out that our calculations of broadening by atoms are only valid provided that the density parameter of dipole – dipole interaction (4.2.8) is small, that is, at the condition

$$\varepsilon_{\text{D}} \sim n\rho_{\text{W}}^3 \sim n r_{\text{min}}^3 \ll 1 . \tag{10.6.10}$$

Otherwise the approximation of paired interaction which has been employed cannot be used.

The theory of broadening in a dense partially ionized plasma can only be developed by using the correlations of the positions of charged particles and atoms. It would be natural to expect that the dependence of γ_{nm} upon concentration is not linear any longer. The influence of correlations on spectral line broadening will be discussed in the next chapter.

In Sects. 10.5, 6 we have investigated the spectral line broadening of atoms in partially ionized plasma for two extreme cases, for ionization close to a hundred per cent and ionization close to zero. For these extremes only one of four terms could be retained in the expression for the spectral density of field fluctuations. In the general case, however, all four terms must be kept. In spectral line broadening the field fluctuations, created by transitions of the particles surrounding the atom from the free state into the bound state, and vice versa, play an important role.

The resonant broadening of spectral lines upon the collisions of atoms is an inelastic process. We shall proceed to analyze the broadening of spectral lines due to elastic collisions. The process which in this case determines the broadening of spectral lines is, in a sense, reciprocal to the process of resonant broadening.

10.7 Spectral Line Broadening upon Elastic Collisions of Atoms

Suppose that the degree of ionization is low. Then of the four processes determining the collision integral $[I_n(P', t)]_1$, only process 4 (p. 212) is left. Accordingly, of four contributions which determine the spectral density of field fluctuations, only the one defined by (8.4.10) needs to be used.

In order to single out the contribution of elastic collisions one must set $n = m$ and $n_1 = m_1$ in the double sum in (8.4.10). With a crisscross replacement $n \to m_1$, $m \to n_1$ we get from (8.4.10) the contribution of resonant interaction (Sect. 10.6). Singling out the elastic processes ($n = m$, $n_1 = m_1$) in the collision integral, obtained from (8.5.2) by substituting $\alpha = n$, $\beta = m$, $\alpha_1 = n_1$, $\beta_1 = m_1$, produces the following expression:

$$I_n(P', t) = \frac{4N}{V} \sum_{n_1} \int \frac{V}{(2\pi\hbar)^3} dP_1' \, dP_1'' \, dP'' \, d\omega \, dk \frac{|P_{nn}(k)|^2 |P_{n_1 n_1}(k)|^2}{k^4 |\varepsilon(\omega, k)|^2}$$

$$\times \delta(\hbar k - (P' - P'')) \, \delta(\hbar\omega - (E_{P'} - E_{P''})) \, \delta(P' + P_1'' - (P_1' + P''))$$

$$\times \delta(E_{P'} + E_{P_1''} - E_{P''} - E_{P_1'}) [f_{n_1}(P_1') f_n(P'') - f_{n_1}(P_1'') f_n(P')] .$$

$$(10.7.1)$$

This equation equals zero in the state of equilibrium, when the distribution function $f_n(P)$ is the Maxwell – Boltzmann distribution.

In the dipole approximation, (8.3.30) can be used for the matrix elements in (10.7.1). In this approximation the interaction is determined by the diagonal matrix elements d_{nn}. If they are equal to zero, then dipole interaction does not exist.

In order to obtain a more general expression for the integral of elastic collisions in perturbation theory approximation with respect to interactions, one must make the following substitution in (10.7.1):

$$\frac{(4\pi)^2}{k^4} \frac{|P_{nn}(k)|^2 |P_{n_1 n_1}(k)|^2}{|\varepsilon(\omega, k)|^2} \rightarrow |U_{nn \atop n_1 n_1}(k)|^2 \tag{10.7.2}$$

where

$$U_{nn \atop n_1 n_1}(k) = \int U(R - R_1, r, r_1) |\Psi_n(r)|^2 |\Psi_{n_1}(r_1)|^2$$

$$\times \exp\left[-ik(R - R_1)\right] dr \, dr_1 \, d(R - R_1) . \tag{10.7.3}$$

Here $U(R - R_1, r, r_1)$ is the potential of interaction of the atoms at points R, R_1. Having carried out (10.7.2) and integrating over ω, k, we get

$$[I_n(P', t)] = \frac{N}{V} \frac{1}{(2\pi)^2 \hbar^4} \sum_{n_1} \int \frac{V}{(2\pi\hbar)^3} dP_1' \, dP_1'' \, dP'' |U_{nn \atop n_1 n_1}(k)|^2_{k=(P'-P'')/\hbar}$$

$$\times \delta(P' + P_1'' - (P_1' + P'')) \, \delta(E_{P'} + E_{P_1''} - E_{P''} - E_{P_1'})$$

$$\times [f_{n_1}(P_1') f_n(P'') - f_{n_1}(P_1'') f_n(P')] . \tag{10.7.4}$$

If the potential of interaction does not depend upon the internal state of the atoms, then

$$U_{nn \atop n_1 n_1}(k) = U(k) . \tag{10.7.5}$$

Using this equation, we derive from (10.7.4) the expression giving the collision integral for a system of structureless atoms. It is connected with the integral $I_n(P', t)$ through the equation

$$I(P, t) = \sum_n I_n(P, t); \quad f(P, t) = \sum_n f_n(P, t) . \tag{10.7.6}$$

For a single-component gas the expression for the collision integral $I(P, t)$ coincides with [Ref. 10.8, Eq. (69.11)].

In order to determine the influence of elastic collisions of atoms on the spectral linewidth, the corresponding expression for the collision integral I_{nm} at $n \neq m$ must be determined.

Following the method in [10.14, 15], the collision integral $I_n(P', t)$ can be written in the form

$$I_n(P, t) = \int A_n(P', P'', t) f_n(P'', t) \frac{V}{(2\pi\hbar)^3} dP'' - v_n(P', t) f_n(P', t) , \quad (10.7.7)$$

where $v_n(P', t)$ is the frequency of elastic collisions of atoms in the state nP'. The expressions for functions $A_n(P', P'')$ and $v_n(P')$ can be found by comparing (10.7.4, 7). Naturally, they are dependent on the distribution functions.

The collision integral I_{nm} in the equation for the function f_{nm} can be expressed with these same functions. To demonstrate this we can start from the expression which connects the collision integral I_{nm} with the two-particle correlation function $g_{nm_1 n_1 m}(P'_1 P''_1, P'', P'_1, t)$,

$$I_{nm}(P', t) = -\frac{i}{\hbar} \frac{N}{V} \sum_{l n_1 m_1} \int \left[\frac{V}{(2\pi\hbar)^3} \right]^2 dP'' \, dP'_1 \, dP''_1$$

$$\times \delta(P' + P''_1 - P'' - P'_1) U_{nl}_{m_1 n_1}(k) g_{ln_1 mm_1}(P'', P'_1 P', P''_1, t)$$

$$- g_{nm_1 ln_1}(P', P''_1, P'', P'_1, t) U_{lm}_{n_1 m_1}(k)]_{k=(P'-P'')/\hbar} . \quad (10.7.8)$$

Instead of the correlation function used in the perturbation theory approximation one may take another equation [Ref. 10.8, Eq. (81.8)]. As a result we get for the collision integral $I_{nm}(P', t)$ (with only the elastic collision accounted for),

$$I_{nm}(P', t) = \int A_{nm}(P'', P', t) f_{nm}(P'', t) \frac{V}{(2\pi\hbar)^3} dP'' - v_{nm}(P', t) f_{nm}(P', t) . \quad (10.7.9)$$

The functions $A_{nm}(P'', P')$, $v_{nm}(P')$ are connected with $A_n(P', P'')$, $v_n(P')$ via equations similar to (10.4.6, 10),

$$A_{mn}(P'', P', t) = \tfrac{1}{2}[A_n(P'', P', t) + A_m(P', P'', t)] ,$$

$$v_{nm}(P', t) = \tfrac{1}{2}[v_n(P', t) + v_m(P', t)] . \quad (10.7.10)$$

The role that elastic collisions play depends significantly on the relative values of the characteristic frequencies: the Doppler width $k v_T$, the rate of elastic collisions v, the linewidth determined by radiational friction, and collisions which directly influence the emission process γ (see Sects. 10.5, 6).

In the above discussion we have assumed that the distribution over velocities is the Maxwell distribution. This is justified if the frequency of collisions v is much greater than γ, i.e., $\gamma \ll v$.

When $k v_T \gg \gamma$, the linewidth (after averaging over velocities) is not determined by γ, but by the Doppler width. In this case, for a nonuniformly broadened line the change in atom's velocity due to elastic collisions is not taken into account in the approximation of preset velocity distribution. This is justified when $v \ll k v_T$.

Thus, the velocity distribution can be assumed to be set in advance in calculating the width and form of spectral lines provided that two conditions are satisfied,

$$k v_T \gg v \gg \gamma . \tag{10.7.11}$$

We conclude that the change in the form of spectral lines which accompanies an increase in pressure is brought about not only by a change in γ, but also by a change in the relative values of $k v_T$ and v, that is, due to the change in the atom's velocity during emission.

Since the elastic collision integral is very complicated, relatively simple model integrals are usually employed in calculations. For instance, the so-called model of strong collisions is used in place of (10.7.7) [10.3, 16],

$$I_n(P, t) = v f(P) \int f_n(P', t) \frac{V}{(2 \pi \hbar)^3} dP' - v f_n(P, t) , \tag{10.7.12}$$

where v is the effective collisions rate, and $f(P)$ is the Maxwell distribution. Another example is the model of weak interactions [10.3, 14, 16, 17]. In this case the Fokker–Planck operator is the collision operator (4.8.10, 5.15.5), and the corresponding collision integral has the form

$$I_n(P, t) = D \frac{\partial^2 f_n(P, t)}{\partial P^2} + \frac{\partial}{\partial P} (v P f_n(P, t)) , \quad D = v M \varkappa T . \tag{10.7.13}$$

The calculations carried out in [10.3, 14, 16, 17] show that the linewidth decreases with an increase in pressure. The form changes also, due to the change in the atom's velocity at elastic collisions. For example, the Doppler width in the model of weak interactions is

$$k v_T \to \frac{k v_T}{1 + \dfrac{2}{3 \sqrt{\pi}} \dfrac{v}{k v_T}} \quad \text{at} \quad v \ll k v_T ; \quad k v_T \to k v_T \frac{k v_T}{v} \quad \text{at} \quad v \gg k v_T . \tag{10.7.14}$$

Qualitatively the picture does not depend upon the choice of elastic collision integral model. Naturally, in order to reveal the general dependence of the width and form of the line on pressure, one must also take the influence of collisions on γ into account.

10.8 Radiation Capture (Imprisonment of Radiation)

In Sect. 10.6 we calculated the spectral line broadening of atom radiation due to the resonant transmission of radiation when similar particles collide. Naturally, this broadening is proportional to the gas pressure. According to (10.6.8), the ratio of $\gamma_{res} \equiv \gamma_{bound}$ to the radiation broadening γ_{rad} is proportional to the optical density parameter $n\lambda^3$, where n is the concentration of atoms.

Let us now consider another effect of pressure, the so-called capture of radiation, which plays an important role in the theory of gas lasers. In the construction of the theory of optical quantum generators (Chaps. 12, 13) it is convenient to single out the two operational levels from the very beginning. We denote them by the indices a and b. Laser radiation is created in the transition $a \rightarrow b$.

Suppose that the upper operating level a is connected with the ground state o by an allowed optical transition. The possible processes are spontaneous emission in the transition $a \rightarrow o$, and resonant absorption of emitted photons (transitions $o \rightarrow a$). The latter results in the transition of atoms back into the excited state a. Naturally, this process of resonant absorption prevents spontaneous radiation from leaving the laser cavity. One could say that a capture of spontaneous radiation occurs.

The extent of this capture depends, of course, on the concentration of atoms in the ground state, and therefore on pressure. If l_o is the free path of a photon, emitted in the transition $a \rightarrow o$, before it is resonantly absorbed in the transition $o \rightarrow a$, then the capture is complete when $l_o \ll L$, where L is the characteristic size of the system.

We shall see that

$$l_o \sim v_o v_T, \quad v_o \sim n_o \lambda_{ao}^3 \gamma_{ao}. \tag{10.8.1}$$

Here n_o is the concentration of atoms in the ground state, λ_{ao} is the wavelength of radiation emitted in the transition $a \rightarrow o$, γ_{ao} is the corresponding radiation broadening, and v_o is the corresponding "collision rate" of photons.

We shall also see that during reabsorption of resonant photons (capture of radiation) changes occur in the function of the distribution of atoms over velocities. Consequently, the process of radiation capture results in the establishment of the Maxwell distribution.

The theory of radiation capture was first developed as a phenomenological theory [10.18 – 20]. The relevant quantum theory was developed in comparatively recent times [10.21 – 24]. Here we shall demonstrate that the capture of radiation is one of the pressure effects described by the collision integral (9.5.2).

Let us again consider the kinetic equation for the distribution function of atoms $f_n(P, t)$. The collision integral is given by (9.5.2), which, as we have seen, can be employed in different approximations. For example, with the condition given in (9.5.3), i.e., neglecting the recoil effect and the Doppler effect, one may use a much simpler expression for the collision integral (9.5.4) in place of (9.5.2). It is then possible to single out the contribution which falls into the transparency region; it is given by (9.5.6, 10).

The principal contribution to the capture of radiation naturally comes from the transparency region, since the resonant reabsorption of radiation takes place there. We shall see, however, that the Doppler effect plays a significant role. Therefore, the contribution from the transparency region should now be singled out in the initial expression (9.5.2), and not just in (9.5.4).

Simplifications commonly employed in the theory of gas lasers are:

1) In the initial kinetic equation we single out the operating levels. For that purpose we set $n = a$; b in (9.5.2).

2) The Doppler and recoil effects are taken into account only for the transitions from the operating levels to the ground level o.

3) The induced transitions to the operating levels from any level other than ground one are neglected. In other words, we take only spontaneous radiation from these transitions into account. The corresponding dissipative coefficients are denoted by $\gamma'_{a,b}$; thus

$$\gamma'_a = \sum_{a>m>o} A^a_m .$$ (10.8.2)

With these assumptions we find, instead of (9.5.2), the following expression for the collision integral in the kinetic equation for the function $f_a(P, t)$:

$$I_a(P, t) = -\gamma'_a (f_a - f_a^{(0)}) + \frac{1}{3\hbar^2} \int d\omega \frac{dk}{(2\pi)^3} \delta(\omega - kV - \omega_{ao}) |d_{ao}|^2$$

$$\times \{(\delta E \, \delta E)_{\omega,k} [f_o(P - \tfrac{1}{2}\hbar k) - f_a(P + \tfrac{1}{2}\hbar k)]$$

$$- A(\omega, k) [f_o(P - \tfrac{1}{2}\hbar k) + f_a(P + \tfrac{1}{2}\hbar k)]\},$$ (10.8.3)

the function $A(\omega, k)$ being determined by (9.3.2).

The pumping introduced into (10.8.3) creates the inverse population of the operating levels; $f_a^{(0)}$ is a preset function. The integral term in (10.8.3) is responsible for the transition $a \to o$. In it we single out the contribution from the transparency region, taking the Doppler effect into account as well. By analogy with (9.5.8) we can introduce the function

$$4\pi^2 \rho(\omega, k) = (\delta E^\perp \delta E^\perp)_{\omega,k} - A(\omega, k) .$$ (10.8.4)

The integral term in (10.8.3) can then be written

$$\frac{4\pi^2}{3\hbar^2} \int d\omega \frac{dk}{(2\pi)^3} |d_{ao}|^2 \delta(\omega - kV - \omega_{ao})$$

$$\times \left\{ \rho(\omega, k) [f_o(P - \tfrac{1}{2}\hbar k) - f_a(P + \tfrac{1}{2}\hbar k)] \right.$$

$$\left. - \frac{1}{2\pi^2} A(\omega, k) f_a(P + \tfrac{1}{2}\hbar k) \right\}.$$ (10.8.5)

We can make further simplifications.

4) We take into account that the populations of levels a and b are much lower than that of the ground level, i.e., f_a, $f_b \ll f_o$.

5) In the term describing spontaneous emission both Doppler and recoil effects can be neglected (9.5.3), and the contribution from the transparency region can be singled out. As a result we get

$$-A_o^a f_a(P, t),\tag{10.8.6}$$

where A_o^a is the Einstein coefficient (9.5.9) for the transition $a \rightarrow o$.

6) In the term describing the induced emission we neglect the recoil effect [by virtue of the first inequality of (9.5.3)]. The Doppler effect at resonant absorption, as we shall see later, is important.

With these assumptions we can derive (10.8.3, 5, 6) the following expression for the collision integral:

$$I_a(P, t) = -\gamma_a'(f_a - f_a^{(0)}) - A_o^a f_a$$

$$+ \frac{1}{6\pi\hbar^2} \int d\omega\, dk\, |d_{ao}|^2 \rho(\omega, k)\, \delta(\omega - kV - \omega_{ao})\, f_o(P). \tag{10.8.7}$$

Now we have to make similar simplifications for the function $\rho(\omega, k)$ in (10.8.4). We substitute the expression for the spectral density of field fluctuations (9.2.4) into (10.8.4), keeping at the same time only the contribution from the transition $a \rightarrow o$. In the same approximation we also write the expression for the imaginary part of dielectric permittivity (9.1.7). As a result, neglecting the recoil effect, we get

$$\rho(\omega, k)$$

$$= \frac{16\pi}{3} \frac{N}{V} |d_{ao}|^2 \frac{\omega^4}{|\omega^2\varepsilon - c^2 k^2|^2} \int \frac{V}{(2\pi\hbar)^3} dP'\, \delta(\omega - kV' - \omega_{ao})\, f_a(P', t). \tag{10.8.8}$$

Note that only the distribution function for the state "a" is included. The contributions determined by the distribution function of the ground state have dropped out.

Finally, we single out the contribution from the transparency region in (10.8.8). By multiplying and dividing (10.8.8) by $\mathrm{Im}\{\varepsilon(\omega, k)\}$, and employing (9.4.1), which separates out of a more general expression (9.3.2), the contribution from the transparency region, so we get instead of (10.8.8) (at $\omega > 0$)

$$\rho(\omega, k)$$

$$= \frac{16\pi^2}{3} \frac{N}{V} |d_{ao}|^2 \frac{\omega^2 \delta(\omega^2 \mathrm{Re}\{\varepsilon\} - c^2 k^2)}{\mathrm{Im}\{\varepsilon(\omega, k)\}} \int \frac{V\, dP'}{(2\pi\hbar)^3} \delta(\omega - kV' - \omega_{ao})\, f_a(P', t). \tag{10.8.9}$$

Here (in the employed approximation)

$$\mathrm{Im}\{\varepsilon(\omega, k)\} = \frac{4\pi^2}{3\hbar} \frac{N}{V} |d_{ao}|^2 \int \frac{V dP''}{(2\pi\hbar)^3} \delta(\omega - kV'' - \omega_{ao}) f_o(P'') . \quad (10.8.10)$$

Now we substitute these expressions into (10.8.7), getting as a result

$$I_a(P, t) = -\gamma_a' (f_a - f_a^{(0)}) + A_o^a \left[\frac{1}{4\pi} \int d\Omega K(k_o V, k_o) f_o(P) - f_a(P, t) \right]. \quad (10.8.11)$$

Here we have introduced the following designation for the kernel of the integral equation,

$$K(k_o V, k_o) = \frac{\int \delta(k_o V - k_o V') f_a(P', t) \dfrac{V dP'}{(2\pi\hbar)^3}}{\int \delta(k_o V - k_o V'') f_o(P'') \dfrac{V dP''}{(2\pi\hbar)^3}} , \quad (10.8.12)$$

where $k_o = k/|k|$ is the unit vector and $d\Omega$ is the element of the solid angle in the space of wave numbers.

Thus, the distribution function $f_a(P, t)$ [for a preset distribution function $f_o(P)$ of atoms in the ground state] is determined by the solution of a linear integral equation. This equation corresponds to [Ref. 10.21, Eq. (17)], provided that the summation in that equation is carried out over Zeeman sublevels.

The radiation capture process leads to redistribution of atoms over velocities, which may result in the establishment of the Maxwell distribution. Indeed, the substitution of the Maxwell distributions

$$\frac{N}{V} \frac{V}{(2\pi\hbar)^3} f_{a;o}(P) = \frac{n_{a;o}}{(2\pi M \mathscr{K} T)^{3/2}} \exp\left(-\frac{P^2}{2M \mathscr{K} T}\right) \quad (10.8.13)$$

into (10.8.12) after integration over P', P'' gives

$$K = n_a/n_o . \quad (10.8.14)$$

Therefore the part of the collision integral (10.8.11) which is determined by the capture of radiation becomes zero and

$$I_a(P, t) = -\gamma_a'(f_a - f_a^{(0)}) . \quad (10.8.15)$$

In this case the reduction in the population of the operating level "a" caused by spontaneous emission in the transition $a \to o$ is completely compensated for by the increase in population due to the resonant absorption in the transition $o \to a$, i.e., the complete capture of radiation takes place.

Notice that the redistribution over velocities proceeds given the resonance of Doppler shifts (10.8.12).

The influence of the capture of radiation upon processes taking place in gas lasers is dealt with in [10.22, 24].

10.9 The Influence of the Static Electric Field upon the Atomic Emission Spectrum

In Sect. 10.5 we discussed the broadening of spectral lines caused by free charged particles in a plasma. We have shown that the quantity γ_{nm} is determined by the spectral density of field fluctuations (10.2.2, 4). Naturally, we took into consideration only those field fluctuations whose correlation times are much smaller than the lifetime of the transition $n \to m$, i.e.,

$$\tau_{\text{cor}} \ll 1/\gamma_{nm} . \tag{10.9.1}$$

Due to great difference in mass between electrons and ions, their characteristic times τ_{cor} differ significantly; this condition is therefore satisfied for electrons and ions in a differing degree.

The broadening and shift of spectral lines can be determined both by diagonal and by off-diagonal matrix elements of atoms' dipole moments.

From quantum theory it is known (see, e.g., [Ref. 10.9, Sect. 37, 76]; [Ref. 10.25, Chaps. 5, 8] and [Ref. 10.26, Sect. 14]) that atoms possess permanent dipole moments (the diagonal elements d_{nn} are nonzero) only in the case of so-called random degeneration, when energy levels do not depend on the quantum number l. Such is the case, e.g., of hydrogen.

Consider a hydrogen plasma. When the polarization of the plasma has little influence on spectral line broadening (10.5.7), the characteristic times of the field fluctuations which caused the broadening are determined only by the parameters of the "collisions" of charged particles with atoms. The quantity γ_{nm}, given by (10.5.9) via the parameters of collisions, is

$$\gamma_{nm} \sim n \frac{C^2}{v_{\text{T}}} \Lambda \sim n \rho_{\text{W}}^2 v_{\text{T}} \Lambda , \quad C \sim \frac{e^2}{\hbar} \sqrt{(r_{nn} - r_{mm})^2} , \tag{10.9.2}$$

where Λ is the parameter which determines the dependence on the mistuning $\Delta \omega_{nm}$ (10.5.9). We also have introduced the designation

$$\rho_{\text{W}_{e,i}} = C/v_{\text{T}_{e,i}} \tag{10.9.3}$$

for the Weisskopf radius for electrons and ions, respectively. According to (10.9.2), the Weisskopf radius acts as the radius of interaction.

The relevant temporal parameters which characterize fast temporal processes in the collisions of electrons and atoms are

$$\tau_{e,i} \equiv 1/\Omega_{e,i} = \rho_{\text{W}_{e,i}}/v_{\text{T}_{e,i}} . \tag{10.9.4}$$

Since the ratio $\tau_e / \tau_i \sim m_e / m_i$ is small, situations may exist when the inequality (10.9.1) is satisfied for electrons while not being satisfied for ions. In order to single out corresponding regions for the values of concentrations and temperatures, we introduce dimensionless density parameters [cf. (4.2.8)]

$$\varepsilon_{W_{e,i}} = n \rho_{W_{e,i}}^3 . \tag{10.9.5}$$

Then the characteristic times for broadening by electrons and by ions are

$$\frac{1}{(\gamma_{nm})_{e,i}} \sim \frac{\tau_{e,i}}{\varepsilon_{W_{e,i}}} \frac{1}{\varLambda_{e,i}} . \tag{10.9.6}$$

Hence it follows that the inequalities (10.9.1) can only be satisfied simultaneously for electrons and for ions provided

$$\varepsilon_{W_{e,i}} \varLambda_{e,i} \ll 1 , \tag{10.9.7}$$

which is only possible at low pressures ($n < 10^{15}$ cm^{-3} at $T \sim 10^2$ K).

Consider the approximation

$$\varepsilon_{W_e} \varLambda_e \ll 1 , \quad \text{but} \quad \varepsilon_{W_i} \varLambda_i \gg 1 . \tag{10.9.8}$$

In this approximation the field fluctuations of ions are slow, so the field of the ions may be considered to be static. For the static field of ions the wave eigenfunctions $\varPsi_n$, which the dipole moments of atoms are expressed in, cannot be determined for free atoms as before [see (7.2.9)], but must take the field of atoms into account. Then the functions $\varPsi_n$ and the eigenvalues of the energy will be determined by the solution of the Schrödinger stationary equation,

$$(\hat{H}_0 - dE) \, \varPsi_n(r) = E_n \varPsi_n , \tag{10.9.9}$$

where E is the static field of ions at the point of the atom's position.

As is known [10.9, 25] (10.9.9) is most conveniently solved in a parabolic coordinate system with z axis directed along the field vector. The state of the atom is then given by a set of quantum numbers n_1, n_2, m or n, $n_1 - n_2$, m, since

$$n_1 + n_2 + |m| + 1 = n , \tag{10.9.10}$$

where n is the principal quantum number.

In this presentation the atom has a permanent dipole moment, directed along z axis,

$$d = -3 \, e a_0 n (n_1 - n_2)/2 . \tag{10.9.11}$$

This reveals that the distribution of charges in an atom is nonsymmetric with respect to z axis. At $n_1 < n_2$ the dipole moment is parallel to E, and at $n_1 > n_2$ antiparallel to E. At $n_1 = n_2$ the dipole moment is zero.

At a fixed n the number $|m|$ takes values from zero to $n-1$. Accordingly,

$$n - |m| - 1 \geqslant n_1 \geqslant 0, \quad n_2 = n - |m| - 1 - n_1. \tag{10.9.12}$$

Hence it follows that at a fixed n the difference $n_1 - n_2$ may assume the values

$$n - 1 \geqslant n_1 - n_2 \geqslant -(n-1), \tag{10.9.13}$$

and therefore in a constant field level n is split into $2n-1$ sublevels.

With the aid of (10.9.11) the shift of the energy level can be described by

$$\Delta E_{n, n_1 - n_1} = -d_{n, n_1 - n_2} E. \tag{10.9.14}$$

Hence it follows that degeneracy remains with respect to the magnetic quantum number.

In accordance with (10.9.14) the transition frequencies are given by

$$\omega_{n, n_1 - n_2; m, m_1 - m_2} - \Delta \omega_{n, n_1 - n_2; m, m_1 - m_2}, \tag{10.9.15}$$

where m, $m_1 - m_2$ are parabolic quantum numbers of the lower state. These designations should not lead to a misinterpretation since magnetic quantum numbers are not used at any other place.

In order to obtain the distribution of radiation over frequencies in transitions between states with the principal quantum numbers n, m, in the equation for the distribution of intensity in a single line at a given field

$$I_{nm}(\omega | E) = 2 \frac{\gamma_{nm}}{\left[\omega - \omega_{nm} + \dfrac{1}{\hbar}(d_n - d_m)E \right]^2 + \gamma_{nm}^2} \tag{10.9.16}$$

the following replacement must be performed,

$$n \to n, \quad n_1 - n_2; \quad m \to m, \quad m_1 - m_2. \tag{10.9.17}$$

Here for the sake of compactness $d_{nn} = d_n$, $d_{mm} = d_m$.

We see that the position and the form of spectral lines depend upon the field strength E. At a random distribution of atoms and ions, the fields acting on different atoms differ. This gives rise to the problem of finding the distribution function of the microfields created by ions. This problem (with certain simplifying assumptions) was solved by Holzmark as far back as 1919 [10.2, 5, 27].

10.10 The Distribution of Microfields Created by Ions. The Holzmark Formula

We denote the unknown distribution function by $W(E)$. If N is the number of ions creating a Coulomb field, then

$$W(E) = \left\langle \delta \left(E - \sum_{1 \leqslant i \leqslant N} \frac{er_i}{r_i^3} \right) \right\rangle$$

$$= \frac{1}{(2\pi)^3} \int d\rho \exp(i\rho E) \left\langle \exp \left(-i \sum_{1 \leqslant i \leqslant N} \rho \frac{er_i}{r_i^3} \right) \right\rangle. \tag{10.10.1}$$

The angular brackets symbolize averaging over the distribution function of the positions of all ions.

If $\varepsilon_{W_i} \gg 1$, the number of ions within a sphere of Debye radius is large (at $r_D \gtrsim \rho_{W_i}$), and so in zero approximation with respect to plasma parameter μ one may employ self-consistent approximation. The distribution function of ions then splits into a product of one-particle distribution functions, i.e., in (10.10.1)

$$\left\langle \exp \left(-i \sum_{1 \leqslant i \leqslant N} \frac{er_i \rho}{r^3} \right) \right\rangle = \left\langle \exp \left(-i \frac{er\rho}{r^3} \right) \right\rangle^N$$

$$= \exp \left[-n \int dr \left(1 - \cos \frac{er\rho}{r^3} \right) \right], \tag{10.10.2}$$

where $n = N/V$ is the mean concentration of ions. Using the integral value

$$\int dr \left(1 - \cos \frac{er\rho}{r^3} \right) = (\lambda e|\rho|)^{3/2}, \quad \lambda = 2\pi \left(\frac{4}{15} \right)^{2/3}, \tag{10.10.3}$$

we write the desired distribution in the form

$$W(E) = \frac{1}{(2\pi)^3} \int d\rho \cos(\rho E) \exp[-n(\lambda e|\rho|)^{3/2}], \tag{10.10.4}$$

which is, of course, symmetric with respect to all components of the field.

Since (10.9.16) only includes the field strength E, we find with the aid of (10.10.4) the corresponding distribution,

$$W(E) = 4\pi E^2 W(E) \equiv \frac{1}{E_0} H \left(\frac{E}{E_0} \right). \tag{10.10.5}$$

Here we have introduced the designation for the Holzmark function

$$H(\beta) = \frac{2}{\pi} \beta \int_0^\infty dx\, x \sin(\beta x) \exp(-x^3/2),$$

$$\int_0^\infty H(\beta)\, d\beta = 1, \quad \beta = \frac{E}{E_0}, \quad E_0 = \lambda e n^{2/3}. \tag{10.10.6}$$

The parameter E_0 determines the strength of the ion field at a distance of the order $r_{av} \sim n^{-1/3}$. The function has its maximum at the point $\beta \approx 1.6$. At large and small values of β,

$$H(\beta) \approx \frac{2\pi}{\lambda^{9/2}} \beta^{-5/2} \quad \text{at} \quad \beta \gg 1 ,$$

$$H(\beta) \approx \frac{4}{3\pi} \beta^2 \quad \text{at} \quad \beta \ll 1 . \tag{10.10.7}$$

Let us now return to (10.9.16), and employ the distribution $W(E)$ obtained here.

10.11 The Atomic Emission Spectrum with the Ion Field Distribution Taken into Account

Due to the nonuniformity of the field created by ions, the positions of stationary energy levels of different atoms differ. For that reason the emission at frequency ω is determined by the summary radiation of atoms having different Stark shifts. The intensity of radiation at frequency ω is thus given by [Ref. 10.5, Eq. (7.1–4)]

$$I_{nm}(\omega) = 2 \int_0^\infty \frac{\gamma_{nm}}{\left[\omega - \omega_{nm} + \frac{1}{\hbar}(d_n - d_m)E\right]^2 + \gamma_{nm}^2} W(E) dE , \tag{10.11.1}$$

where $W(E)$ is the distribution (10.10.5), and n, m are sets of parabolic quantum numbers [see (10.9.17)]; γ_{nm} is the overall linewidth, determined by all the processes [see (10.2.2) and the calculations in Sect. 10.5].

The function $W(E)$ determines the distribution of Stark shifts. These shifts are known to be nondissipative since they only determine the positions of stationary energy levels in the presence of a static field.

Dissipation is determined by small-scale field fluctuations [see inequalities (10.9.1)]. In (10.11.1) they are accounted for through γ_{nm}. At $\gamma_{nm} = 0$ [in the initial equation, not in (10.11.1)], that is, with the small-scale fluctuations left out, the intensity of radiation is zero ($I_{nm} = 0$). This also follows from the fact that at $\gamma_{nm} = 0$ the imaginary part of permittivity becomes zero.

The term "Stark broadening of hydrogen lines" is in this respect not entirely correct since the distribution of the field in compliance with the Holzmark formula does not by itself determine the spectral linewidth. It determines only the distribution of the positions of stationary energy levels of atoms in the static field created by ions. This is the basis for the remark made by Pauli: "Holzmark attempted to calculate the broadening of spectral lines in connection with gas pressure as a Stark effect due to intermolecular fields. His results are ill based, since the intermolecular fields introduced by him cannot be considered uniform and, what is more, independent of time" (as quoted in [10.5]). This applies equally to the Doppler effect, which determines only the shift of levels (the real part of permittivity), while the linewidth (the imaginary part of permittivity) is determined by small-scale fluctuations.

It is convenient to introduce in (10.11.1) a designation for the Stark shift,

$$\Delta \omega_{St} = \frac{1}{\hbar} |d_n - d_m| E \equiv \frac{CE}{e}, \quad \Delta \omega_0 = \frac{CE_0}{e} = \lambda C n^{2/3}, \tag{10.11.2}$$

where C is the Stark constant, which, of course, depends upon parabolic quantum numbers of the relevant energy levels. Instead of (10.10.5) one can then employ the distribution

$$W(\Delta \omega_{St}) = \frac{e}{C} W(E) = \frac{1}{\Delta \omega_0} H \left(\frac{\Delta \omega_{St}}{\Delta \omega_0} \right), \quad \int_0^\infty W(\Delta \omega_{St}) d\Delta \omega_{ft} = 1. \tag{10.11.3}$$

In this case, (10.11.1) takes the form

$$I_{nm}(\omega) = 2 \int_0^\infty \frac{\gamma_{nm}}{[\omega - \omega_{nm} + \text{sign}\,(d_n - d_m)\,\Delta \omega_{St}]^2 + \gamma_{nm}^2} W(\Delta \omega_{St}) d\Delta \omega_{St}. \tag{10.11.4}$$

With the aid of the distributions given by (10.10.6, 11.3) we can find the mean Stark shift,

$$\langle \Delta \omega_{St} \rangle = \Delta \omega_0 \int_0^\infty \beta H(\beta) d\beta \approx \Delta \omega_0 = \frac{CE_0}{e}. \tag{10.11.5}$$

Using the definition of E_0 [see (10.10.6)] we get

$$\langle \Delta \omega_{St} \rangle \sim C n^{2/3}. \tag{10.11.6}$$

The distribution (10.11.4) thus includes two parameters γ_{nm}, $\langle \Delta \omega_{St} \rangle$, which have the dimension of frequency. According to (10.9.2, 11.6),

$$\frac{\gamma_{nm}}{\Delta \omega_0} \sim n^{1/3} C \frac{\Lambda_e}{v_{T_e}} \sim \frac{\rho w_e}{r_{av}} \Lambda_e. \tag{10.11.7}$$

Let us now consider the distribution (10.11.4) for several special cases.
1) Zero mistuning with respect to an unshifted line ($\omega - \omega_{nm} = 0$):

$$I_{nm}(\omega_{nm}) = 2 \int_0^\infty \frac{\gamma_{nm} W(\Delta \omega_{St})}{(\Delta \omega_{St})^2 + \gamma_{nm}^2} d\Delta \omega_{St} \approx \frac{\gamma_{nm}}{(\Delta \omega_0)^2 + \gamma_{nm}^2}. \tag{10.11.8}$$

To be concrete, we assume the lower level to be the ground one ($m = 1$, $m_1 = m_2 = 0$), and the upper level to have the principal quantum number 2. In this case there is no Stark splitting at the lower level, while the upper one is split into three sublevels with the transition frequencies ω_{nm}, $\omega_{nm} \pm \Delta \omega_{St}$. The contribution from the transition $\omega_{nm} (\Delta \omega_{St} = 0)$ to (10.11.8) equals $2/\gamma_{nm}$. Here γ_{nm} is determined by (10.2.2) at $d_{nn} = d_{mm} = 0$.

2) Weak fields ($\Delta \omega_0/\gamma_{nm} \ll 1$). In zero approximation with respect to this parameter, the Stark shift equals zero, and distribution over frequencies has the shape of the Lorentz line.

3) Strong fields. In zero approximation with respect to parameter $\gamma_{nm}/\Delta \omega_0$, the Lorentz line under the integral can be replaced, e.g., for the transition from the upper sublevel, by the δ function,

$$\pi \delta(|\omega - \omega_{21}| - \Delta \omega_{St}) .$$

As a result,

$$I_{21}(\omega) = \pi W(|\omega - \omega_{21}|) = \frac{\pi}{\Delta \omega_0} H \left(\frac{|\omega - \omega_{21}|}{\Delta \omega_0} \right),$$

$$\frac{1}{\pi} \int_0^\infty I_{21} d\omega = 1 , \quad \omega - \omega_{21} > 0 . \tag{10.11.9}$$

The function W is determined by (10.10.6, 11.3).

At the wings of the line, i.e., at $|\omega - \omega_{21}| \gg \Delta \omega_0$, this expression takes the form

$$I_{21}(\omega) = \frac{2\pi^2 C^{3/2} n}{|\omega - \omega_{21}|^{5/2}} . \tag{10.11.10}$$

Here we have taken the behavior of the function $H(\beta)$ at $\beta \gg 1$ (10.10.7) and the definition $\Delta \omega_0 = \lambda C n^{2/3}$ (10.11.2) into account. Since (10.11.9, 10) are obtained in zero approximation over $\gamma_{nm}/\Delta \omega_0$, (10.11.10) only determines the contributions of ions at the wings of the line.

What then is the contribution of electrons to the intensity distribution at the wings of the line? From the general formula (10.11.1) it follows that the contribution of electrons at large values of $|\omega - \omega_{21}|/\gamma_{21}$; $\Delta \omega_0$ is given by

$$I_{21}(\omega) = 2\gamma_{21} |\omega - \omega_{21}|^{-2} . \tag{10.11.11}$$

This therefore seems to the most important factor in determinating the behavior of the intensity at the wings of the line since the contribution given by ions, according to (10.11.10), decreases more abruptly (as $|\omega - \omega_{21}|^{-5/2}$). Let us investigate this matter in greater detail.

10.12 The Influence of an Electron Field on the Intensity Distribution at the Wings of the Spectral Line

As before we shall assume that the density parameter for electrons $n\rho_{W_e}^3$ is small. Then, according to (10.9.6, 7), the inequality given in (10.9.1) is satisfied. This gave us reason to consider the fluctuations of electrons to be fast. This separation of fluctuations into "fast" and "slow" is based on the employment of the parameter γ_{nm}, which determines the width of the line at "level $\frac{1}{2}$".

In investigating the intensity distribution at the wing of the line, $1/\gamma_{21}$ in the right-hand side of (10.9.1) must be replaced by time parameter $1/|\omega - \omega_{21}|$, which at the wings of the line is much smaller than $1/\gamma_{21}$. Hence it follows that at

$$|\omega - \omega_{21}| \gg \frac{\rho_{W_e}}{v_{T_e}} \equiv \Omega_e \qquad (10.12.1)$$

[see the definition (10.9.4)] the fluctuations of electrons ought to be considered slow. Therefore in analyzing the behavior of the radiation intensity at the wings of the line (provided this condition is satisfied), the field of electrons as well as of ions may be considered to be static in zero approximation with respect to parameter $\Omega_e/|\omega - \omega_{21}|$.

The contribution of ions to $I_{21}(\omega)$ at the wings of the line is proportional to the concentration of ions (10.11.10). This means that the interaction between ions and atoms at the wings of the line is binary in spite of the condition that $\varepsilon_{W_e} = n\rho_{W_e}^3 \gg 1$, that is, an atom does not simultaneously interact with two or more ions.

Due to the binary nature of the interaction between electrons and atoms (with accuracy to a logarithmic term), the contribution of electrons at the wing of the line is also proportional to their concentration (10.11.11). This means that an atom cannot (in the binary approximation) interact simultaneously with two or more electrons. This also means that an atom cannot interact with an electron and an ion at the same time.

Since the Stark constant is proportional to the square of the charge, the Holzmark distribution at the condition given in (10.12.1) is equally applicable to ions and electrons. As they mutually exclude one another, we get a doubled result (10.11.10) when calculating the intensity of radiation at the wings of the line, the statistical contribution of electrons being taken into account.

Consistent calculations turn out to be rather complicated. In order to obtain a single formula for the intensity of radiation which describes the actual behavior at both small and large mistunings qualitatively, correctly we substitute in (10.11.4) [or (10.11.1)]

$$\gamma_{nm} \to \tilde{\gamma}_{nm} = \frac{\gamma_{nm}\gamma_{\Delta\omega_{nm}}}{\gamma_{nm} + \gamma_{\Delta\omega_{nm}}}, \qquad \gamma_{\Delta\omega_{nm}} = \frac{\pi^2 n C^{3/2}}{|\omega - \omega_{nm}|^{1/2}}. \qquad (10.12.2)$$

Hence it follows that at the mistunings

$$|\omega - \omega_{nm}| \ll \frac{\gamma_{nm}}{n\rho_{W_e}^3} , \quad \text{when} \quad \tilde{\gamma}_{nm} \approx \gamma_{nm} , \tag{10.12.3}$$

(10.11.1) remains the same. In contrast, at large mistunings

$$|\omega - \omega_{nm}| \gg \frac{\gamma_{nm}}{n\rho_{W_e}^3} , \quad \text{when} \quad \tilde{\gamma}_{nm} \approx \gamma_{\Delta\omega_{nm}} , \tag{10.12.4}$$

we get the expression for the intensity

$$I_{nm}(\omega) = 4\pi^2 n \frac{C^{3/2}}{|\omega - \omega_{nm}|^{5/2}} . \tag{10.12.5}$$

This result differs from (10.11.10) by exactly a factor of 2. Therefore, the contributions of electrons and ions at the wings of the line are equal.

10.13 Taking Strong Short-Range Interactions and Collective Long-Range Interactions into Account Simultaneously

To calculate spectral line broadening due both to strong interactions at small distances and to collective interactions, one can employ the technique presented in [Ref. 10.8, Sect. 56], which we shall now describe.

For a completely ionized plasma in the approximation of pair interactions, the chain of equations developed by *Bogolyubov* [1.2], *Born* and *Green* [1.28], *Kirkwood* [1.29], and *Yvon* [1.30], can be replaced by a closed set of equations for the one-particle distribution functions f_a and the two-particle correlation functions g_{ab}.

In order to take the collective effects in the equations for functions g_{ab} into account, the potential of interaction of two particles is replaced by the effective potential,

$$\Phi_{ab}(r) \rightarrow \tilde{\Phi}_{ab}(r) . \tag{10.13.1}$$

The Fourier component of the effective potential is defined as, cf. (10.5.2):

$$\tilde{\Phi}_{ab}(k) = \Phi_{ab}(k) \frac{(\delta E \, \delta E)_k}{(\delta E \, \delta E)_k^0} , \tag{10.13.2}$$

where $(\delta E \, \delta E)_k$ is the spatial spectral density of field fluctuations with collective interactions taken into account, and $(\delta E \, \delta E)_k^0$ is the same but without taking the collective interactions into account, i.e., not accounting for the polarization.

In particular, for the equilibrium state for totally ionized plasma, it follows from (10.13.2) that

$$\tilde{\Phi}_{ab}(r) = \frac{e_a e_b}{r} e^{-r/r_\mathrm{D}}. \tag{10.13.3}$$

In this case the effective potential coincides with the effective Debye potential.

In the same approximation we get the following expression for the correlation function:

$$f_{ab}(r) = 1 + g_{ab} = \exp\left[-\tilde{\Phi}(r)/\mathcal{X}T\right] = \exp\left[-\frac{e_a e_b}{\mathcal{X}Tr}\exp\left(-\frac{r}{r_\mathrm{D}}\right)\right].$$
$$\tag{10.13.4}$$

Thus, the set of equations for the functions f_a, f_{ab} (taking the effective potential account) can be written as

$$\frac{\partial f_a}{\partial t} = \sum_b n_b \int \frac{\partial \Phi_{ab}}{\partial r}\,\frac{\partial f_{ab}}{\partial p}\,dr'\,dp' \equiv I_a(p,t),$$

$$\left(\frac{\partial}{\partial t} + v\,\frac{\partial}{\partial r} + v'\,\frac{\partial}{\partial r'} - \frac{\partial \tilde{\Phi}_{ab}}{\partial r}\,\frac{\partial}{\partial p} - \frac{\partial \tilde{\Phi}_{ab}}{\partial r'}\,\frac{\partial}{\partial p'}\right) f_{ab}$$

$$= \frac{\partial}{\partial t}(f_a(p,t)\,f_b(p',t)). \tag{10.13.5}$$

In particular, the expression for the Boltzmann collision integral with the collective interactions of charged particles taken into account follows from this equation. The integral therefore does not divergence at either large or small distances.

From the above definitions it follows that the introduction of the effective potential amounts to taking the averaged dynamic polarization into account. It is therefore a more general operation than the introduction of static Debye screening [10.28, 29].

Such a technique for considering the collective interactions can also be employed in the theory of the spectral line broadening of atoms in partially ionized plasma (see [10.17]. In the polarization approximation one then gets, of course, the results presented in Sects. 10.4−6. It becomes possible to take both strong short-range interactions and collective interactions into consideration, and to establish a correspondence with the well-known results obtained by *Voslamber* [10.30, 31].

10.14 Some Problems of the Kinetic Theory
of Spectral Line Broadening

The above discussion, naturally, does not pretend to give an exhaustive analysis of all the problems of the kinetic theory of broadening. Some of the problems which remain beyond the scope of our discussion are:

The influence of dynamic polarization on spectral line broadening. In the concrete calculations (Sect. 10.5) we took the polarization into account only approximately. Namely, we employed the averaged dynamic polarization in place of the dynamic polarization. This allows taking the screening effects into account, but excludes the resonance effects from consideration. The latter, however, become significant, e.g., when the frequency of plasma oscillations is close to the transition frequencies between Stark sublevels.

The influence of external fields upon the shape and broadening of spectral lines. In the past few years a new branch of plasma spectroscopy has been successfully developing: the investigation of phenomena observed in strong fields. At first these phenomena were analyzed on the basis of approximative kinetic equations with constant relaxation characteristics, with the field accounted for, as a rule, in the dipole approximation. In many cases such an approach is not exact enough. The explanation of some phenomena requires that the action of the field be taken into account in describing the collisions of atoms and free charged particles with radiating atoms and ions. The dependence upon the field is then observed in the collision integrals themselves. Certain results in this connection were obtained in [10.32, 33].

The influence of strong fields on collisions has been analyzed in detail for completely ionized plasma (see [Ref. 10.8, Chap. 8] and references cited there) and for semiconductors.

According to the approach adopted in the present book, this problem demands the calculation of the nonequilibrium small-scale fluctuations which determine the collision integral with the mean field taken into account. To be more precise, one must take the influence of the mean field into account in (4.6.5, 7) when calculating the fluctuations in a system of free charged particles and a field, as well as in (10.1.8, 9). As a result, both the spectral densities of small-scale fluctuations and the collision integrals determined by them are functions of the field. This scheme is carried out in [Ref. 10.8, Chap. 8] for completely ionized plasma.

It must also mentioned that the investigation of the influence of the field on spectral line broadening is connected with other problems, e.g., with the aim of narrowing gamma resonances [10.34].

The influence of correlations of free charged particles and atoms on spectral line broadening. This domain is a very vast one. It will suffice to note that in calculating the ions' field we did not consider the static or dynamic polarization of ions. A large body of literature exists on this subject; we refer here to just two works, by *Kogan* [10.35] and by *Ecker* [10.28]. In calculations of the spectra in dense systems, the difference between microscopic and local fields becomes important. As we have shown in Sects. 6.4 – 7, taking this difference into account brings about significant alterations in the equation for the polarization vector, and therefore significant changes in the shape of the spectral line. In this regard there are still many unresolved questions. The corresponding quantum results will be discussed in the following chapter.

Spectral line broadening in nonequilibrium plasma. The theory developed above enables performing calculations of spectral line broadening in nonequilibrium plasma, as well. This is possible by employing the distribution functions for non-equilibrium states in the expressions for the spectral densities of field fluctuations and in the formulae for γ, $\Delta \omega$. These distribution functions are obtained as solutions of the relevant kinetic equations. Problems of this kind are very compli-cated, though they can be simplified by using model collision integrals [10.7].

The influence of turbulence upon spectral line broadening. The investigation of the influence of turbulence on spectral line broadening is one of the tasks of the kinetic theory of fluctuations in highly nonequilibrium plasma. The name "kinetic", (Sect. 4.8 and Chaps. 5, 6) is given to fluctuations whose correlation times are of the order of relaxation times in the kinetic equations. In other words, they are the fluctuations of the distribution functions themselves as well as of the corresponding field fluctuations.

In a turbulent plasma all the kinetic characteristics are drastically changed. Naturally, this leads to marked changes in the shape of spectral lines. The growing concern with this problem is connected with achievements in the experi-mental investigation of turbulent plasma. Of the more general works in this field we mention [10.36], where the approach using the kinetic theory of fluctuations is employed which is akin to the one employed here and in [Ref. 10.8, Chaps. 4, 11].

A study concerned with a more specific problem is [10.37]. It deals with the influence of the Langmuir turbulence on the shapes of Stark components. The same problem is analyzed in [10.38] without using perturbation theory. This spreads the limits on the applicability of the theory, and permits a more reliable interpretation of the experimental results.

The tasks listed above for the theory of the spectral line broadening of atoms in a plasma are connected with the influence of collective processes in nonequilib-rium and of nonideal plasma on the atoms' emission of radiation. The solution of these complicated problems is important not only for the kinetic theory of plasma, but for many practical applications as well.

11. Fluctuations and Kinetic Processes in Systems Composed of Strongly Interacting Particles

This chapter deals with the solution of two problems of the statistical theory of strongly interacting systems: 1) the influence of the correlations of atoms' positions on the characteristics of atomic transitions; and 2) the kinetic basis of the fluctuation-dissipation theorem (FDT). The kinetic approach allows formulating the FDT in its most general form, as a relationship between the spectral density of the many-particle distribution function and the imaginary part of the respective response. We define the scope of nonequilibrium processes for which the FDT can be formulated. In this case, the spectral density of fluctuations depends on the nonequilibrium distribution functions, for which a relevant kinetic equation has been established.

11.1 The Influence of the Correlations of Atoms' Positions on Their Spontaneous Radiation

Among the basic elements of kinetic theory we have described the microscopic equations which do not distinguish between a local (acting) field and a microscopic field created by all particles. The only exception is in Sects. 6.4 – 7, where we saw that the correlations which determine the difference between the mean acting field and the Maxwell field are very important in dense systems. In particular, they give rise to three effects: the appearance of the Lorentz correction (Lorentz field), the broadening of the emission lines of the atoms belonging to medium, and the shift of eigenfrequencies (the transition frequencies) of atoms.

In the first three sections of this chapter we shall discuss the corresponding quantum results. The remaining sections are devoted to the kinetic description of processes in systems of strongly interacting atoms. We will also generalize the Callen – Welton and Kubo equations for nonequilibrium states.

The kinetic equation for the distribution function of a system composed of strongly interacting particles is, of course, very complex. Nevertheless, it attracts our interest for several reasons. For one thing, it provides an example of an equation which describes the process of relaxation towards the Gibbs distribution. Then the spectral density, e. g., of field fluctuations, relaxes toward the spectral distribution described by the Callen – Welton equation.

Proceeding to the first of the named problems, we recall that (10.1.17, 18) relate the collision integrals I_{nm} and I_n to the correlations of fluctuations δN_{nm} and δE. Now, however, we can express these integrals through the correlators of the operator matrix N_{nm} and the operator of the acting field E_a, e. g.,

$$I_{nm} = \frac{i}{\hbar N} \sum_{n_1} [d_{nn_1} \langle N_{n_1 m} E_a \rangle_{R,P,t,R,t} - \langle N_{nn_1} E_a \rangle d_{n_1 m}] .$$ (11.1.1)

We shall employ an approach similar to that in Sect. 6.3. The mean acting field will again be split into two parts: the mean Maxwell field, and the field created by particles confined within a physically infinitesimal volume centered around the position of the chosen atom. We use the relationship between the polarization vector and the matrix N_{nm},

$$P(R, t) = \frac{1}{V} \sum_{n_1 m_1} \int d_{m_1 m_1} N_{n_1 m_1}(R, P, t) \frac{V}{(2\pi\hbar)^3} dP ,$$ (11.1.2)

and employ the solution of the field equation in the form of (3.6.4).

As a result we obtain the expression for the correlation between matrix δN_{nm} and the field created by atoms which are within the limits of the physically infinitesimal volume l_{ph}^3 [cf. the second term in the right-hand side of (6.3.4)],

$$\langle \delta N_{nm}(R, P, t) \, \delta E(R, t) \rangle$$

$$= \int_{V_{ph}} \text{rot}_R \, \text{rot}_R \frac{\langle \delta N_{nm} \delta P \rangle_{R,t,R'_1 t} - \frac{|R-R'|}{c}}{|R-R'|} dR' - 4\pi \langle \delta N_{nm} \delta P \rangle_{R,P,t,R,t} .$$ (11.1.3)

In the future one must bear in mind that the operator $\text{rot}_R \, \text{rot}_R$ does not act upon the function δN_{nm}.

As in Sect. 6.3, we single out in (11.1.3) the contribution not connected with the correlation of particles. It determines the radiational friction of a separate atom in the collision integral and the corresponding frequency shift. As a result, (11.1.3) becomes [cf. the second term in the right-hand side of (6.3.7)]

$$\langle \delta N_{nm} \delta E \rangle$$

$$= \frac{N^2}{V} \int_{V_{ph}} \text{rot}_R \, \text{rot}_R \frac{\sum_{n_1 m_1} d_{m_1 n_1} g_{nmn_1 m_1} \left(R, t, R', t - \frac{|R-R'|}{c} \right)}{|R-R'|} dR'$$

$$- \frac{4\pi N^2}{V} \sum_{n_1 m_1} d_{m_1 n_1} g_{nmn_1 m_1}(R, P, R, t) .$$ (11.1.4)

Substituting this correlator into the expressions for the collision integrals I_{nm} and I_n, we obtain the additional contributions due to the correlation of particles.

Recall that integration in (11.1.4) is performed over the physically infinitesimal volume V_{ph}. This restriction, however, is unimportant provided the correlation radius of atoms' positions is smaller than the physically infinitesimal length l_{ph} (6.3.9).

11.2 The Effective Lorentz Field

The equation for the correlator of fluctuations δN_{nm}, δE (11.1.4) is thus determined by the correlation function $g_{nmn_1m_1}(R, R', t)$. The latter is connected with a simpler correlation function of atoms' positions by the relation

$$g(R, R', t) = \sum_{nn_1} g_{nnn_1n_1}(R, R', t) \,. \tag{11.2.1}$$

Again as in Sect. 6.3, the correlation function (11.2.1) is considered to be set in advance. A more general correlation function which appears in (11.1.4) can be presented in the form, cf. (6.3.11),

$$g_{nmn_1m_1}(R, P, R', t) = f_{nm}(R, P, t)\, f_{n_1m_1}(R', t)\, g(R, R', t) \,. \tag{11.2.2}$$

In this approximation the total correlation (accounting for the mean field) can be presented as

$$\langle N_{nm}(R, P, t)E^m(R, t)\rangle$$

$$= \left[E(R, t) + \int \mathrm{rot}_R\, \mathrm{rot}_R \left(\frac{P\left(R', t - \dfrac{|R-R'|}{c}\right)}{|R-R'|} \right) \right.$$

$$\left. \times\, g(R, R')\,dR' - 4\pi P(R, t)\, g(R, R) \right] N f_{nm}(R, P, t) \,. \tag{11.2.3}$$

Here we have employed the connection between the mean polarization vector with the function f_{nm}.

In the quantum approach we employ the inequalities in (6.3.9) as well. Under this condition one can expand the integrand in (11.2.3) with respect to lagging and to the parameter of nonuniformity, since in both cases the expansion is actually carried out over one and the same parameter r_{cor}/λ (cf. Sect. 6.4).

In zero approximation (11.2.3) becomes (the calculation is similar to that performed in Sect. 6.4)

$$\langle N_{nm}(R, t)E^m(R, t)\rangle = \left[E - \frac{4\pi}{3} g(0)P \right] N f_{nm} \equiv E_{eff} N f_{nm} \,. \tag{11.2.4}$$

Here we have used the designation given in (6.4.7) for the mean field. If the correlation function is such that $g(0) = -1$, then we get the Lorentz result for the magnitude of the mean field.

Let us point out once more that (11.2.4) holds only at the condition given in (11.2.2), which implies that the distribution of the positions of atom centers and the internal motion of atoms are statistically independent. Naturally, this condition only has limited plausibility. It presents a good approximation in cases when the approximation of immobile atoms is good. The situation is radically changed when transitions from the discrete spectrum into the continuous one become sufficiently probable. In particular, in the extreme case of completely ionized plasma the effective field has an entirely different structure [11.1–4].

Indeed, the correlational addendum to the mean field in this case is proportional to the plasma parameter μ, and is therefore small in the case of slightly nonideal plasma where $\mu \ll 1$.

It would, of course, be desirable to have a general expression for the effective field, which would apply for any arbitrary degree of ionization and for nonideal systems. This makes it necessary to separate the correlation

$$\langle N_{\alpha\beta}(R, P, t) E^{\mathrm{m}}(R, t) \rangle \tag{11.2.5}$$

for both discrete and continuous parts of the spectrum. The exact solution of this problem is hardly possible. The situation is simplified, however, in those cases when the characteristic times of the radiation processes are significantly different from those of processes determined by the correlation of atoms.

11.3 The Influence of the Correlations of Atoms' Positions on the Coefficient of Spontaneous Emission

If higher terms of expansion over r_{cor}/λ are taken into account, one can, as in Sect. 6.5, single out the terms describing the influence of the correlations of atoms' positions on the dissipative processes.

First we separate the contribution proportional to $(r_0/\lambda)^3$ in the right-hand side of (11.2.3). Having carried out transformations similar to those performed in Sect. 6.5, we get

$$\langle \delta N_{nm} \delta E \rangle (=) \frac{2}{3c^3} \int g(|r|) \, dr \frac{\partial^3 P}{\partial t^3} N f_{nm} \,. \tag{11.3.1}$$

The sign $(=)$ warns that only the dissipative contribution is singled out.

In order to find the corresponding contribution to the collision integral, we substitute (11.3.1) into (11.1.1) and single out the resonance terms. In this approximation,

$$P(R, t) = n d_{mn} f_{nm}, \quad \frac{\partial^3 P}{\partial t^3} = (-\mathrm{i})^3 \omega_{nm}^3 n d_{mn} f_{nm}, \tag{11.3.2}$$

and in the relevant correlation expressions the functions $f_{nm} \to \delta_{n_1 m} f_m$ and $f_{nn_1} \to f_n \delta_{nn_1}$. As a result we get the following equation for the dissipative contribution:

$$I_{nm}(=) - \frac{2|d_{nm}|^3}{3\hbar c^3} \omega_{nm}^3 (f_m - f_n) n \int g(|r|) \, dr f_{nm} \,. \tag{11.3.3}$$

Taking both this dissipative contribution determined by the correlation of particles and the effective field into account, we can present the equation for the function f_{nm} in the form

$$\left[\frac{\partial}{\partial t} + (\gamma_{nm})_{\text{eff}} + i\omega_{nm}\right]f_{nm} = \frac{i}{\hbar}\sum_{n_1}(d_{nn_1}\,f_{n_1m}E_{\text{eff}} - f_{nn_1}d_{n_1m}E_{\text{eff}}),\qquad(11.3.4)$$

where

$$(\gamma_{nm})_{\text{eff}} = \gamma_{nm} + \tfrac{1}{2}A^n_m(f_m-f_n)\,n\int g(|r|)\,dr\,;\qquad(11.3.5)$$

here A^n_m is the Einstein coefficient (9.5.9) (at $\text{Re}\{\varepsilon\} = 1$), and γ_{nm} is the coefficient of radiational friction, given by (10.4.10). In the latter it is possible to single out the contribution resulting from the transition $n \to m$ (10.4.11).

To facilitate the comparison between the newly obtained expression for γ_{eff} with the classical result (6.5.7), we single out in (11.3.5) the resonance contribution (10.4.11), employ the equality in (6.5.5), and take m as the ground state ($f_m \approx 1$). Then

$$(\gamma_{nm})_{\text{eff}} = \frac{1}{2}A^n_m\frac{\langle(\delta N_{V_{\text{ph}}})^2\rangle}{\langle N_{V_{\text{ph}}}\rangle}.\qquad(11.3.6)$$

This equation coincides with the classical result (6.5.8) with the only difference being that the coefficient of the radiational friction is now determined by the Einstein coefficient. For this reason the arguments developed in Sect. 6.5 apply equally to the quantum theory.

The complete coincidence of classical and quantum results is observed in a special case. A more general expression (11.3.5) can be presented in the form

$$(\gamma_{nm})_{\text{eff}} = \frac{1}{2}A^n_m\left[1 + (f_m-f_n)\left(\frac{\langle(\delta N_{V_{\text{ph}}})^2\rangle}{\langle N_{V_{\text{ph}}}\rangle} - 1\right)\right].\qquad(11.3.7)$$

If the surrounding atoms are positioned in a completely ordered way, i.e., if $\langle(N_{V_{\text{ph}}})^2\rangle = 0$, then

$$(\gamma_{nm})_{\text{eff}} = \tfrac{1}{2}A^n_m[1 - (f_m-f_n)]\,.\qquad(11.3.8)$$

Hence the radiational friction is entirely compensated for only if the chosen atom is in the ground state. In contrast, if the inversion is total ($f_m = 0$, $f_n = 1$), then

$$(\gamma_{nm})_{\text{eff}} = A^n_m\,,\qquad(11.3.9)$$

and the coefficient of the radiational friction is therefore doubled.

The marked increase in radiational friction approaching the critical point is equally true for both classical and quantum theories.

We shall now elaborate on the method for calculating electromagnetic fluctuations in a nonequilibrium nonideal system of atoms. First we shall derive the well-known Callen – Welton equation (fluctuation – dissipation theorem). It will help us to comprehend better the later discussion.

11.4 The Kubo and the Callen – Welton Equations

Consider a macroscopic system with the Hamiltonian

$$\hat{H} = \hat{H}_0 + \hat{H}_1, \quad \hat{H}_1 = - \sum_i X_i F_i. \tag{11.4.1}$$

Here $\hat{H}_0$ is the Hamiltonian of the macroscopic system comprising N particles, however strongly they interact; F_i are the external forces; and X_i are the corresponding internal parameters. In general they are functions of all the variables. The external forces are assumed to depend only on time. Later we shall explain the meanings of functions F_i and X_i.

We denote by Ψ_n the eigenfunctions of the Hamiltonian operator $\hat{H}_0$, i.e.,

$$\hat{H}_0 \Psi_n = E_n \Psi_n, \quad \int \Psi_m^*(X) \, \Psi_n(X) \, \frac{dX}{V^{2N}} = \delta_{nm},$$

$$X = (R_1, \ldots, R_N, r_1 \ldots, r_N). \tag{11.4.2}$$

For a system of noninteracting atoms,

$$H_0 = \sum_{1 \leqslant i \leqslant N} H_i, \quad E_n = \sum_{1 \leqslant i \leqslant N} E_{n_i}, \quad \Psi_n = \prod_i \Psi_{n_i}, \tag{11.4.3}$$

where Ψ_{n_i} are the eigenfunctions of the Hamiltonian operator of a single atom (Sect. 7.2).

The density matrix of the system in question is labelled f_{nm}. In the state of equilibrium, which is established in the absence of external forces ($F_i = 0$), the density matrix is diagonal,

$$f_{nm} = \delta_{nm} f_n. \tag{11.4.4}$$

Here f_n is the Gibbs distribution,

$$f_n = \frac{1}{Z} \exp\left(- \frac{E_n}{\varkappa T}\right), \quad Z = \sum_n \exp\left(- \frac{E_n}{\varkappa T}\right). \tag{11.4.5}$$

Many-particle functions are written here with f_{nm} and Ψ_n, the same designations which were employed for one-particle functions. The introduction of new designations is hardly justified; where it might lead to misinterpretation, we shall supply the necessary explanations.

The Equation for the density matrix of a system with the Hamiltonian $\hat{H} = \hat{H}_0 + \hat{H}_1$ is [cf., e.g., (7.5.9)]

$$\left(\frac{\partial}{\partial t} + i\omega_{nm}\right) f_{nm} = \frac{i}{\hbar} \sum_{n_1} [(X_i)_{nn_1} f_{n_1 m} - f_{nn_1} (X_i)_{n_1 m}] F_i(t) , \qquad (11.4.6)$$

where $(X_i)_{nm}$ is the matrix element of the operator X_i. Using this equation we can find the mean response of the system to a weak external field $F_i(t)$. We assume $\langle X_i \rangle = 0$ in the absence of an external field.

In linear approximation with respect to F_i the density matrix in the right-hand side of (11.4.6) has the form of (11.4.4) as a result for the deviation from the equilibrium function

$$f^1_{nm} = f_{nm} - \delta_{nm} f_n , \qquad (11.4.7)$$

so we get the following equation:

$$\left(\frac{\partial}{\partial t} + \Delta + i\omega_{nm}\right) f^1_{nm} = \frac{i}{\hbar} (X_i)_{nm} (f_m - f_n) F_i(t) . \qquad (11.4.8)$$

Here we have introduced the attenuation Δf^1_{nm}. In the final results $\Delta \to 0$.

Using this equation we proceed to find the Fourier component of the mean response,

$$\langle X_i \rangle_\omega = \sum_{nm} (X_i)_{mn} f^1_{nm}(\omega) = \alpha_{ij}(\omega) F_j(\omega) , \qquad (11.4.9)$$

where the designation for the tensor of linear susceptibility is introduced,

$$\alpha_{ij}(\omega) = \frac{i}{\hbar} \sum_{nm} \int_0^\infty d\tau \exp[-\Delta \tau + i(\omega + \omega_{mn})\tau] (X_i)_{mn}(X_j)_{nm} (f_m - f_n) . \qquad (11.4.10)$$

The tensor of susceptibility determines both dissipative and nondissipative components of response. The dissipative part is given by the tensor

$$i[\alpha^*_{ij}(\omega) - \alpha_{ij}(\omega)] = \frac{2\pi}{\hbar} \sum_{mn} \delta(\omega + \omega_{mn})(X_i)_{nm}(X_j)_{mn}(f_m - f_n) . \qquad (11.4.11)$$

If the internal parameter X_i only depends on the coordinates, and is therefore invariant with respect to the reversal of time, then the corresponding matrix element is real, i.e.,

$$(X_i)_{nm} = (X_i)^*_{nm} = (X_i)_{mn} . \qquad (11.4.12)$$

Under this condition the tensor of susceptibility is symmetric,

$$\alpha_{ij}(\omega) = \alpha_{ji}(\omega) . \tag{11.4.13}$$

Therefore (11.4.11) determines in this case the doubled imaginary part of the tensor of susceptibility $\alpha_{ij}(\omega)$.

Consider now the expression for the spectral density of fluctuations,

$$\delta X_i = X_i - \langle X_i \rangle \equiv X_i , \quad \text{since } \langle X_i \rangle = 0 . \tag{11.4.14}$$

We shall show that there exists a connection between the functions $(X_iX_j)_\omega$ and $\alpha_{ij}(\omega)$ in equilibrium. The correlation of operators at different moments of time is given by the expression

$$\langle X_i(t)X_j(t-\tau)\rangle = \sum_{nm} [X_i(t)X_j(t-\tau)]_{mn} f_{nm}$$

$$= \sum_{nmn_1} [X_i(t)]_{mn_1} (X_i(t-\tau))_{n_1n} f_{nm}$$

$$= \sum_{nm} [X_i(t)]_{nm}[X_j(t-\tau)]_{mn} f_n . \tag{11.4.15}$$

Here we have used the definition of the mean value in terms of the density matrix f_{nm}, replaced the matrix $(X_iX_j)_{mn}$ by the product of the matrices $\sum_{n_1} (X_i)_{mn_1} (X_j)_{n_1n}$, and employed (11.4.4), which is valid for the state of equilibrium.

The time-dependent matrix element can be presented in the form

$$X_{nm}(t) = X_{nm} \exp(-i\omega_{nm}t) . \tag{11.4.16}$$

Then the expression for the correlator takes the form

$$\langle X_i(t)X_j(t-\tau)\rangle = \sum_{nm} (X_i)_{nm}(X_j)_{mn} \exp(-i\omega_{nm}\tau) f_n . \tag{11.4.17}$$

Hence the equation for the unknown spectral density is

$$(X_iX_j)_\omega = 2\pi \sum_{nm} (X_i)_{nm}(X_j)_{mn} \delta(\omega - \omega_{nm}) f_n . \tag{11.4.18}$$

Symmetrizing this expression consistent with the condition of stationarity,

$$\langle X_i(t)X_j(t-\tau)\rangle = \langle X_j(t)X_i(t+\tau)\rangle, \quad (X_iX_j)_\omega = (X_jX_i)_{-\omega}, \tag{11.4.19}$$

we get, instead of (11.4.18),

$$(X_iX_j)_\omega = \pi \sum_{nm} [(X_i)_{nm}(X_j)_{mn} \delta(\omega - \omega_{nm}) + (X_i)_{mn}(X_j)_{nm} \delta(\omega - \omega_{mn})] f_n . \tag{11.4.20}$$

Finally, in the first term in the right-hand side we substitute $n \rightleftarrows m$. As a result we arrive at the final expression,

$$(X_i X_j)_\omega = \pi \sum_{nm} (X_i)_{nm} (X_j)_{mn} \delta(\omega - \omega_{nm})(f_m + f_n) . \tag{11.4.21}$$

To compare (11.4.11) and (11.4.21), we first multiply and divide the summand in (11.4.21) by $f_m - f_n$, and employ the equality

$$\delta(\omega - \omega_{nm}) \frac{f_m + f_n}{f_m - f_n} = \delta(\omega - \omega_{nm}) \coth\left(\frac{\hbar\omega}{2\mathcal{K}T}\right)$$

$$\equiv 2\delta(\omega - \omega_{nm}) \left(\frac{1}{2} + \frac{1}{\exp(\hbar\omega/\mathcal{K}T) - 1}\right). \tag{11.4.22}$$

This results in the following relation:

$$(X_i X_j)_\omega = \frac{\hbar i}{2}(\alpha_{ij}^* - \alpha_{ji}) \coth\left(\frac{\hbar\omega}{2\mathcal{K}T}\right) \tag{11.4.23}$$

which is the well-known Callen – Welton equation expressing the fluctuation-dissipation theorem (FDT) [Ref. 11.5, Eq. (125.10); 11.6 – 9].

The Callen – Welton equation connects the spectral density of fluctuations of the internal parameters X_i with the dissipative part of the susceptibility tensor α_{ij}. The reciprocal relation

$$i(\alpha_{ij}^* - \alpha_{ji}) = \frac{2}{2}(X_i X_j)_\omega \tanh\left(\frac{\hbar\omega}{2\mathcal{K}T}\right) \tag{11.4.24}$$

is called the Kubo formula. The latter permits determining the dissipative part of the susceptibility tensor using the spectral density (for more details see [11.8, 10]).

In these equations only the temporal dispersion is taken into account. Naturally, it is possible to obtain more general relationships which would connect space and time spectral densities with the corresponding susceptibility tensor [11.9, 11].

In the next section we shall generalize the fluctuation – dissipation relationships to include the nonequilibrium states as well. This is the case when the distribution function of a system with the Hamiltonian $\hat{H}_0$ is no longer a Gibbs distribution, but satisfies a certain kinetic equation. One of our immediate tasks will be to establish the form of this particular kinetic equation.

We shall see that under certain conditions, (11.4.11, 21) also hold for non-equilibrium states. The Callen – Welton equation, however, does not hold any longer since (11.4.22) is true only for the equilibrium state.

The fluctuation – dissipation relation remains true for nonequilibrium states, although only for separate transitions. Indeed, from (11.4.11, 21) we get for separate transitions $n \to m$ the following expression:

$$(X_i X_j)_\omega^{n,m} = \frac{i\hbar}{2}(\alpha_{ij}^* - \alpha_{ji})_{n,m} \frac{f_m + f_n}{f_m - f_n}. \qquad (11.4.25)$$

It even remains true for the inverse population of levels n and m. Both multiplicands in (11.4.25) are then negative [see (11.3.4)].

We see that FDT in the form of (11.4.23) is only true provided that (11.4.22) holds. The latter equation contains two time parameters, one of which ($\tau_\hbar \equiv \tau_0 = 1/\omega_{nm}$) enters explicitly, while the other characterizes the "width" of the function $\delta(\omega - \omega_{nm})$ and enters implicitly. Let us designate this width by $\Delta \omega$. Finally, there exists one more time parameter, which determines the amount of time needed to establish the Gibbs distribution. This symbol for relaxation time is τ_{rel}. It also characterizes the correlation time of the corresponding kinetic fluctuations. Fulfilling (11.4.22) is only possible provided that two strong inequalities are satisfied,

$$\tau_\hbar \ll \frac{1}{\Delta \omega} \ll \tau_{rel}. \qquad (11.4.26)$$

Equation (11.4.22) is stringent and corresponds to a zero-order approximation with respect to the parameter $\tau_\hbar / \tau_{rel}$.

This parameter can equal zero in two physically different situations:

1) Time $\tau_\hbar$ is finite, and τ_{rel} is infinitely large. By analogy with the theory of gases and plasmas we call this the "collisionless" approximation, which finds application in both classical and quantum theories. Special forms of FDT have been used several times, e.g., the expressions for partially ionized plasma (8.4.1, 2).

2) The correlation time of the kinetic fluctuations, which is determined by the relaxation time τ_{rel}, is finite, whereas $\tau_\hbar = 0$.

It is important that the FDT can be derived in the classical approximation without using equalities such as (11.4.22), i.e., that FDT holds for the "collisions" region as well. In other words, it can be derived when the width of the spectral lines corresponding to separate eigenfrequencies (in quantum theory, to the transition frequencies) is finite. An example of this is provided by the well-known Nyquist formula for the spectral density of emf $\mathscr{E}(t)$ in the analysis of thermal oscillations in an electric circuit. It has the form

$$(\mathscr{E}^2)_\omega = 2R \mathscr{K}T; \quad (J^2)_\omega = \frac{(\mathscr{E}^2)_\omega}{R^2 + (L\omega - 1/C\omega)^2}. \qquad (11.4.27)$$

Here R is the ohmic resistance, and $(J^2)_\omega$ is the spectral density of current.

The Nyquist formula is usually generalized for the quantum case in the following way:

$$(\mathscr{E}^2)_\omega = 2R\,\mathscr{K}T_\omega, \quad \mathscr{K}T_\omega = \frac{\hbar\omega}{2}\coth\left(\frac{\hbar\omega}{2\,\mathscr{K}T}\right). \tag{11.4.28}$$

It is thus assumed that the mean energy of the oscillator does not depend on the eigenfrequency $\omega_0 = 1/\sqrt{LC}$, but on the running frequency ω. This is only justified in zero approximation with respect to parameters $\tau_0 = 1/\omega_0$ and $\tau_{rel} \sim L/R$, that is, in the approximation $R \to 0$. Otherwise it is necessary to substitute

$$\mathscr{K}T \to \mathscr{K}T_{\omega_0} \tag{11.4.29}$$

in the quantum generalization of the Nyquist formula.

Then the mean energy of the oscillator depends on the eigenfrequency ω_0, and not on the running frequency ω.

Equation (11.4.25) is derived without using (11.4.22), and is therefore true for the equilibrium state and at finite widths of the spectral lines corresponding to separate transitions. Instead of the Callen – Welton equation we then have the following expression:

$$(X_i X_j)_\omega = \frac{\hbar i}{2}\sum_{nm}[\alpha^*_{ij}(\omega) - \alpha_{ji}(\omega)]_{nm}\coth\left(\frac{\hbar\omega_{nm}}{2\,\mathscr{K}T}\right). \tag{11.4.30}$$

This coincides with (11.4.23) under the additional assumption of infinitely narrow resonances, i.e., under the condition given by (11.4.22).

In all the equations in this section, the matrix elements $(X_i)_{nm}$ are determined by the wave functions Ψ_n, which are the eigenfunctions of the Hamiltonian $\hat{H}_0$. An attempt to generalize these eigenfunctions (for a macroscopic system of particles, however strongly they interact) seems futile. So the fluctuation – dissipation theorem is actually only applied to rather simple systems, i.e., when one can carry out the calculations by using small parameters such as the plasma parameter. In such cases FDT has been applied for nonequilibrium states as well; two examples are the theory of fluctuations in lasers [11.12, 13], where (11.4.25) has been employed for two operating levels, and the theory of nonequilibrium fluctuations in rarefied plasma [11.3, 14, 15]. In both cases the distribution functions entering the fluctuation – dissipation relationships satisfy the relevant kinetic equations for one-particle distribution functions.

In order to derive the kinetic equations for a many-particle distribution function entering (11.4.11, 21), another approach to the establishment of fluctuation – dissipation relationships must be taken. Namely, the susceptibility tensor ought to be determined from the response to the internal fluctuation forces, and not to the external forces. This possibility has been discussed [Ref. 11.5, Sects. 124, 125], though its ultimate development has been hindered by the unsettled question regarding the nature of these random forces. Naturally, they must be determined by the structure of the system under consideration. Thus, before one proceeds to construct the kinetic equation for the many-particle distribution function f_n, one must give a self-consistent formulation of FDT, where not only the spectral density of fluctuations X_i and the tensor α_{ij}, but also the spectral density of random forces F_i are determined by the properties of the analyzed system.

The task of constructing an equation for the distribution function of states of macroscopic systems has drawn the attention of research workers for a long time [11.10, 16]. Of the more recent work we refer to the article by *Kukharenko* [11.17], see also [11.22, Chap. 24]. In contrast to that article, we do not stress the weak interaction of these systems with the environment; the emphasis is on the interaction of particles within the system. To be specific, we further assume that H_1 includes weak (in the dipole approximation) interaction through a fluctuational electromagnetic field.

In the future we shall call a system with the Hamiltonian $\hat{H}_0$ basic, and a system with the Hamiltonian $\hat{H}_0 + \hat{H}_1$ extended.

11.5 Fluctuations of the Distribution Function of the Density Matrix. The Random Source in the Liouville Equation

The density matrix $f_{nm}(t)$ gives a complete statistical description of the basic system. Its counterpart in classical theory is the distribution function $f_N(x, t)$, which obeys the Liouville equation. The two-time distribution function $f_N(x, t, x', t')$ also obeys the Liouville equation since the variables x' and t' enter as parameters. Here x is the set of coordinates and momenta of the basic system.

The solution of the Liouville equation is necessary at the initial $(t = t')$ condition,

$$f_N(x, t, x', t')|_{t=t'} = \delta(x-x') f_N(x, t) , \quad \int f_N(x, t) dx = 1 , \tag{11.5.1}$$

in order to determine the two-time distribution function. The designation for the deviation of the two-time distribution function from the product of one-time distribution functions at each of the two instants is

$$\langle \delta f_N \delta f_N \rangle_{x, t, x'; t'} = f_N(x, t; x', t') - f_N(x, t) f_N(x', t') . \tag{11.5.2}$$

For coinciding moments of time, taking (11.5.1) into account, we hence find

$$\langle \delta f_N \delta f_N \rangle_{x, x'; t} = \delta(x-x') f_n(x, t) - f_N(x, t) f_N(x', t) . \tag{11.5.3}$$

The utility of these relations will be appreciated when we turn to the extended system. Describing the system with the aid of the function $f_N(x, t)$ (or f_{nm}) is then no longer complete, since in this case the state of the system is determined by the variables of the field as well as of the particles. Thus function $f_N(x, t)$ is a random function (see [11.23] and [11.22, Chap. 24]).

As a consequence, for the extended system the expressions in the left-hand sides of (11.5.2, 3) can be interpreted as two-time and one-time correlators of fluctuations of the distribution function, respectively. The functions f_N in the right-hand sides must then be treated as the corresponding averaged functions; so in the future $\langle f_N \rangle \equiv f_N$ (the angular brackets $\langle \rangle$ mean averaging over the whole ensemble of extended systems).

Among the corresponding definitions for quantum systems, the equation which is the counterpart to (11.5.3) is

$$\langle \delta f_{nm}(t) \delta f^*_{n_1 m_1}(t) \rangle = \delta_{nn_1} f_{m_1 m} - f_{nm} f_{m_1 n_1} . \tag{11.5.4}$$

If the density matrix is nonequilibrium although diagonal, this definition is simplified, and after symmetrization becomes

$$\langle \delta f_{nm}(t) \delta f^*_{n_1 m_1}(t) \rangle = \tfrac{1}{2} \delta_{nn_1} \delta_{mm_1} (f_m + f_n) - \delta_{nm} \delta_{n_1 m_1} f_n f_{n_1} . \tag{11.5.5}$$

Let us treat this equation as the initial ($t = t'$) condition for the equation for the two-time correlator [11.18],

$$\left(\frac{\partial}{\partial t} + \Delta + \mathrm{i}\,\omega_{nm} \right) < \delta f_{nm}(t) \delta f^*_{n_1 m_1}(t') > = 0 \quad (\Delta \to 0) . \tag{11.5.6}$$

Recall that, e. g., for a system of free charged particles, the correlator of phase density fluctuations has the structure of (8.2.1). Then first term in the right-hand side is determined by the correlation function. For a single-component system this contribution is proportional to $N(N-1)$, where N is the number of similar objects in the system under consideration. The second term (in brackets) is determined by one-particle distribution functions. It can be interpreted as the "source" of the fluctuations when calculating the spectral densities (Chaps. 8 – 10).

Now we are dealing with fluctuations of the distribution function of the whole basic system; there is therefore just one object present. Since the correlation function characterizes the statistical connection between different objects, the contribution to the one-time correlator δf_N determined by the correlation function is absent. This leaves only one term in the expression for the correlator; this is the "source", determined by the distribution function itself. As we shall see, the name "source" here is also worthwhile. The source determines the contribution of the fluctuations of the basic system in the expressions for the correlators of the extended system.

Consider now the corresponding expression for the spectral density. Jumping ahead a little, we assume that the relaxation time of the distribution function $f_n(t)$ for the extended system is large compared to the characteristic time of the two-time correlator (τ_{cor}). In zero approximation with respect to the ratio of these times ("collisionless" approximation in calculating fluctuations), we find with the aid of (11.5.6) the following expression for the spectral density of fluctuations of the density matrix of the basic system [11.18]:

$$(\delta f_{nm} \delta f_{n_1 m_1})_\omega = \pi \delta(\omega - \omega_{nm}) \delta_{nn_1} \delta_{mm_1} (f_m + f_n) , \quad \omega \neq 0 . \tag{11.5.7}$$

This spectral density depends implicitly (via the distribution function f_n) on time. The condition that $\omega \neq 0$ excludes the term with $\delta(\omega)$.

Note that the spectral density (11.4.21) can now be presented as

$$(X_i X_j)_\omega = \sum_{nm} \sum_{n_1 m_1} (X_i)_{nm} (X_j^*)_{n_1 m_1} (\delta f_{nm} \delta f_{n_1 m_1})_\omega , \tag{11.5.8}$$

that is, as a quantum-mechanical average with the spectral density (11.5.7). Naturally, (11.5.7) is more general than (11.4.21).

Instead of using (11.5.6) with the initial condition given in (11.5.5), it is possible to write the corresponding Langevin equation, which is the Liouville equation (in classical theory) or the corresponding equation for the density matrix with a random (Langevin) source. In the quantum-mechanical case it can be written as

$$\left(\frac{\partial}{\partial t} + \Delta + i \omega_{nm} \right) \delta f_{nm}^{\text{source}} = y_{nm} , \quad \langle y_{nm} \rangle = 0 . \tag{11.5.9}$$

The correlator of the random source is determined by

$$\langle y_{nm}(t) y_{n_1 m_1}^*(t') \rangle = A_{nmn_1 m_1} \delta(t - t') ,$$

$$A_{nmn_1 m_1} = 2 \Delta \delta_{nn_1} \delta_{mm_1} \frac{f_m + f_n}{2} . \tag{11.5.10}$$

The equation for intensity includes an auxiliary parameter Δ; in the final results $\Delta \to 0$.

The solution of (11.5.10) brings us back to the expression for the spectral density of fluctuations $\delta f_{nm}^{\text{source}}$ (11.5.7). The superscript "source" reminds us that for the extended system the function $\delta f_{nm}^{\text{source}}$ is the source of fluctuations. This is sufficient reason to put the same subscript in the left-hand side of (11.5.7).

11.6 Fluctuations in an Extended System. The Polarization Approximation

Let us assume once more that the Hamiltonian is given by (11.4.1), but that force F_i is not external. Moreover, the mean value of force $\langle F_i \rangle$ is zero, and its fluctuations are determined by the distribution of atoms in the system under consideration. An example of this case will be given in the next section.

We assume the interaction to be small. Then in the equation for the density matrix fluctuation it is only possible to keep the linear term with respect to δF_i in the terms which are determined by the interaction H_1. As a result we get, instead of (11.5.9), a more general Langevin equation,

$$\left(\frac{\partial}{\partial t} + \Delta + i\omega_{nm}\right)\delta f_{nm} = \frac{i}{\hbar}(X_i)_{nm}(f_m - f_n)\,\delta F_i(t) + y_{nm}(t)\,. \tag{11.6.1}$$

Using (11.5.9), it is possible to write (11.6.1) in another form,

$$\left(\frac{\partial}{\partial t} + \Delta + i\omega_{nm}\right)(\delta f_{nm} - \delta f_{nm}^{\text{source}}) = \frac{i}{\hbar}(X_i)_{nm}(f_m - f_n)\,\delta F_i(t)\,. \tag{11.6.2}$$

This equation is similar, e. g., to the equation (8.1.5) for the fluctuation of the operator density matrix of pairs of charged particles (i. e., atoms) in the polarization approximation.

From (11.6.2) we find the Fourier component of the fluctuation of the density matrix,

$$\delta f_{nm}(\omega) = \delta f_{nm}^{\text{source}} - \frac{(X_i)_{nm}(f_m - f_n)}{\hbar(\omega - \omega_{nm} + i\Delta)}\,\delta F_i(\omega)\,. \tag{11.6.3}$$

We see that the fluctuation of the force $\delta F_i(\omega)$ determines the deviation of fluctuation in an extended system with the Hamiltonian $H = H_0 + H_1$ from the fluctuations in a system with the Hamiltonian H_0.

With the aid of the last equation we can find the Fourier component $\delta X_i(\omega)$,

$$\delta X_i(\omega) = \delta X_i^{\text{source}}(\omega) + \alpha_{ij}(\omega)\,\delta F_j(\omega)\,. \tag{11.6.4}$$

Here we have employed a designation for the corresponding function $\delta X_i^{\text{source}}$ in a system with the Hamiltonian $\hat{H}_0$, i. e., at $\hat{H}_1 = 0$,

$$\delta X_i^{\text{source}}(\omega) = \sum_{nm}(X_i)_{mn}\delta f_{nm}^{\text{source}}(\omega)\,. \tag{11.6.5}$$

The expression for the polarizability tensor coincides with (11.4.10). For that reason the expression for the tensor $i(\alpha_{ij}^* - \alpha_{ji})$ coincides with (11.4.11).

Comparing the expression for the spectral density $\delta X_i^{\text{source}}$ and (11.4.10), we arrive again at the Callen–Welton equation. Now, however, we can write it in the form

$$(\delta X_i\delta X_j)_\omega^{\text{source}} = \frac{\hbar i}{2}[\alpha_{ij}^*(\omega) - \alpha_{ji}(\omega)]\coth\frac{\hbar\omega}{2\mathcal{K}T}\,, \tag{11.6.6}$$

thus changing its meaning. It connects the spectral density of fluctuations in the system with the Hamiltonian $\hat{H}_0$ (at $H_1 = 0$) with the dissipative part of the susceptibility tensor. The latter determines the connection between the difference of fluctuations $\delta X_i(\omega) - \delta X_i^{\text{source}}(\omega)$ in an equilibrium system and the Hamiltonian $\hat{H} = \hat{H}_0 + \hat{H}_1$ with fluctuation of the force δF_i.

The difference in interpretations is by no means formal. The fluctuations $\delta F_i(t)$ do, in fact, occur within the system under consideration, and are therefore themselves determined by the fluctuations of the density matrix, described by (11.6.2). As a result we get a closed set of equations for fluctuations δf_{nm} and δF_i.

Naturally, equations which enable establishing the connection between the force fluctuations and the density matrix fluctuations depend on the concrete form of the interaction operator $\hat{H}_1$. In the next section we shall determine the energy H_1 in such a way that the desired equations will be the Maxwell equations.

Based on the set of equations for fluctuations δf_{nm} and δF_i, it is not only possible to calculate (or express via the distribution function f_n) the spectral densities of the source fluctuations (11.6.6). It is also possible to determine the spectral density of fluctuations δF_i, as well as the more general spectral densities, e. g., the function $(\delta f_{nm} \delta F_i)_\omega$. This approach allows us to generalize easily all the results for the case which features spatial dispersion as well as temporal.

Finally, one can obtain the corresponding results for nonequilibrium states, as well. In particular, one can find the kinetic equation describing the evolution of the distribution function $f_n(t)$ towards the equilibrium (Gibbs) distribution for a system with the Hamiltonian $\hat{H}_0$. The relaxation time is determined by the interaction H_1. In order to make this discussion less formal, and to facilitate the comparison with our earlier results, we shall now specify the type of interaction H_1.

11.7 The Fluctuations of Polarization and Field. The Callen – Welton Equation for Nonequilibrium States

Let us assume that the energy of the interaction between atoms in a system can be split into two parts. One of them (H_1) represents the interaction via the electromagnetic field in the dipole approximation,

$$H_1 = - \int P^m(R, t) E^m(R, t) dR . \tag{11.7.1}$$

Here P^m and E^m are the vector of polarization and the strength of the electric field. The superscript m denotes them as microscopic (operator) characteristics. The remaining interaction between atoms is included in the Hamiltonian H_0; this interaction can be any strength. Our principal assumption is that the interaction H_1 is small.

The polarization vector $P(R, t)$ is defined as (3.3.32)

$$P^m(R, t) = e \sum_{1 \leqslant i \leqslant N} r_i \delta(R - R_i) = e \int r N_{ab}(R, P, r, p, t) \, dr \, dp \, dP ; \tag{11.7.2}$$

hence the matrix element of H_1 is

$$(H_1)_{nm} = - \int \Psi_n^*(X) \left[\int \sum_i e r_i \delta(R - R_i) E^m(R, t) dR \right] \Psi_m(X) \frac{dX}{V^{2N}} . \tag{11.7.3}$$

Here $X = (r_1, \ldots, r_N, R_1, \ldots, R_N)$, and Ψ_n is the eigenfunction of the Hamiltonian H_0 (11.4.2).

The equation for the density matrix of the system with the Hamiltonian $\hat{H} = \hat{H}_0 + \hat{H}_1$ is

$$\left(\frac{\partial}{\partial t} + i\omega_{nm}\right)f_{nm} = -\frac{i}{\hbar}\sum_{n_1}[(H_1)_{nn_1}f_{n_1m} - f_{nn_1}(H_1)_{n_1m}]. \tag{11.7.4}$$

It is a direct corollary of the Schrödinger equation,

$$i\hbar\frac{\partial\Psi}{\partial t} = (\hat{H}_0 + \hat{H}_1)\,\Psi(X, t). \tag{11.7.5}$$

Averaging over the whole ensemble in (11.7.4) we consider those states in which the mean value of the density matrix can be presented as [cf. (8.1.1)]

$$\langle f_{nm}\rangle = \delta_{nm}\langle f_n\rangle \equiv \delta_{nm}f_n. \tag{11.7.6}$$

Here $f_n(t)$ is a nonequilibrium distribution function which, as we shall see, relaxes to the Gibbs distribution because of the interaction H_1.

Naturally, the kinetic stage of evolution exists for the off-diagonal elements of the density matrix as well. However, at small interaction H_1 the contribution from off-diagonal elements to the collision integral is small. In this approximation the dissipative matrix in the equation for the function $f_{nm}(t)$ is expressed through the transition probabilities which determine the collision integral $I_n(t)$ (cf. the results of Sect. 10.4).

After averaging we set $n = m$ in (11.7.4). Under condition (11.7.6), only the correlator of fluctuations δf_{nm}, $(\delta H_1)_{nm}$ is left in the right-hand side, while the terms with f_n drop out.

From (11.7.3) it follows that

$$(\delta H_1)_{nm} = -\int D_{nm}(R)\,\delta E(R, t)\,dR. \tag{11.7.7}$$

Here we have introduced the designation

$$D_{nm}(R) = \int \Psi_n^*(X) \sum_{1\leqslant i\leqslant N} er_i\delta(R - R_i)\,\Psi_m(X)\,\frac{dX}{V^{2N}}. \tag{11.7.8}$$

Thus, the equation for the distribution function $\langle f_n\rangle \equiv f_n$ can be written [cf. (10.1.18)]

$$\frac{\partial f_n}{\partial t} = \frac{2}{\hbar}\sum_m \int \text{Im}\{\langle\delta f_{nm}\delta E(R, t)\rangle\}D_{mn}(R)\,dR \equiv I_n(t). \tag{11.7.9}$$

Here we use the designation $I_n(t)$ for the corresponding collision integral.

In (11.7.3, 7) it is necessary to heed the distinction between the microscopic and the local fields (see Sect. 2.4). Strictly speaking, (11.7.7) only holds when the mean local, not the mean microscopic, field equals zero. This distinction is important, e. g., in the determination of the effective Lorentz field (see Sects. 6.4, 11.2). Later we shall raise this subject again.

The equation for the fluctuation δf_{nm} follows from (11.6.2),

$$\left(\frac{\partial}{\partial t} + \Delta + i\omega_{nm}\right)(\delta f_{nm} - \delta f_{nm}^{\text{source}})$$

$$= \frac{i}{\hbar}(f_m - f_n)\int D_{nm}(R)\,\delta E(R, t)\,dR\,. \tag{11.7.10}$$

Hence we find the expression for the Fourier component,

$$\delta f_{nm}(\omega) = \delta f_{nm}^{\text{source}}(\omega) - \frac{\int D_{nm}(R')\,\delta E(R', \omega)\,dR'}{\hbar[(\omega + i\Delta) - \omega_{nm}]}(f_m - f_n)\,. \tag{11.7.11}$$

The Fourier component of the polarization vector is linked to the function $\delta f_{nm}(\omega)$ by

$$\delta P(R, \omega) = \sum_{nm} D_{nm}(R)\,\delta f_{nm}(\omega)\,. \tag{11.7.12}$$

The connection between fluctuations of polarization and field follows from the last two equations,

$$\delta P_i(R, \omega) = \int \alpha_{ij}(R, R', \omega)\,\delta E_j(R', \omega)\,dR' + \delta P_i^{\text{source}}(R, \omega)\,. \tag{11.7.13}$$

The polarizability tensor and the function $\delta P_i^{\text{source}}$ are given by

$$\alpha_{ij}(R, R', \omega) = \sum_{nm}[D_i(R)]_{mn}[D_j(R')]_{nm}\frac{f_n - f_m}{\hbar(\omega + i\Delta - \omega_{nm})}\,, \tag{11.7.14}$$

$$\delta P_i^{\text{source}}(R, \omega) = \sum_{nm}[D_i(R)]_{mn}\,\delta f_{nm}^{\text{source}}(\omega)\,. \tag{11.7.15}$$

In (11.7.13 – 15) one can carry out expansions in the Fourier integrals over R, R'. To single out the contribution corresponding to spatially uniform distribution of atoms, we define the susceptibility tensor $\alpha_{ij}(\omega, k)$ as

$$\alpha_{ij}(\omega, k)$$

$$= \frac{1}{V}\sum_{nm}[(D_i(R))_{mn}(D_j(R'))]_{nm}\exp[-ik(R - R')]\,dR\,dR'\,\frac{f_n - f_m}{\hbar(\omega + i\Delta - \omega_{nm})}\,, \tag{11.7.16}$$

which reflects averaging over the volume of the system. Consequently we obtain the connection between the Fourier components of the fluctuations of polarization and field vectors,

$$\delta P_i(\omega, k) = \alpha_{ij}(\omega, k) \, \delta E_j(\omega, k) + \delta P_i^{\text{source}}(\omega, k) \,, \tag{11.7.17}$$

$$\delta P_i^{\text{source}}(\omega, k) = \sum_{nm} [D_i(k)]_{mn} \delta f_{nm}^{\text{source}}(\omega) \,. \tag{11.7.18}$$

The permittivity tensor for a homogenous isotropic medium has the structure

$$\varepsilon_{ij} = \frac{k_i k_j}{k^2} \varepsilon^{\|}(\omega, k) + \left(\delta_{ij} - \frac{k_i k_j}{k^2} \right) \varepsilon^{\perp}(\omega, k) \,. \tag{11.7.19}$$

The expressions for the functions $\varepsilon^{\|}$ and $\varepsilon^{\perp}$ can be found by juxtaposing (11.7.16) and (11.7.19),

$$\varepsilon^{\|}(\omega, k) = 1 + \frac{4\pi}{\hbar V} \sum_{nm} \frac{|(k \cdot D_{nm}(-k))|^2}{k^2} \frac{f_n - f_m}{\omega + i\Delta - \omega_{nm}} \,, \tag{11.7.20}$$

$$\varepsilon^{\perp}(\omega, k) = 1 + \frac{2\pi}{\hbar V} \sum_{nm} \frac{|[k \times D_{nm}(-k)]|^2}{k^2} \frac{f_n - f_m}{\omega + i\Delta - \omega_{nm}} \,. \tag{11.7.21}$$

Using these results, we can find the expression for the spectral density of the field fluctuations from the field equations. The overall structure of this function has, as earlier, the form of (10.1.1). Both the structure of the spectral densities of the fluctuations of longitudinal and transverse fields [see (4.4.14, 5.1)] and (6.1.26) remain the same. Only the expression for the spectral density of fluctuations of polarization δP^{source} needs to be obtained.

From (11.7.18) and the equation for the spectral density of fluctuations $\delta f_{nm}^{\text{source}}$ (11.5.7) we find

$$(\delta P \delta P)_{\omega, k}^{\text{source}} = \frac{\pi}{V} \sum_{nm} \delta(\omega - \omega_{nm}) |D_{nm}(-k)|^2 (f_n + f_m) \,. \tag{11.7.22}$$

In order to single out the contributions in (6.1.26) for the spectral densities of fluctuations of the transverse and longitudinal fields, one must employ the identity

$$|D_{nm}(k)|^2 = \frac{1}{k^2} |(k \cdot D_{nm})|^2 + \frac{1}{k^2} |[k \times D_{nm}]|^2 \,. \tag{11.7.23}$$

The final expressions for the spectral densities of the field fluctuations are

$$(\delta E^{\|} \delta E^{\|})_{\omega, k} = \frac{16\pi^3}{V} \sum_{nm} \frac{|(k \cdot D_{nm}(-k))|^2}{k^2} \frac{\delta(\omega - \omega_{nm})(f_n + f_m)}{|\varepsilon^{\|}(\omega, k)|^2} \,, \tag{11.7.24}$$

$$(\delta E^{\perp}\delta E^{\perp})_{\omega,k} = \frac{16\pi^3}{V} \sum_{nm} \frac{|[k \times D_{nm}(-k)]|^2}{k^2} \frac{\delta(\omega - \omega_{nm})(f_n + f_m)\omega^4}{|\omega^2\varepsilon^{\perp}(\omega,k) - c^2k^2|^2}.$$

$$(11.7.25)$$

These spectral distributions [over (ω, k)] depend on the form of the distribution function. If the function f_n is the Gibbs distribution, then the Callen – Welton equations for the fluctuations of longitudinal and transverse fields follow from (11.4.22, 7.20, 21, 24, 25),

$$(\delta E^{\|}\delta E^{\|})_{\omega,k} = \frac{4\pi\hbar \, \mathrm{Im}\{\varepsilon^{\|}(\omega,k)\}}{|\varepsilon^{\|}(\omega,k)|^2} \coth\left(\frac{\hbar\omega}{2\mathscr{K}T}\right),$$

$$(11.7.26)$$

$$(\delta E^{\perp}\delta E^{\perp})_{\omega,k} = \frac{8\pi\hbar\omega^4 \, \mathrm{Im}\{\varepsilon^{\perp}(\omega,k)\}}{|\omega^2\varepsilon^{\perp}(\omega,k) - c^2k^2|^2} \coth\left(\frac{\hbar\omega}{2\mathscr{K}T}\right).$$

$$(11.7.27)$$

These expressions correspond to [Ref. 11.11, Eq. (9.40)]. Equations (11.7.24, 25) are more general however, since they can also be employed for nonequilibrium states.

Indeed, from (11.7.9) the relaxation time of the distribution function f_n towards the equilibrium distribution is determined by the interaction in (11.7.1), which is small compared to the energy H_0. The relaxation time can therefore be considered to be large with respect to the characteristic times of the system with the Hamiltonian $\hat{H}_0$.

This again allows distinguishing between small- and large-scale fluctuations, and treating (11.7.24, 25) as the spectral densities of small-scale fluctuations. In the jargon of plasma theory one can say that (11.7.24, 25) correspond to the "collisionless" region. However, they themselves determine the collision integral. Thus, we encounter the same situation as in the kinetic theory of gases and plasmas.

11.8 The Kinetic Equation Giving the Distribution Function for the States of a System of Interacting Atoms

In order to find the collision integral in (11.7.9), one has to calculate the correlation of fluctuations δf_{nm} and δE. It suffices to employ (11.7.10) for the function δf_{nm}, and the field equations and (11.5.7) for the spectral density of fluctuations $\delta f_{nm}^{\mathrm{source}}$. Since all the calculations are similar to those carried out earlier in searching for the collision integrals for a plasma, we shall just list the final results here.

In Chaps. 8, 9 we separately found the expressions for the collision integrals of partially ionized plasma with Coulomb interaction (Chap. 8) and with electromagnetic interaction (Chap. 9). For the present system the collision integral can also be presented as a sum of two contributions; it is determined by the fluctuations of the longitudinal and transverse fields, i.e., [11.18]

$$I_n(t) = I_n^{\parallel} + I_n^{\perp} . \tag{11.8.1}$$

The expression for $I_n^{\parallel}$ can be written

$$I_n^{\parallel} = \frac{1}{3\hbar^2(2\pi)^3} \sum_m \int d\omega \, dk \, |D_{nm}(-k)|^2 \, \delta(\omega - \omega_{nm})$$

$$\times \left[(\delta E^{\parallel} \delta E^{\parallel})_{\omega,k}(f_m - f_n) - \frac{4\pi\hbar \, \mathrm{Im}\{\varepsilon^{\parallel}(\omega,k)\}}{|\varepsilon^{\parallel}(\omega,k)|^2}(f_m - f_n) \right], \tag{11.8.2}$$

which must be supplemented by (11.7.20, 24) for the permittivity and for the spectral density of fluctuations of the longitudinal field. In obtaining (11.8.2) we have taken into account that in a homogeneous and isotropic medium the tensor

$$[D_\alpha(k)]_{nm}[D_\beta(-k)]_{nm} = \tfrac{1}{3}\delta_{\alpha\beta}|D_{nm}(k)|^2 . \tag{11.8.3}$$

The collision integral $I_n^{\perp}$ is given by a similar expression,

$$I_n^{\perp} = \frac{1}{3\hbar^2(2\pi)^3} \sum_m \int d\omega \, dk \, \delta(\omega - \omega_{nm})|D_{nm}(-k)|^2$$

$$\times \left[(\delta E^{\perp} \delta E^{\perp})_{\omega,k}(f_m - f_n) - \frac{8\pi\hbar\omega^4 \, \mathrm{Im}\{\varepsilon^{\perp}(\omega,k)\}}{|\omega^2\varepsilon^{\perp}(\omega,k) - c^2k^2|^2}(f_m + f_n) \right]. \tag{11.8.4}$$

It must be supplemented by (11.7.21, 25).

This equation formally coincides with (9.5.4), although they differ in essence. Indeed, (9.5.4) determines the collision integral in the kinetic equation for one-particle distribution function of atoms $f_n(P, t)$. Here n is the set of quantum numbers of a separate atom, and P is the atom's momentum. Equation (11.8.4) determines the collision integral in the kinetic equation for the distribution function f_n of a set of N particles, however strongly interacting they are. The number N can be large enough for the system under consideration to be a macroscopic body. Only the interaction through the electromagnetic field, the interaction H_1, is considered small. All the information referring to the interaction of atoms which is not related to dipole interaction through the field is stored in the matrix elements D_{nm} (11.7.8).

From (11.8.2, 4) with (11.4.22) taken into account, it follows that

$$I_n^{\parallel} = 0 , \qquad I_n^{\perp} = 0 , \tag{11.8.5}$$

if the function f_n is the Gibbs distribution, and the spectral densities of fluctuations of the longitudinal and transverse fields are determined by the Callen – Welton equations (11.7.26, 27).

As in the case of a plasma (8.5.1, 2), two forms of presenting the collision integral (11.8.1) are possible here. One of them is determined (11.8.2, 4), which corresponds to the form that (8.5.1) for partially ionized plasma was presented in.

Writing the collision integral I_n in the form of the Boltzmann collision integral, cf. (8.5.2), we substitute into (11.8.2, 4) the expressions (11.7.24, 25) for the spectral densities of fluctuations of the longitudinal and transverse fields and the expressions for the functions $\mathrm{Im}\{\varepsilon^{\|}\}$ and $\mathrm{Im}\{\varepsilon^{\perp}\}$. The latter follow from (11.7.20, 21). As a result we get the following expression:

$$
I_n = \frac{4}{9Vh} \sum_{mn_1 m_1} \int d\omega\, dk\, |D_{nm}(-k)|^2 |D_{n_1 m_1}(-k)|^2 \delta(\omega - \omega_{nm})
$$

$$
\times \delta(E_n + E_{m_1} - E_{n_1} - E_m) \left[\frac{1}{|\varepsilon^{\|}(\omega, k)|^2} + \frac{2\omega^4}{|\omega^2 \varepsilon^{\perp}(\omega, k) - c^2 k^2|^2} \right]
$$

$$
\times (f_m f_{n_1} - f_{m_1} f_n) .
\tag{11.8.6}
$$

As in Sect. 8.5, it is possible to investigate the general properties of this collision integral. Here we discuss just one of these. We multiply (11.8.6) by $-\mathscr{K} \ln f_n$, carry out summation over n, and perform double symmetrization: 1) $n \rightleftarrows m_1$, $m \rightleftarrows n_1$; and 2) $n \rightleftarrows m$, $n_1 \rightleftarrows m_1$. Then it becomes evident that

$$
-\mathscr{K} \sum_n \ln f_n(t) I_n(t) \geqslant 0 .
\tag{11.8.7}
$$

This property ensures fulfillment of the increase-of-entropy law,

$$
S(t) = -\mathscr{K} \sum_n \ln f_n \times f_n , \quad \frac{dS(t)}{dt} \geqslant 0 ,
\tag{11.8.8}
$$

for the system of interacting atoms with the Hamiltonian $\hat{H}_0$. The equals sign corresponds to the state of equilibrium.

Let us make three remarks.

1) The kinetic equation for the distribution function f_n with the collision integrals (11.8.2, 4, 6) describes only those dissipative processes that are determined by the interaction H_1, which is assumed to be small. The nature of this interaction can be quite diverse. In our example H_1 is the interaction of atoms via the field in the dipole approximation. Accordingly, the increase of the entropy (11.8.8) is due only to those dissipative processes which are determined by this equation.

2) In obtaining the collision integrals (11.8.2, 4), we took into account only the correlations of small-scale fluctuations, i.e., of those fluctuations whose correlation times are smaller than the relaxation times in the kinetic equation with the collision integrals (11.8.2, 4). The relaxation time is determined by the weak interaction H_1 and is therefore large. As in the theory of plasmas and gases (see Sects. 4.8, 8.7 and [Ref. 11.3, Chaps. 4, 11]), one can also develop a kinetic theory of large-scale fluctuations whose characteristic times are determined by the interaction H_1.

3) The kinetic equation with the collision integral (11.8.6) leads to the law of conservation of mean energy $\langle \hat{H}_0 \rangle$, though not the mean energy of the extended

system $\langle \hat{H}_0 + \hat{H}_1 \rangle$. This implies in our approximation that the weak interaction H_1 determines only the dissipative processes, and does not contribute to the law of conservation of mean energy. This situation is similar to that encountered in the kinetic theory of gases and plasmas [11.3]. For example, in the Boltzmann kinetic equation the interaction of atoms determines only the dissipative process. The contribution of this interaction to the thermodynamic functions is not accounted for. For a more complete account of the interaction, the time lag must be taken into account when obtaining the kinetic equations. Thus, the contribution of the interaction H_1 is not accounted for in the expression for the entropy (11.8.8), which is clear if only from the fact that the latter does not depend upon the energy H_1 in the state of equilibrium. To take the interaction H_1 into account, the expression for the entropy must be changed (cf. the results in [Ref. 11.3, Sect. 14]).

That the entropy reaches its maximum in the end state, when f_n is the Gibbs distribution, can be proved without employing the kinetic equation provided the mean energy $\langle \hat{H}_0 \rangle$ (not the total energy $\langle \hat{H}_0 + \hat{H}_1 \rangle$) is conserved in the relaxation process.

Indeed, let us compare the entropy values calculated with (11.8.8) for two distribution functions: the Gibbs distribution function f_n and an arbitrary distribution f_n^1. For the latter one can use a nonequilibrium solution of the kinetic equation for a certain moment of time t, e. g., $f_n^1 = f_n(t)$. We shall compare these two values at two conditions,

$$\sum_n f_n = \sum_n f_n^1 = 1 , \quad \sum_n E_n f_n^1 = \sum_n E_n f_n \equiv \langle \hat{H}_0 \rangle , \tag{11.8.9}$$

i. e., given normalization and given conservation of the mean energy of the basic system in the process of evolution toward the state of equilibrium.

Employing the method proposed by *Gibbs* himself, one can prove that [11.19]

$$S \geqslant S^1 . \tag{11.8.10}$$

The equality corresponds to the case when $f_n = f_n^1$, i. e., to the arrival at the state of equilibrium.

Finally, let us estimate the time relaxation of the function $f_n(t)$ towards the Gibbs distribution, e. g., for the logarithm of the distribution function. From (11.7.20, 21) and the definition of the matrix element D_{nm} (11.7.8) it follows that — in the thermodynamic limit with the condition that the correlation radii of quantum-mechanical distributions are finite — the squares of the moduli of matrix elements $|D_{nm}|^2$ increase as N. Consequently, the spectral densities of field fluctuations (11.7.24, 25) and the polarizabilities $\alpha^{\parallel}$ and $\alpha^{\perp}$ are proportional to the concentration $n = N/V$.

The dependence of the collision integrals (11.8.2, 4) on the number of particles in the transition to the thermodynamic limit is therefore determined by the dependence of $|D_{nm}|^2$ on the number of particles. Since $|D_{mn}|^2$ increases as N, then $I_n \propto N$. The relaxation time is estimated by

$$I_n \sim - \tau_{\text{rel}}^{-1} ; \quad \text{therefore} \quad \tau_{\text{rel}}^{-1} \propto N . \tag{11.8.11}$$

Hence the time derivative of the logarithm of the distribution function, per particle, is determined only by the concentration of particles.

The relaxation time determined by the interaction H_1 characterizes the establishment of local equilibrium (local Gibbs distribution) on the kinetic stage of evolution. The time necessary to arrive at total equilibrium is determined by the "hydrodynamic" stage, and depends, naturally, on the dimensions of the system.

11.9 The Transition to the Kinetic Equation for One-Particle Distribution Functions of Atoms

The kinetic equation giving the distribution function of the states of a system made up of strongly interacting particles can serve as a basic one for obtaining simpler kinetic equations. The simplest of these is the kinetic equation for one-particle distribution functions of the states of separate atoms. Naturally, the transition to the kinetic equation for one-particle distribution functions of the states of separate atoms (equations of this kind were discussed in Chaps. 8 – 10) is only possible under additional simplifying assumptions.

The first of these is the assumption that the internal motion and the motion of an atom as a whole are statistically independent. This is possible if the potential energy of an atom in the Hamiltonian H_0 only depends on the positions of atoms $R_1, \ldots, R_N$, and not on the internal variables. Then

$$H_0 = \sum_{1 \leqslant i \leqslant N} \left[\frac{p_i^2}{2\mu} + \Phi(r_i) \right] + \sum_{1 \leqslant i \leqslant N} \frac{P_i^2}{2M} + V(R_1, \ldots R_N) ; \tag{11.9.1}$$

therefore the eigenfunctions of the operator H_0 can be presented in the form

$$\Psi_n(X) = \prod_{1 \leqslant i \leqslant N} \Psi_{n_i}(r_i) \, \Psi_{P_1, \ldots P_N}(R_1, \ldots, R_N) . \tag{11.9.2}$$

If the potential energy U equals zero, and thus all the interactions of atoms are included in H_1, then the structure of the function Ψ_n is further simplified,

$$\Psi_n(X) = \prod_{1 \leqslant i \leqslant N} \Psi_{n_i}(r_i) \exp\left(i \frac{P_i R_i}{\hbar} \right). \tag{11.9.3}$$

The matrix element $D_{nm}(-k)$, see (11.7.8), can be presented as

$$D_{nm}(-k)$$

$$= \sum_{1 \leqslant i \leqslant N} d_{n_i m_i} \frac{(2\pi\hbar)^3}{V} \delta(\hbar k - (P_i' - P_i'')) \prod_{l \neq i} \delta_{n_l m_l} \frac{(2\pi\hbar)^3}{V} \delta(P_l' - P_l'') . \tag{11.9.4}$$

The square of the modulus of this matrix element is determined by the double sum $\sum\limits_{ij}$. However, because correlations of the directions of vectors d_{nm} for different atoms are absent in the adopted approximation, the double sum only includes the diagonal elements after averaging over angles. As a result, the square of the modulus of the matrix element is

$$|D_{nm}(-k)|^2 = \sum_{1 \leqslant i \leqslant N} |d_{n_i m_i}|^2 \frac{(2\pi\hbar)^3}{V} \delta(\hbar k - (P_i' - P_i'')) \prod_{l \neq i} \delta_{n_l m_l}$$

$$\times \frac{(2\pi\hbar)^3}{V} \delta(P_i' - P_i'') . \tag{11.9.5}$$

Let us substitute this equation into the equations for permittivity $\varepsilon^\perp$ (10.7.21) and spectral density $(\delta E^\perp \delta E^\perp)_{\omega,k}$ (10.7.25). All the terms of the sum $\sum\limits_i$ are equivalent with respect to the sum $\sum\limits_{nm}$, so one can set, e.g., $i = 1$, and multiply the sum $\sum\limits_{nm}$ by N, that is, by the number of equivalent terms of the sum over i.

Since according to (11.9.5) the states $n_l P_l'$ and $m_l P_l''$ coincide for all the atoms except $i = 1$, we can carry out convolution (summation over $n_2, \ldots, n_N$, $m_2, \ldots, m_N$, and integration over $P_2', \ldots, P_N', P_2'', \ldots, P_N''$). As a result (11.7.21) becomes (the subscript 1 is dropped),

$$\varepsilon^\perp(\omega, k) = 1 + \frac{2\pi}{V} \frac{N}{\hbar} \sum_{nm} \int \frac{V}{(2\pi\hbar)^3} dP' \, dP'' \, \delta(\hbar k - (P' - P''))$$

$$\times \frac{f_n(P', t) - f_m(P'', t)}{\omega + i\Delta - (E_n + E_{P'} - E_m - E_{P''})/\hbar} . \tag{11.9.6}$$

After integration over $(P' - P'')$, this expression is reduced to (9.1.7). Similar transformations can be carried out in (11.7.25), reducing it to (9.2.4). After similar transformations the collision integral (11.8.4) is reduced to the form of (9.5.2).

In the course of these transformations one has to substitute (11.9.5) into (11.8.4), and carry out summation and integration over all the variables but n_1, P_1', taking into account that all the terms in (11.9.5) with $i \neq 1$ give a zero contribution. Consequently, dropping the subscript 1, we arrive at (9.5.2).

Thus, for the special case of a Hamiltonian when all the interaction takes place through the field and dipole approximation is employed, the kinetic equation for the many-particle distribution function of a system of atoms is reduced to the kinetic equation for the one-particle distribution function of the states of a single atom [the function $f_n(P, t)$].

Recall that in Sects. 10.6, 7 we have discussed the influence of atomic interaction on spectral line broadening, and in Sect. 10.6 considered resonance collisions. The equation for the many-particle distribution function provides the basis to describe more complicated resonance interactions, when a group of atoms are involved in a single collision.

11.10 The Transparency Region. Probabilities of Transition

Let us single out the contribution from the values of ω and k, which are linked through the dispersional equations (10.1.2), in the collision integrals (11.8.2, 4). This time, however, the permittivities $\varepsilon^\parallel$ and $\varepsilon^\perp$ will be given by more general expressions (11.7.20, 21). In our example we shall also consider the collision integral $I_n^\perp$ [see (11.8.4)].

If we restrict ourselves to the case with only temporal dispersion [set $k = 0$ in the equations for $\varepsilon^\perp(\omega, k)$ and $|D_{nm}(k)|^2$], then integration can be carried out with respect to k. By employing the value of (9.5.5), we get the following expression for the collision integral:

$$
I_n^\perp = \frac{1}{3\hbar^2} \sum_m |D_{nm}(0)|^2 \left\{ (\delta E^\perp \delta E^\perp)_{\omega_{nm}} [f_m(t) - f_n(t)] \right.
$$
$$
\left. - \frac{2\hbar\omega_{nm}^3}{c^3} \sqrt{\mathrm{Re}\{\varepsilon(\omega_{nm})\}} [f_n(t) + f_m(t)] \right\}.
\tag{11.10.1}
$$

Although it coincides in form with (9.5.6), it is far more complicated since f_n and f_m here are the distribution functions of the states of a system composed of interacting atoms, regardless how strongly they interact.

With the aid of (11.7.8), the square of the matrix element in (11.10.1) can be presented as

$$
|D_{nm}(0)|^2 = |\textstyle\int P_{nm}(R)\,dR|^2,
\tag{11.10.2}
$$

where $P(R)$ is the polarization operator of the whole system with the Hamiltonian $\hat{H}$.

The Einstein coefficients for the system under consideration are

$$
B_m^n = \frac{4\pi^2 |D_{nm}(0)|^2}{3\hbar^2}, \qquad A_m^n = \frac{4|D_{nm}(0)|^2 \omega_{nm}^3}{3\hbar c^3} \sqrt{\mathrm{Re}\{\varepsilon\}}.
\tag{11.10.3}
$$

They differ from those introduced earlier [see (9.5.9)] in that the matrix elements of dipole moments have been substituted for the matrix elements of the polarization vector of the whole system of atoms, i.e.,

$$
|d_{nm}|^2 \rightarrow |D_{nm}(0)|^2.
\tag{11.10.4}
$$

Thus, the processes of spontaneous and induced emission of radiation are now given by the matrix elements of the polarization of a system of atoms.

Making use of (11.10.3) and the definition of the electromagnetic energy distribution over frequencies for nonequilibrium states (9.5.8), the collision integral (11.10.1) can be written in the form of (9.5.10). As in (9.5.10), the collision integral is split into two parts. The first covers the transition with $\omega_{nm} > 0$, while the other those with $\omega_{nm} < 0$.

Finally, notice that the collision integral for our system can be written [cf. (10.4.2)]

$$I_n = \sum_m [W_{mn} f_m(t) - W_{nm} f_n(t)] . \tag{11.10.5}$$

Here we have introduced the relevant probabilities of transition [cf. definitions (10.4.2 – 5)]. The dissipative matrix can then be expressed as

$$\gamma_{nm} = \frac{1}{2} \sum_{n_1} (W_{nn_1} + W_{mn_1}) , \tag{11.10.6}$$

which determines the spectral linewidths for the transitions between states for a system of atoms, although not for a single atom.

Consider a simple example. Let the temperature be zero ($\hbar \omega_{nm} \gg \mathcal{K} T$), and a resonant transition between two multiparticle states be singled out. Then for γ_{nm} we get an equation similar to (10.4.11),

$$\gamma_{nm} = \frac{1}{2} A^n_m = \frac{2}{3} \frac{|D_{nm}(0)|^2 \omega^3_{nm}}{\hbar c^3} \sqrt{Re\{\varepsilon\}} . \tag{11.10.7}$$

In contrast to its counterpart (10.4.11), this transition is given by the matrix element of the dipole moment of the whole system. In the special case when the square of modulus of the matrix element is represented in the form of (11.9.5) at $\hbar k = 0$, the expression for the transition between states of a single atom, e.g., with $i = 1$, coincides with (10.4.11) [from (11.10.7)].

The calculation of the fluctuations and kinetic processes in Sects. 11.4 – 10 was carried out in the polarization approximation. This approximation corresponds to perturbation theory with respect to interaction H_1, though with the polarization of the medium also being taken into account. It has been assumed that the mean values of the polarization vector and of the field were zero, and thus that only the second moments of fluctuations δP_n and δE were nonzero. Now we shall turn to a totally different approximation, in which the first moments of the vectors P and E are nonzero and fluctuations are not taken into account. Such an approximation can be called the first-moments approximation (cf. Sect. 4.3), or the approximation of the self-consistent field.

11.11 The Distribution Function of a System of Atoms and Mean Field. The First-Moments Approximation

In the first-moments approximation the equation for the mean matrix of density follows from (11.7.4),

$$\left(\frac{\partial}{\partial t} + i \omega_{nm} \right) \langle f_{nm} \rangle = \frac{i}{\hbar} \sum_{n_1} \int [D_{nn_1}(R) \langle f_{n_1 m} \rangle - \langle f_{nn_1} \rangle D_{n_1 m}(R)] E(R, t) dR . \tag{11.11.1}$$

Here we used (11.7.3) and the definition of the matrix element (11.7.8); $E(R, t)$ is the mean field.

The mean polarization vector is linked to the density matrix by, cf. (11.7.12),

$$P(R, t) = \sum_{nm} D_{mn}(R) \langle f_{nm}(t) \rangle. \tag{11.11.2}$$

The charge and current densities in the dipole approximation are given, using the polarization vector, by

$$q(R, t) = -\operatorname{div} P, \quad j(R, t) = \frac{\partial P}{\partial t}. \tag{11.11.3}$$

Equations (11.11.1) together with the Maxwell equations form a closed set.

Consider a special solution of the set of equations of the self-consistent field,

$$\langle f_{nm} \rangle = \delta_{nm} f_n, \quad P = 0, \quad E = 0, \tag{11.11.4}$$

where f_n is a stationary distribution, in particular the Gibbs distribution. The equations giving the linear approximation with respect to deviations from the special solution (11.11.4) provide the wave properties of the system. The connection between ω and k is given by the dispersional equations (10.1.2). This time, however, (11.7.20, 21), where f_n and f_m are stationary distribution, must be employed for the functions $\varepsilon^{\parallel}(\omega, k)$ and $\varepsilon^{\perp}(\omega, k)$.

In this chapter we have just been able to touch on a few of the great variety of problems in the kinetic theory of electromagnetic processes in strongly interacting systems. The most interesting of these concern the kinetic theory of equilibrium and nonequilibrium coherent states in such systems, e. g., the kinetics of phase transitions and superradiation. Some problems of this kind will be the subject of our discussion in the following chapters. To conclude this chapter, we shall consider the influence of the correlations of atoms' positions on the scattering of light.

11.12 The Influence of the Correlations of Atoms' Positions on the Absorption and Scattering of Electromagnetic Waves

In Sects. 6.4, 5 we have shown for the example of a classic system of atom oscillators that three important effects are defined by the correlation of the atoms' positions: the Lorentz field (Sect. 6.4), the dependence of the time of spontaneous emission on the thermodynamic state of the system (Sect. 6.5), and frequency shift. In Sects. 11.1 – 3 we obtained the corresponding results on the basis of quantum theory.

In this section we shall consider the absorption and scattering processes on the basis of classical theory. From (6.7.1) we get the equation for dielectric permittivity (6.7.4) and hence for the complex coefficient of refraction,

$$n + i\varkappa = \sqrt{\varepsilon(\omega)} \ . \tag{11.12.1}$$

The extinction coefficient is connected with the absorption coefficient by the equations

$$h = \frac{4\pi}{\lambda_0}\varkappa, \quad \lambda_0 = \frac{\lambda}{n} \ . \tag{11.12.2}$$

The set of equations for the refraction index n and the absorption coefficient $\varkappa$, from (6.7.4, 11.12.1), is

$$n^2 - \varkappa^2 = 1 + \frac{4\pi e^2 N}{\mu V} \frac{\tilde{\omega}_0^2 - \omega^2}{(\tilde{\omega}_0^2 - \omega^2)^2 + \omega^2 \gamma_{\text{eff}}^2} , \tag{11.12.3}$$

$$n\varkappa = \frac{2\pi e^2 N}{\mu V} \frac{\omega \gamma_{\text{eff}}}{(\tilde{\omega}_0^2 - \omega^2)^2 + \omega^2 \gamma_{\text{eff}}^2} . \tag{11.12.4}$$

According to (6.5.6), in these equations

$$\gamma_{\text{eff}} = \frac{2e^2\omega^2}{3\mu c^3} \frac{\overline{(\delta N_{v_{\text{ph}}})^2}}{\bar{N}_{v_{\text{ph}}}} . \tag{11.12.5}$$

When the frequency of the incident wave is such that

$$\tilde{\omega}_0^2 - \omega^2 \gg \gamma_{\text{eff}}^2 ,$$

then it is far from resonant. Since in this case $\varkappa \ll n$, $\varkappa^2$ can be neglected in comparison with n^2 in the first equation. Excluding in this case the difference $\tilde{\omega}_0^2 - \omega^2$ from (11.12.3, 4), we obtain the following equation for the extinction coefficient:

$$h = \frac{(n^2 - 1)^2}{n\lambda_0} \frac{V}{2e^2 N} \omega \gamma_{\text{eff}} . \tag{11.12.6}$$

Assuming the refraction index to be close to 1, one can simplify this formula. Using the equation for the dissipative coefficient γ_{eff} (11.12.5) and observing that the wavelength in vacuum is $\lambda_0 = 2\pi c/\omega$, we get the following expression for the extinction coefficient [11.20]:

$$h = \frac{32\pi^3}{3} \frac{(n-1)^2}{\lambda^4 N/V} \frac{\overline{(\delta N_{v_{\text{ph}}})^2}}{\bar{N}_{v_{\text{ph}}}} . \tag{11.12.7a}$$

For the ideal gas

$$\frac{\overline{(\delta N_{v_{\mathrm{ph}}})^2}}{\overline{N}_{v_{\mathrm{ph}}}} = 1, \quad \text{so} \quad h = \frac{32\,\pi^3(n-1)^2}{3\,\lambda^4 N/V}, \tag{11.12.7b}$$

the result first obtained by Rayleigh [Ref. 11.21, Eq. (96.3)].

According to (6.5.9), the extinction coefficient can usually be expressed in terms of isothermal compressibility. Thus

$$h = \frac{32\,\pi^3(n-1)^2}{3\,\lambda^4} \frac{\mathcal{K}T}{\rho} \left(\frac{\partial\rho}{\partial p}\right)_{\mathrm{T}}, \tag{11.12.8}$$

where ρ is the mass density. This result coincides with the well-known Einstein equation for the extinction coefficient [Ref. 11.21, Eq. (96.2)] if the temperature dependence of the permittivity is insignificant, while

$$\rho\left(\frac{\partial\varepsilon}{\partial\rho}\right)_{\mathrm{T}} \approx \varepsilon - 1 \approx 2(n-1). \tag{11.12.9}$$

The extinction coefficient characterizes the scattering of light as well. Indeed, the extinction coefficient is determined as a ratio of the total (sum over all directions) intensity of scattered light to the flux density of incident light (per volume unit of scattering medium) (see [Ref. 11.21, Sect. 95]). From this definition it follows that (11.12.7) characterizes the integral (over all angles) intensity of the scattered light. Naturally, we are referring here to scattering which brings about little change in frequency, the so-called Rayleigh scattering.

From the Rayleigh formula (11.12.7b) it follows in particular, that for an ideal gas the intensity of scattered light is inversely proportional to the fourth power of the wavelength of the incident light. On approaching the critical point of the gas – liquid transition, the intensity of scattered light according to (11.12.7a) markedly increases, which explains the phenomenon of so-called critical opalescence [Ref. 11.21, Sect. 97].

Similar results can also be obtained on the basis of quantum theory, presented here in Sects. 11.1 – 3. It is important to bear in mind that the substitution (11.3.2) is not possible, in handling the problem of the frequency-dependent extinction coefficient h. In other words, the instant frequency ω cannot be replaced by the transition frequency ω_{nm} in the dissipative terms.

12. Fluctuations in Quantum Self-Oscillatory Systems

The temporal and spatial correlations in a quantum autooscillatory system are analyzed using a quantum optical generator as an example. The fluctuations in amplitude, phase, and energy are treated as examples of Brownian motion at a nonequilibrium phase transition (the crossing of generation threshold).

12.1 A System Composed of Two-Level Atoms and a Field

In Sects. 5.10 – 14 we discussed several subjects which belong to the theory of fluctuations in classical self-oscillatory systems. Our next task is to describe fluctuations in quantum self-oscillatory systems. This can be done for the examples of a quantum optical generator (laser) and a molecular generator.

To clarify which are the general properties, we shall start our discussion with the simplest case, i.e., when the atoms of the active medium can be considered immobile. Such a model can be applied to solid-state lasers as well as to gas lasers with uniformly broadened lines.

As in the case of classical generators, the fluctuations in lasers can be classified into technical and natural. Technical fluctuations are due to comparatively slow changes of parameters, e.g., thermal fluctuations of the cavity length. They can be reduced by various improvements of the device. Natural fluctuations are due to the molecular structure of the active medium and the walls of the cavity, and are therefore irreducible. They determine the limitations to the generator's stability, which is measured by the ratio $\Delta\omega/\omega$, where $\Delta\omega$ is the emission linewidth of the quantum generator. The spectrum of natural fluctuations is as a rule wider than that of technical fluctuations, the width of which is of the order $10^3 - 10^4$ Hz. This allows detecting natural fluctuations on the background of stronger technical fluctuations.

Two sources of natural fluctuations can be named: first, thermal fluctuations in the cavity, and second, fluctuations of polarization of the active medium. They are determined by the atomic structure of the active medium, and by the spontaneous emission of atoms. Of course, fluctuations of the active medium polarization are nonequilibrium. It is specifically one of the principal tasks of the theory of natural fluctuations in lasers to calculate their spectral characteristics.

In our calculations of fluctuations in lasers, pumping, which creates an inverse population in atoms belonging to the active medium, is assumed to be given in advance. In this approximation it is sufficient to investigate the evolution of the population of only two levels a and b, of which a is the upper one. This is the so-called two-level approximation.

The state of the system is determined by the four functions f_a, f_b, f_{ab}, f_{ba}. The equations for these functions follow from the kinetic equations for the functions f_n and f_{nm}. For the collision integrals in the kinetic equations we employ the simplest model representations,

$$I_a = -\gamma(f_a - f_a^0), \quad I_b = -\gamma(f_b - f_b^0), \quad I_{ab} = -\gamma_{ab} f_{ab}. \tag{12.1.1}$$

Here γ and γ_{ab} are the dissipative constants, and f_a^0 and f_b^0 are the given distributions, determined by the pumping. In the absence of pumping f_a^0 and f_b^0 are the equilibrium distributions.

In place of the functions f_a and f_b it is more practical to employ their combinations,

$$D = f_a - f_b, \quad R = f_a + f_b. \tag{12.1.2}$$

The quantity D is called the difference in populations.

The equations for the functions $f_a, f_b, f_{ab} = f_{ba}^*$ can be obtained, e.g., from (10.1.5), with the following simplifying assumptions:

1) dipole approximation (the term with $\partial/\partial P$ drops out);
2) the approximation of immobile atoms (the term with $V \partial/\partial R$ drops out); and
3) the diagonal elements d_{nn} equal zero.

Setting successively in (10.1.5) $n = m = a$; $n = m = b$; $n = a$, $m = b$; $n = b$, $m = a$; and using the equations for the collision integrals (12.1.1), we get the following equations:

$$\frac{\partial D}{\partial t} = \frac{2i}{\hbar} (d_{ab} f_{ba} - f_{ab} d_{ba}) E(R, t) - \gamma(D - D^0), \tag{12.1.3}$$

$$\frac{\partial R}{\partial t} = -\gamma(R - R^0), \tag{12.1.4}$$

$$\left(\frac{\partial}{\partial t} + \gamma_{ab} + i\omega_{ab}\right) f_{ab} = -\frac{i}{\hbar} d_{ab} D E(R, t), \tag{12.1.5}$$

$$f_{ba} = f_{ab}^*. \tag{12.1.6}$$

Notice that the dependence on the coordinates enters these equations only through the function $E(R, t)$. Since the function R does not enter the equations for D, f_{ab}, f_{ba}, the equation for R can be solved separately,

$$\partial R/\partial t = 0, \quad R = R^0. \tag{12.1.7}$$

Recall that the function f_n is normalized in the following manner:

$$\sum_n \int f_n(R, t) \frac{dR}{V} = 1. \tag{12.1.8}$$

The polarization vector is linked to the functions D, f_{ab}, R by the equation,

$$P(R, t) = \frac{N}{V}(d_{ba} f_{ab} + d_{ab} f_{ba}) + \frac{1}{2}(d_{aa} - d_{bb})D + \frac{1}{2}(d_{aa} + d_{bb})R . \quad (12.1.9)$$

If only off-diagonal matrix elements d_{nm} are nonzero, then this expression is simplified,

$$P(R, t) = \frac{N}{V}(d_{ba} f_{ab} + d_{ab} f_{ba}) . \quad (12.1.10)$$

Sometimes it would be more convenient to employ the equation for the polarization vector,

$$\frac{\partial^2 P}{\partial t^2} + 2\gamma_{ab} \frac{\partial P}{\partial t} + \omega_{ab}^2 P = -\frac{2}{3} \frac{|d_{ab}|^2}{\hbar} n \omega_{ab} DE , \quad n = \frac{N}{V} , \quad (12.1.11)$$

instead of (12.1.5, 6).

The first term in the right-hand side of (12.1.3) can also be expressed in terms of the vector P. Then it becomes

$$\frac{\partial D}{\partial t} + \gamma(D - D^0) = \frac{2}{\hbar \omega_{ab} n} \left(\frac{\partial}{\partial t} + \gamma_{ab} \right) P \cdot E . \quad (12.1.12)$$

If $\gamma_{ab} \ll \omega_{ab}$, the term with γ_{ab} can be neglected.

Thus, calculating the polarization is reduced to finding the solution of the system of two equations for the functions D and P. For the description of fluctuations to be complete, one must also include the equation for E. Considering only the transverse field, the equation for E has the form

$$\frac{\partial^2 E}{\partial t^2} + \gamma_E \frac{\partial E}{\partial t} - c^2 \Delta E = -4\pi \frac{\partial^2 P^\perp}{\partial t^2} ,$$

$$\text{div } E = 0 , \quad \gamma_E \equiv \Delta \omega_{\text{r}} , \quad (12.1.13)$$

where $\Delta \omega_{\text{r}}$ is the halfwidth of the resonator band, which is determined by thermal loss in the cavity and escape of radiation.

In obtaining (12.1.11) we employed the equality

$$(d_{ab})_i (d_{ba})_j = \tfrac{1}{3} \delta_{ij} |d_{ab}|^2 , \quad (12.1.14)$$

true for a homogeneous and isotropic medium [cf. (11.8.3)]. Recall that the collision integrals I_a and I_{ab} are determined by small-scale fluctuations, whose correlation times are much smaller than the relaxation times, determined by the collision integrals themselves (times $1/\gamma_a$, $1/\gamma_{ab}$).

The fluctuations which determine the statistical properties of laser radiation are described by (12.1.3 – 6) and are therefore large-scale (kinetic) fluctuations. The outline of the kinetic theory of fluctuations is given in Sect. 4.8 (for more details see [Ref. 12.1, Chaps. 4, 11]). Here we shall employ the results of this theory in the approximation in which the collision integrals are given by the simple expressions (12.1.1).

In the kinetic theory of fluctuations in quantum generators, new characteristic parameters appear. One of them, $\Delta \omega_a$, determines the spectrum width of the fluctuations in amplitude, and the other, $\Delta \omega = D_\varphi$, the spectrum width of the radiation set free by phase diffusion; D_φ is the corresponding diffusion coefficient.

The relative values of the dissipative constants vary with different types of quantum generators. For instance, for gas lasers with a nonuniformly broadened lines we have the inequalities

$$ku \gg \gamma, \quad \gamma_{ab} \gg \Delta \omega_r \gtrsim \Delta \omega_a \gg \Delta \omega, \tag{12.1.15}$$

where ku is the Doppler broadening. For gas lasers with a uniformly broadened line one can employ the approximation using immobile atoms, and set $ku = 0$.

In a molecular generator [12.2, 3]

$$\Delta \omega_a \gg \gamma \gg \Delta \omega, \tag{12.1.16}$$

and therefore the spectrum of fluctuations in amplitude is the widest.

In solid-state lasers the spectral density of the amplitude fluctuations is a nonmonotonic function of frequency. It can be approximately considered to be combined of two lines, the wide line $\Delta \omega_a$ and the narrow line $\Delta \omega_{a1}$. The relative values of the dissipative parameters are given by the inequalities [12.3]

$$\gamma_{ab} \gg \Delta \omega_a \gg \Delta \omega_{a_1} \sim \gamma \gg \Delta \omega. \tag{12.1.17}$$

Here we shall calculate the fluctuations under the conditions given in (12.1.15), first for a system of immobile atoms ($ku = 0$). Then we shall discuss the changes brought about by the motion of atoms.

12.2 Stationary Generation Regime, Without Taking Fluctuations into Account

The equations for the mean values $\langle D \rangle$, $\langle f_{ab} \rangle$, $\langle E \rangle$ (fluctuations not taken into account) coincide with (12.1.3 – 6, 13). The field $\langle E \rangle$ is set in the form of a running wave,

$$\langle E(R, t) \rangle = \tfrac{1}{2}[E(\omega_0, k_0) \exp [-i(\omega_0 t - k_0 R)] + \text{c.c.}] . \tag{12.2.1}$$

To calculate the polarization, we seek the solution for functions $\langle D \rangle$, $\langle f_{ab} \rangle$ in the form

$$\langle D \rangle = D = \text{const} , \tag{12.2.2}$$

$$\langle f_{ab}(R, t) \rangle = f_{ab}(\omega_0, k_0) \exp[-i(\omega_0 t - k_0 R)] , \quad f_{ba} = f_{ab}^* , \tag{12.2.3}$$

$$\langle P(R, t) \rangle = \tfrac{1}{2}[P(\omega_0, k_0) \exp[-i(\omega_0 t - k_0 R)] + \text{c.c.}] . \tag{12.2.4}$$

From (12.1.10, 2.2, 3) follows the expression

$$\tfrac{1}{2}P(\omega_0, k_0) = n d_{ba} f_{ab}(\omega_0, k_0) . \tag{12.2.5}$$

Substituting (12.2.2 – 4) into (12.1.3 – 6) and retaining the resonant terms, we get a set of algebraic equations which connect the functions D and f_{ab} with the function $E(\omega_0, k_0)$,

$$\gamma(D - D^0) = \frac{i}{\hbar}[d_{ab} f_{ba} E(\omega_0, k_0) - d_{ba} f_{ab} E^*] , \tag{12.2.6}$$

$$i(\omega_{ab} - \omega_0 - i\gamma_{ab}) f_{ab} = -\frac{i}{2\hbar} D d_{ab} E , \quad f_{ba} - f_{ab}^* . \tag{12.2.7}$$

Hence the expression for the difference of populations is

$$D = D^0 \frac{(\omega_{ab} - \omega_0)^2 + \gamma_{ab}^2}{(\omega_{ab} - \omega_0)^2 + \gamma_{ab}^2(1 + a_s|E|^2)} . \tag{12.2.8}$$

Here we have used the designation for the "saturation parameter" a_s, which characterizes the influence of the field on the difference in populations,

$$a_s = \frac{|d_{ab}e|^2}{\hbar^2 \gamma \gamma_{ab}} = \frac{1}{3} \frac{|d_{ab}|^2}{\hbar^2 \gamma \gamma_{ab}} . \tag{12.2.9}$$

The connection between the complex amplitudes of the polarization vectors and the field is given by

$$P(\omega_0, k_0) = \alpha(\omega_0, k_0) E(\omega_0, k_0) , \tag{12.2.10}$$

and the dielectric permittivity

$$\varepsilon(\omega_0, k_0) = 1 + \frac{8\pi|d_{ab}|^2 n}{3\hbar} \frac{D}{\omega_0^2 - \omega_{ab}^2 + i2\gamma_{ab}\omega_0} . \tag{12.2.11}$$

From (12.2.8, 11) it follows that the permittivity depends on the field intensity $|E|^2$. The dependence on k_0 enters implicitly via $|E(\omega_0, k_0)|^2$. In order to find the generation frequency and the field amplitude, we substitute (12.2.1, 4) in the field equation and employ (12.2.10). As a result we get the dispersional equation,

$$\omega_0^2 \varepsilon(\omega_0, k_0) + i \omega_r \Delta \omega - c^2 k_0^2 = 0 . \tag{12.2.12}$$

Hence, equating the real and the imaginary parts to zero separately, we get two equations,

$$\omega_0^2 \, \mathrm{Re}\{\varepsilon(\omega_0, k_0)\} - c^2 k_0^2 = 0 , \tag{12.2.13}$$

$$\mathrm{Im}\{\varepsilon(\omega_0, k_0)\} + \frac{1}{Q} = 0, \quad \omega_0 > 0 ; \tag{12.2.14}$$

where $Q = \omega_0 / \Delta \omega_r$ is the quality of the resonator.

From (12.2.8, 11) it follows that the imaginary part of permittivity in the resonance approximation $(\gamma_{ab} \ll \omega_{ab})$

$$\mathrm{Im}\{\varepsilon(\omega_0, k_0)\} = - \frac{4\pi |d_{ab}|^2 n}{3\hbar} \frac{D_0 \gamma_{ab}}{(\omega_0 - \omega_{ab})^2 + \gamma_{ab}^2 (1 + a|E|^2)} \tag{12.2.15}$$

is negative at $D^0 > 0$, when the population of the upper level is greater than that of the lower one. Under this condition the equation for the determination of $|E|^2$ follows from (12.2.14). Since the real part of permittivity for gas lasers is close to one, (12.2.13) is simplified,

$$\omega_0^2 = c^2 k_0^2 . \tag{12.2.16}$$

Thus, we have determined the frequency of generation ω_0 (via k_0), and the field amplitude. Like in classical self-oscillatory systems, the phase of oscillations cannot be found from these equations either, but is determined by the initial distribution of fluctuations.

12.3 Sources of Fluctuations in a Quantum Generator

Now we proceed to find the spectral characteristics of the sources of fluctuations in the equations for the polarization vector and field (12.1.11 – 13), or in the corresponding equations for the functions D, f_{ab}, E. The fluctuations described by these equations are large-scale (kinetic) ones, since their correlation times are determined by the characteristic relaxation times $1/\gamma$ and $1/\gamma_{ab}$ of the kinetic equations for the distribution functions f_a and f_{ab}.

As in the kinetic theory of fluctuations for a system made up of free charged particles and a field (Sects. 4.8, 9), we also have two sources of fluctuations here. One of them is determined by the atomic structure of the active medium of the quantum generator, and the other by the structure of the field as a system of field oscillators.

The equation giving the fluctuation of the polarization vector follows from (12.1.11) and is similar to the classical equation (6.8.1).

The equations for the correlations of fluctuations δP^{source} and $\delta J^{\text{source}} = \partial \delta P^{\text{source}}/\partial t$ [cf. (6.8.2)] are

$$\frac{\partial}{\partial \tau}\overline{(\delta P\, \delta P)}^{\text{source}}_{\tau,R,R'} = \overline{(\delta J\, \delta P)}^{\text{source}}_{\tau,R,R'}, \tag{12.3.1}$$

$$\left(\frac{\partial}{\partial \tau} + 2\gamma_{ab}\right)\overline{(\delta J\, \delta P)}^{\text{source}}_{\tau,R,R'} + \omega^2_{ab}\overline{(\delta P\, \delta P)}^{\text{source}}_{\tau,R,R'} = 0. \tag{12.3.2}$$

These equations must be supplemented by the initial ($\tau = 0$) conditions, cf. (6.8.3). These follow from the equation giving the correlation of fluctuations of the density matrix [cf. (8.2.11, 14, 11.5.4)],

$$\overline{(\delta P\, \delta P)}^{\text{source}}_{\tau=0,R,R'} = n|d_{ab}|^2\,\delta(R-R')(f_a+f_b),$$

$$\overline{(\delta J\, \delta P)}^{\text{source}}_{\tau=0,R,R'} = 0. \tag{12.3.3}$$

Here we employ (12.1.10) for the fluctuations δP and δf_{ab}.

The solution of (12.3.1), given the initial conditions in (12.3.2), brings us to the following expression for the spectral density of fluctuations δP^{source}:

$$(\delta P^\perp \delta P^\perp)^{\text{source}}_{\omega,k} = \frac{2}{3}n|d_{ab}|^2\frac{\gamma_{ab}}{(\omega-\omega_{ab})^2+\gamma^2_{ab}}(f_a+f_b), \quad \omega>0. \tag{12.3.4}$$

In the limit $\gamma_{ab}\to 0$ this expression corresponds to (11.4.21).

Using (12.2.8, 15) for the difference in populations D and the imaginary part of permittivity, one can present (12.3.4) in the form

$$(\delta P^\perp \delta P^\perp)^{\text{source}}_{\omega,k} = \frac{\hbar}{2\pi}\text{Im}\{\varepsilon(\omega,k)\}\frac{f_b+f_a}{f_b-f_a}, \quad \omega>0. \tag{12.3.5}$$

For the regime of generation both multiplicands are negative. Indeed, in the generation regime $\text{Im}\{\varepsilon\} < 0$, while the difference in populations $D = f_a - f_b > 0$. Then, on the basis (12.2.14), (12.3.5) can be written as

$$(\delta P^\perp \delta P^\perp)^{\text{source}}_{\omega,k} = \frac{\hbar}{2\pi Q}\frac{f_a+f_b}{f_a-f_b}, \quad D = f_a-f_b > 0. \tag{12.3.6}$$

In the state of equilibrium both multiplicands in (12.3.5) are positive; by virtue of (11.4.22), (12.3.5) can be written as

$$(\delta P^\perp \delta P^\perp)^{\text{source}}_{\omega,k} = \frac{\hbar}{2\pi}\text{Im}\{\varepsilon(\omega,k)\}\coth\frac{\hbar\omega_{ab}}{2\mathcal{K}T}, \quad \omega>0. \tag{12.3.7}$$

In the classical limit this expression corresponds to (6.8.4).

In contrast to (11.7.27), ω_{ab} has replaced ω, in the argument of the coth. This is due to the fact the width of resonance is finite here, and that therefore ω_{ab} in (12.3.6) can be replaced by ω provided that $\hbar \gamma_{ab} \ll \mathscr{K}T$. At $\gamma_{ab} \sim 10^{11}$ to $10^{12} \, s^{-1}$ (solid-state lasers) this condition is not satisfied even at temperatures $T = 1$ to $10 \, K$.

In zero approximation with respect to γ_{ab}/ω_{ab} it is possible to carry out a limiting transition $\gamma_{ab} \to 0$ in (12.2.15) (at $a|E|^2 = 0$). It is also possible to employ (11.4.22), with the aid of which we get from (12.3.6) an equation which corresponds to the Callen – Welton equation (11.4.23). When comparing these expressions one must bear in mind that $\mathrm{Im}\{\varepsilon\} = 4\pi \, \mathrm{Im}\{\alpha\}$, and that for the fluctuations of the transverse field the tensor spur $\alpha_{\alpha\alpha}^\perp = 2\alpha^\perp$ [see (11.7.19)].

In the classical limit, (12.3.7) coincides with (6.8.4). In juxtaposing them, (6.8.6) must be taken into account. Notice that in the classical limit the structure of the equations is the same whether $\gamma_{ab} = 0$ or $\gamma_{ab} \neq 0$, a consequence of the fact that the "width" of distribution of the function $\mathscr{K}T$ over frequencies is infinite.

Considering now the spectral density of the source of field fluctuations, we introduce the source of the field fluctuations in the same way as in Chap. 4 [see (4.7.9, 17)],

$$(y_f y_f)_{\omega,k} = \frac{16\pi}{c^2 k^2} \gamma_k \hbar c k \left(\bar{n}_k + \frac{1}{2} \right) = \frac{16\pi}{Q} \hbar \left(\bar{n}_k + \frac{1}{2} \right)$$

$$= \frac{8\pi h}{Q} \cot \frac{\hbar c k}{2 \mathscr{K}T} . \tag{12.3.8}$$

Here we have used (4.7.1) for $\bar{n}_k$ (the Planck distribution) and the equality $\gamma_k = ck/Q$.

Thus we have found the equations giving the spectral densities of the sources of fluctuations in the equations for the polarization vectors and field, when the relative values of dissipative constants satisfy the inequalities in (12.1.15). The time required to establish both the polarization and the difference in populations D is much smaller than the characteristic correlation times of the radiation field. The field $E(R, t)$ can therefore be considered as nonfluctuating in calculating the fluctuations of polarization.

For these reasons the polarization vector of the active molecules of the laser medium can be presented as a sum of two parts,

$$P(R, t) = P^{\mathrm{ind}} + P^{\mathrm{source}} . \tag{12.3.9}$$

Here $P^{\mathrm{ind}}(E)$ is the induced part of polarization – the responce of the system to the total field $E = \langle E \rangle + \delta E$; $\delta P^{\mathrm{source}}$ is the source of polarization fluctuations, and its spectral density is given by (12.3.6).

The connection between the Fourier components of the polarization vector and the field is described by the common equation,

$$P^{\text{ind}}(\omega, k) = \frac{\varepsilon(\omega, k) - 1}{4\pi} E(\omega, k) . \qquad (12.3.10)$$

The permittivity $\varepsilon(\omega, k)$ entering this equation is given by (12.2.11) at $\omega_0 \to \omega$, $k_1 \to k$. From (12.2.8, 11) it follows that the permittivity is a nonlinear function of the square of the modulus of the complex field amplitude $|E|^2$.

Let us substitute the equation for the polarization vector (12.3.9) into the field equation (12.1.13), including also the source of field fluctuations y_f, whose spectral density is given by (12.3.8). The spectral density of the total source $y = 4\pi\delta P^{\text{source}} + y_f$ can be written as, by virtue of (12.3.6, 8), [12.3 – 6]

$$(y \cdot y)_{\omega, k} = \frac{16\pi\hbar}{Q} \left(\bar{n}_k + \frac{1}{2} + \frac{1}{2} \frac{f_a + f_b}{f_a - f_b} \right) . \qquad (12.3.11)$$

Now let us examine the fluctuations of the field, created by the source y.

12.4 Field Equations with Fluctuations Taken into Account

Substituting (12.3.10) into the field equation, we get a closed equation for the electric field strenght with a random source. The spectral density of the random source is given by (12.3.11).

To single out the slow movements in this equation, we introduce two new functions to replace the functions $E(R, t)$, $\dot{E}(R, t)$ [cf. (5.10.4)],

$$E(R, t) = \hat{e}[E_c \cos(\omega_0 t - k_0 R) + E_s \sin(\omega_0 t - k_0 R)] ,$$
$$\dot{E}(R, t) = \hat{e}[E_s \omega_0 \cos(\omega_0 t - k_0 R) - E_c \omega_0 \sin(\omega_0 t - k_0 R)] . \qquad (12.4.1)$$

Here $\hat{e}$ is the unit vector along $E(R, t)$. Assuming that our new functions only depend on time, their dependence on the coordinates is completely described by (12.4.1). This assumption is justified provided that the lifetime of a photon in the cavity L/c is much smaller than the characteristic times, determined by the spectrum width of amplitude fluctuations $\Delta\omega_a$ and by the width of the emission line $\Delta\omega$ (see the following section). Since $\Delta\omega_a \gg \Delta\omega$ according to (12.1.15), this condition can be written

$$\frac{L}{c} \ll \frac{1}{\Delta\omega_a} . \qquad (12.4.2)$$

Naturally, in calculating the spectral characteristics of spatial correlations the slow changes over coordinates must be taken into account as well (see Sect. 12.5). From the field equations after averaging over the volume of the cavity, V, (or over the period $2\pi/\omega_0$) we get the following equation for slowly changing functions E_c and E_s [cf. (5.10.7)]:

$$\frac{dE_c}{dt} + \frac{\omega_0}{2}\left(\frac{1}{Q} + \text{Im}\{\varepsilon\}\right)E_c = \omega_0 y_c(t),$$

$$\frac{dE_s}{dt} + \frac{\omega_0}{2}\left(\frac{1}{Q} + \text{Im}\{\varepsilon\}\right)E_s = \omega_0 y_s(t). \tag{12.4.3}$$

Here we have introduced the designations for the Langevin sources y_c and y_s,

$$y_{c,s} = \mp\frac{1}{V\omega_0^2}\int\frac{\partial^2}{\partial t^2}(e\cdot y(t))\begin{Bmatrix}\sin(\omega_0 t - k_0 R)\\\cos(\omega_0 t - k_0 R)\end{Bmatrix}dR, \quad \langle y_{c,s}\rangle = 0. \tag{12.4.4}$$

The imaginary part of permittivity is given by (12.2.15), while

$$|E|^2 = E_c^2 + E_s^2 \tag{12.4.5}$$

and therefore the imaginary part $\text{Im}\{\varepsilon\}$ is determined by the functions (12.4.3).
The correlations of the Langevin sources y_c and y_s follow from (12.3.11, 4.4),

$$\langle y_c(t)y_c(t')\rangle = \langle y_s(t)y_s(t')\rangle = D\delta(t-t'),$$

$$\langle y_c(t)y_s(t')\rangle = \langle y_s(t)y_c(t')\rangle = 0. \tag{12.4.6}$$

Both sources are δ correlated and have the same intensity, which is given by

$$D = D_c = D_s = \frac{(ey)^2\omega_0 k_0}{2V} = \frac{4\pi\hbar}{QV}\left(\bar{n}_{k_0} + \frac{1}{2} + \frac{1}{2}\frac{f_a+f_b}{f_a-f_b}\right). \tag{12.4.7}$$

Here we have employed (12.3.11). The additional factor $\frac{1}{2}$ appears here because (12.4.7) only accounts for the contribution of one of the two possible polarizations.

In Sect. 12.3 the fluctuations of polarization were calculated without the field being taken into consideration, see (12.3.1). Therefore (12.3.6) and consequently (12.4.7) are only true for a weak field, when $a_s|E|^2 \ll 1$. The field can be accounted for according to the same scheme [Ref. 3, Chap. 17]. The intensities D_c and D_s are then however no longer the same; they are given by

$$D_c = \frac{4\pi\hbar}{QV}\left(\bar{n}_{k_0} + \frac{1}{2} + \frac{1}{2}\frac{f_a+f_b}{f_a-f_b}\right),$$

$$D_s = \frac{4\pi\hbar}{QV}\left\{\bar{n}_{k_0} + \frac{1}{2} + \frac{1}{2}\frac{f_a+f_b}{f_a-f_b}[1 + a_s(E_c^2 + E_s^2)]\right\}. \tag{12.4.8}$$

Thus calculating the fluctuations of laser radiation under the condition given by (12.4.2) has been reduced to finding the solution of the Langevin equations

(12.4.3). In a weak field these equations coincide with the corresponding Langevin equations (5.10.7), which were employed to describe fluctuations in a classical self-oscillatory system.

Calculating the fluctuations of laser radiation given (11.1.15, 4.2) is similar to calculating the fluctuations in a classical generator of the Thomson type. The quantum nature of radiation only shows up when calculating the intensity of fluctuations of the Langevin sources. This permits us to take advantage of the results obtained in Sects. 5.10–14, which we shall discuss in the following section.

12.5 Fluctuations of Radiation in a Quantum Optical Generator

Similar to Sect. 5.12, we can here go from (12.4.3) to the amplitude and phase equations. The quantities A and φ are given by the relations

$$E_c = A \cos \varphi, \quad E_s = -A \sin \varphi, \quad A = \sqrt{E_c^2 + E_s^2} \equiv \sqrt{|E|^2}. \tag{12.5.1}$$

The amplitude and phase equations can be presented in the form given in (5.12.5). In a weak field ($a_s A^2 \ll 1$), the constants a and b used there are given by

$$a = \omega_0 \left(\frac{1}{Q} + \mathrm{Im}\{\varepsilon_{E=0}\} \right), \quad b = \omega_0 \left(\frac{\partial \mathrm{Im}\{\varepsilon\}}{\partial |E|^2/2} \right)_{E=0}. \tag{12.5.2}$$

The latter of these can be written in a more comprehensive way by dint of (12.2.11). At zero mistuning ($\omega_0 = \omega_{ab}$) we get the following expression:

$$b = 2\omega_0 |\mathrm{Im}\{\varepsilon_{E=0}\}|a_s = 2\omega_0 \frac{a_s}{Q}. \tag{12.5.3}$$

The saturation constant (12.2.9) here is denoted a_s to avoid being mixed up with a in (12.5.2).

The strength of the Langevin sources in the equations for amplitude and phase fluctuations is given by (12.4.7) since [cf. (5.10.10, 12.15)]

$$D_A = D_\varphi = D_c = D_s = D. \tag{12.5.4}$$

The spectrum of the amplitude fluctuations is given by (5.12.11), while according to (12.5.2, 3)

$$\Delta \omega_s = \omega_0 \left(|\mathrm{Im}\{\varepsilon_{E=0}\}| - \frac{1}{Q} \right) = \Delta \omega_{res} a_s |E|^2. \tag{12.5.5}$$

The second equality employs the designation for the resonator linewidth $\Delta \omega_{res} = \omega_0/Q$ and the equality $|a| = b|E^2|/2$.

In zero-order approximation with respect to the contribution of the amplitude fluctuations, the halfwidth of the spectrum of laser radiation is determined by the coefficient of phase diffusion D_φ. According to (5.13.7),

$$\Delta\omega = D_\varphi = \frac{D}{2\langle A\rangle_0^2}\omega_0^2; \quad \langle A^2\rangle_0 = 2\frac{|a|}{b}.$$

(12.5.6)

Using the defintion for the intensity of fluctuations (12.4.7) and the definition of power

$$P = \Delta\omega_{res}\frac{|E|^2}{8\pi}V,$$

(12.5.7)

one can express the full width $(2\Delta\omega)$ of the emission line in the form

$$2\Delta\omega = 2D_\varphi = \frac{\hbar\omega_0}{2P}(\Delta\omega_{res})^2\left(\bar{n}_{\omega_0} + \frac{1}{2} + \frac{1}{2}\frac{f_a+f_b}{f_a-f_b}\right).$$

(12.5.8)

This result has been obtained by *Haken* [12.5] and *Lax* [12.4]. In an earlier work by *Schawlow* and *Townes* [12.7] only the thermal fluctuations in the cavity were taken into account, which gave the following result:

$$2\Delta\omega = \frac{\hbar\omega_0}{2P}(\Delta\omega_{res})^2.$$

(12.5.9)

The juxtaposition of (11.5.8) and (11.5.9) reveals that under the same conditions (neglecting polarization fluctuations, and with $\bar{n}_\omega = 0$ by dint of $\hbar\omega_0/\mathcal{K}T \gg 1$) the latter yields a linewidth twice as great as (12.5.8) does.

On the basis of the results obtained in Sect. 5.13, one can find the spectral distribution of the mean energy as well as the upper limit of the linewidth, with amplitude fluctuations taken into account. In accordance with the said results, the spectral distributions for arbitrarily crossing the generation threshold can be approximated (in the vicinity of the threshold) as a Lorentz line (5.13.24). The full linewidth is then determined according to (5.13.25) by the expression (in Chaps. 5, 10, $D \leftrightarrow D\omega_0^2$)

$$2\Delta\omega = \frac{D\omega_0^2}{\langle E\rangle} = \frac{D\omega_0^2}{\langle A^2\rangle/2},$$

(12.5.10)

where $\langle E\rangle$ is the mean energy per swing. Hence for the region of well-developed generation, and employing the definition of the phase diffusion coefficient, we get the following estimate for the upper limit of the full linewidth:

$$2\Delta\omega = 4D_\varphi = \frac{\hbar\omega_0}{P}(\Delta\omega_{res})^2\left(\bar{n}_{\omega_0} + \frac{1}{2} + \frac{1}{2}\frac{f_a+f_b}{f_a-f_b}\right).$$

(12.5.11)

The Schawlow – Townes result thus follows if the fluctuations of polarization are neglected and if $\bar{n}_{\omega_0} = 0$.

This calculation of the fluctuations in a quantum optical generator is based on semiclassical theory since we do not take the quantization of the field into account. Such an approximation is justified if the number of quanta in the whole region of generation (including the region next to the threshold) is sufficiently high.

The average number of quanta $\langle n_{ph} \rangle$ at $a_s |E|^2 \ll 1$ can be estimated as

$$\langle n_{ph} \rangle = \frac{\langle |E|^2 \rangle}{8\pi\hbar\omega_0} V. \tag{12.5.12}$$

Employing (5.11.7) determines the mean energy per oscillation to be

$$\langle n_{ph} \rangle \sim \sqrt{\frac{D\omega_0^2}{b}} \frac{V}{\hbar\omega_0} \sim \sqrt{\frac{V}{\hbar\omega_0 a_s}}. \tag{12.5.13}$$

Hence for a cavity of volume $V = 1\,\text{cm}^3$, the frequency $\omega_0 = 10^{15}\,\text{s}^{-1}$, and the value of the saturation parameter $a_s = 10^2$ in CGS units,

$$\langle n_{ph} \rangle \sim 10^5. \tag{12.5.14}$$

Thus, even at the generation threshold the average number of photons is much greater than one.

The width of the emission line $\Delta\omega$ of a laser at commonly employed powers is of the order of a fraction of one c/s.

We have discussed here the simplest case, namely when the field is set in the form of a running wave. This is the case, e. g., of a ring laser, with suppression of one of the two opposing waves. A large number of studies deal with the calculation of fluctuations in more complex cases. One can get acquainted with them in [12.3, 4] as well as in [12.8, 9]. Experimental investigations of natural fluctuations in lasers are reported in [12.10 – 12] [1].

Note that other sources of natural fluctuations exist besides those discussed above. For instance, in a He – Ne laser the pumping of the operating levels of Ne atoms is determined by two processes. First, the helium atoms are excited by the impact of electrons. Then, in the collisions of He and Ne atoms there is an exchange of excited states: the helium atoms change to the ground state, and the neon atoms to excited state. Natural fluctuations in the electronic component then bring about fluctuations in pumping, which in turn produce additional fluctuations in radiation.

---

[1] A wider coverage of experimental investigations regarding fluctuations in lasers can be found in the reviews [12.3, 5, 6, 12].

12.6 Spatial and Temporal Correlations
of a Field Below the Generation Threshold

In calculating temporal correlations for states above the generation threshold we have employed the set of equations for slowly changing functions of time E_c and E_s (12.4.3). If we take into account the slow changes over coordinates as well, we obtain in first approximation the following set of equations instead of (12.4.3) [12.13, 14]:

$$\left\{\frac{\partial}{\partial t} + \frac{c^2}{\omega_0} k_0 \frac{\partial}{\partial R} + \frac{\omega_0}{2} \left[\frac{1}{Q} + \text{Im}\{\varepsilon(\omega)\}\right]\right\} \left\{\begin{matrix} E_c(r, t) \\ E_s(r, t) \end{matrix}\right\}$$

$$= \omega_0 \left\{\begin{matrix} y_c(t, R) \\ y_s(t, R) \end{matrix}\right\} = \mp \frac{1}{\omega_0^2} \frac{\partial^2}{\partial t^2} (e \cdot y(R, t)) \left\{\begin{matrix} \sin(\omega_0 t - k_0 R) \\ \cos(\omega_0 t - k_0 R) \end{matrix}\right\}. \qquad \begin{matrix}(12.6.1) \\ (12.6.2)\end{matrix}$$

These equations have been derived assuming the field to be transverse and spatial dispersion absent.

In (12.6.1, 2) the spatial changes of the field are only accounted for in the direction of propagation, i.e., in the direction of the vector k_0. Such an approximation naturally does not account for the confined nature of the laser beam. In other words, no provisions are made for the nonuniformity of the field in the plane normal to the direction of propagation.

In order to make the calculation of spatial correlations more complete, equations have to be employed which also include the operator $\Delta_\perp$, the Laplace operator in the plane normal to the direction of propagation [12.13, 14].

To calculate spatial and temporal field correlations below the generation threshold one can, naturally, employ the linear equations for transverse field fluctuations, without using the equations for slowly changing functions. For an isotropic medium we get the following equation for the Fourier component $\delta E(\omega, k)$ from (12.1.13, 2.10), with the random source (12.3.11):

$$[\omega^2 \varepsilon_{E=0}(\omega) + i\omega\gamma_k - c^2 k^2] \, \delta E(\omega, k) = \omega^2 y(\omega, k) \, , \quad \gamma_k = \frac{ck}{Q}. \qquad (12.6.3)$$

Hence the expression for the spectral density of transverse field fluctuations is

$$(\delta E \, \delta E)_{\omega, k} = \frac{\omega^4 (y \cdot y)_{\omega, k}}{|\omega^2 \varepsilon_{E=0}(\omega) + i\omega\gamma_k - c^2 k^2|^2}. \qquad (12.6.4)$$

For the states above the generation threshold the spectral density of fluctuations y is given by (12.3.11). For states below the generation threshold an alternative form of presentation is more practical where the spectral density of the source of the polarization fluctuations is given by (12.3.5). With the aid of (12.3.8), the spectral density $(y \cdot y)_{\omega, k}$ can be expressed as

$$(y \cdot y)_{\omega,k} = 8 \pi \hbar \left[\frac{1}{Q} \coth \left(\frac{\hbar c k}{\varkappa T} \right) + \mathrm{Im}\{\varepsilon(\omega)\} \frac{f_a + f_b}{f_b - f_a} \right], \quad \omega > 0. \quad (12.6.5)$$

Recall that in the regime of generation both multiplicands in the second term in parentheses are negative, while below the threshold both of them are positive.

To demonstrate the manner of obtaining the expression for the spectral density of the field fluctuations in the state of equilibrium from (12.6.4, 5), we employ the equation for the spectral density of the source of the field fluctuations (12.3.7). Observe that the resonance band on the function $\mathrm{Im}\{\varepsilon(\omega)\} \gamma \ll \varkappa T/\hbar$. In zero approximation the following substitution is possible:

$$\coth \left(\frac{\hbar \omega_{ab}}{2 \varkappa T} \right) \to \coth \left(\frac{\hbar \omega}{2 \varkappa T} \right). \quad (12.6.6)$$

Bearing in mind that we are interested in the contribution coming from the transparency region, we carry out a substitution in the first term of the right-hand side of (12.6.5),

$$\coth \left(\frac{\hbar c k}{2 \varkappa T} \right) \to \coth \left(\frac{\hbar \omega}{2 \varkappa T} \right). \quad (12.6.7)$$

Then from (12.6.4, 5) we get the following expression for the equilibrium spectral density of the field fluctuations:

$$(\delta E \, \delta E)_{\omega,k} = \frac{8 \pi \hbar \omega^4 [\gamma_k/\omega + \mathrm{Im}\{\varepsilon(\omega)\}]}{|\omega^2 \varepsilon(\omega) + i \omega \gamma_k - c^2 k^2|^2} \coth \left(\frac{\hbar \omega}{2 \varkappa T} \right). \quad (12.6.8)$$

In the classical limit this expression is transformed to (6.8.7). As in that equation, the integration here can also be carried out over k, and the contribution from the transparency region can be singled out. Then we arrive at the Planck formula

$$(\delta E \, \delta E)_{\omega} = \frac{2 \pi \hbar \omega^3}{c^3} \sqrt{\mathrm{Re}\{\varepsilon\}} \coth \left(\frac{\hbar \omega}{2 \varkappa T} \right), \quad (12.6.9)$$

where $\sqrt{\mathrm{Re}\{\varepsilon\}}$ is the refraction index of the medium.

In order to perform the corresponding transformations for the state of non-equilibrium, we multiply and divide (12.6.4) by

$$\frac{1}{Q} + \mathrm{Im}\{\varepsilon(\omega)\} = \frac{1}{Q} - |\mathrm{Im}\{\varepsilon(\omega)\}|, \quad (12.6.10)$$

where $\omega \approx \omega_{ab}$, and ω_{ab} is the frequency of transition between the inversely populated operating levels of a laser. For states below the generation threshold the quantity given by (12.6.10) is positive.

We can single out the contribution from the transparency region by using the relationship ($\omega \approx \omega_{ab}$)

$$\frac{\omega^2 \left[\dfrac{1}{Q} - \mathrm{Im}\{\varepsilon(\omega)\} \right]}{(\omega^2 \mathrm{Re}\{\varepsilon\} - c^2 k^2)^2 + \omega^4 \left[\dfrac{1}{Q} - \mathrm{Im}\{\varepsilon(\omega)\} \right]^2} \rightarrow \pi \delta(\omega^2 \mathrm{Re}\{\varepsilon(\omega)\} - c^2 k^2)$$

(12.6.11)

and integrating over k. This results in the expression for the spectral density of fluctuations for the frequencies $\omega \sim \omega_{ab}$ in the band of the order γ_{ab}, cf. (12.6.9),

$$(\delta E\, \delta E)_\omega = \frac{2 \pi \hbar \omega^3 \sqrt{\mathrm{Re}\{\varepsilon\}}}{c^3} \cdot \frac{\dfrac{1}{Q} \coth\left(\dfrac{\hbar \omega_{ab}}{2 \mathscr{K} T} \right) + |\mathrm{Im}\{\varepsilon(\omega)\}| \dfrac{f_a + f_b}{f_a - f_b}}{\dfrac{1}{Q} - |\mathrm{Im}\{\varepsilon(\omega)\}|}.$$

(12.6.12)

Thus, approaching the generation threshold the strength of the fluctuations at frequencies $\omega \approx \omega_{ab}$ (within the band γ_{ab}) increases. The behavior of this increase in fluctuations formally obeys Curie's law if

$$T - T_c \leftrightarrow \frac{1}{Q} - |\mathrm{Im}\{\varepsilon(\omega)\}|,$$

(12.6.13)

that is, if approaching the generation threshold corresponds to approaching the critical point.

12.7 Spatial and Temporal Correlations of the Fluctuations of Laser Radiation

In calculating the spectral characteristics of laser radiation one must take into account the spatial anisotropy created by the resonator [12.15, 16]. The presence of a resonator creates a preferred direction along the vector k_0 in the space of wave numbers, i.e., along the axis of the resonator. This was accounted for in the "shortened equations" for slowly changing functions $\delta E_{c,s}(R, t)$ (12.6.1, 2). Let us employ these equations to calculate the fluctuations above the generation threshold.

Using (12.5.2, 3) we present (in a weak field) the dissipative term in (12.6.1, 2) in the form

$$\frac{\omega_0}{2}\left(\frac{1}{Q} + \text{Im}\{\varepsilon\}\right)E_{c,s} = \frac{1}{2}(a + bE)E_{c,s}, \tag{12.7.1}$$

with the designation for energy E [cf. (5.10.8)]

$$E = \tfrac{1}{2}(E_c^2 + E_s^2).$$

Like in the classical theory of fluctuations in self-oscillatory systems (Sects. 5.10 – 14), (12.6.1, 2) form the basis for describing both fast and slow fluctuations.

As is known in the regime of well-developed generation the fast fluctuations are those of amplitude δA and energy δE. The latter is more interesting since it is directly related to the fluctuations in the intensity of radiation. Let us first discuss a simpler case, when the functions $\delta E_{c,s}$ only depend on the coordinate along the axis of the resonator. Then $\Delta_{\perp}\delta E_{c,s} = 0$, and the equation for the fluctuation of energy $\delta E(x, t)$ in the linear approximation is [cf. (5.10.19)]

$$\left[\frac{\partial}{\partial t} + \frac{c^2}{\omega_0}k_0\frac{\partial}{\partial x} + \frac{1}{2}(a + 2b\langle E\rangle)\right]\delta E(x, t) = \delta y_E(x, t). \tag{12.7.2}$$

The correlation of fluctuations of the Langevin source is given by an expression similar to (5.10.20),

$$\langle\delta y_E(x, t)\,\delta y_E(x', t')\rangle = 2LD\omega_0^2\langle E\rangle\delta(x-x')\delta(t-t'). \tag{12.7.3}$$

The diffusion coefficient is given by (12.4.7).

In this approximation ($\Delta_{\perp}\delta E_{c,s} = 0$), the Fourier components of the function $\delta E(x, t)$ only depend on the longitudinal component of the wave vector. The corresponding equation for the spectral density of fluctuations of energy is

$$(\delta E)^2_{\omega, k_{\parallel}} = \frac{2LD\langle E\rangle\omega_0^2}{(\omega - ck_{\parallel})^2 + (\Delta\omega_E)^2}. \tag{12.7.4}$$

At $k_{\parallel} = 0$ this formula (if divided by L) coincides with (5.11.9), since $(\delta E)^2_{\omega} = (\delta E)^2_{\omega, k_{\parallel}=0}/L$ and $D\omega_0^2 \to D$.

In order to estimate the radius of longitudinal correlations, let us examine (12.7.4) at $\omega = 0$. This means that (at zero mistuning) we can single out the value of the spectral density at the frequency of the operational transition ω_{ab}. From (12.7.4) we find

$$(\delta E)^2_{\omega=0, k_{\parallel}} = \frac{2LD\langle E\rangle\omega_0^2}{(ck_{\parallel})^2 + (\Delta\omega_E)^2}. \tag{12.7.5}$$

Hence the radius of correlation of longitudinal (depending but upon x) energy fluctuations is

$$r_{E, \parallel} \sim \frac{c}{\Delta \omega_E} \sim \frac{c}{\Delta \omega_A} \sim \frac{c}{|a|} . \tag{12.7.6}$$

Here we have taken into account that the linewidth $\Delta \omega_E$ is of the same order as the linewidth of amplitude fluctuations, $\Delta \omega_A$ see (5.11.13, 12.12).

The inequality in (12.4.2) being satisfied, the correlation radius $r_{E, \parallel}$ is much greater than the resonator length L. This substantiates our calculation of the fluctuations of laser radiation performed in Sects. 12.4, 5.

Finally, we have to investigate the behavior of spatial correlations in the plane normal to the resonator axis. For states far below the generation threshold, when nonlinear terms can be neglected [$b = 0$ in (12.7.1, 2)], we can immediately employ the expression for the spectral density of field fluctuations (12.6.4), without using the equations for slowly changing functions. In (12.6.4) we set

$$\omega = \omega_0 + \Omega, \quad k^2 = (k_0 + K)^2 = k_0^2 + 2 k_0 K + K^2 \tag{12.7.7}$$

and retain the principal terms with respect to parameters Ω / ω_0 and $k_\parallel / k_0$. Here ω_0 and k_0 are, respectively, the frequency and the wave number of the mode which is formed after crossing the generation threshold.

Using the dispersional equation [assuming that $\mathrm{Re}\{\varepsilon(\omega_{ab})\} = 1$], we get the following expression:

$$(\delta E \, \delta E)_{\Omega, k_\parallel k_\perp} = \frac{\omega_0^4 (y \cdot y)_{\omega_0, k_0}}{|2 \omega_0 (\Omega - c k_\parallel) + i \omega_0 a - c^2 k_\perp^2|^2} . \tag{12.7.8}$$

Here we set $\Omega = 0$, $k_\parallel = 0$. This amounts to singling out the spectral component at the frequency of the transition, and to singling out the principal longitudinal mode, which is justified by (12.4.2). Consequently we get

$$(\delta E \, \delta E)_{\Omega = 0, k_\parallel = 0, k_\perp} = \frac{\omega_0^2 (y \cdot y)_{\omega_0, k_0}}{a^2 + \dfrac{c^4}{\omega_0^2} k_\perp^4} . \tag{12.7.9}$$

Hence the correlation radius of spatial fluctuations in the plane normal to the resonator axis is

$$r_\perp \sim \frac{c}{\sqrt{\omega_0 |a|}} \sim \sqrt{\frac{|a|}{\omega_0}} \, r_{E, \parallel} . \tag{12.7.10}$$

Approaching the generation threshold, the correlation radius $r_\perp$ increases less rapidly than the correlation radius $r_\parallel$. Notice that the dependence of the correlation radius $r_\perp$ upon a, that is, upon the distance from the generation threshold, corresponds to the dependence of fluctuations of the parameter of order upon $T - T_c$ in the Landau theory of phase transitions, cf. [12.15 – 17]. This analogy, however, should not be overestimated. In fact, despite the similarities in the behavior of equilibrium and nonequilibrium phase transitions, there exists a fundamental difference between them. Namely, all the peculiarities observed when the critial point at nonequilibrium transitions in lasers is crossed are determined by the behavior of the dissipative part of permittivity (response), while at equilibrium transitions they are determined by the nondissipative part [see, in particular, (12.6.13)]. A more detailed comparison will be made in the next chapter, which deals with the theory of phase transitions in systems composed of atoms and a field.

13. Phase Transitions in a System Composed of Atoms and a Field

Various types of phase transitions in a system of two-level atoms and a field are considered here. In addition to the equilibrium phase transition, a nonequilibrium phase transition induced by laser irradiation is considered. We also discuss the possibility of equilibrium and nonequilibrium (laser) phase transitions taking place simultaneously in certain media (ferroelectrics with dope ions and liquid crystals with dope organic molecules). The influence of fluctuations which accompany the conventional phase transition on the parameters of laser radiation are also analyzed.

13.1 A Phase Transition in a System Composed of Two-Level Atoms and a Field

We have seen in Sects. 6.3 – 9 that distinguishing between the mean Maxwell field and the effective field leads to changes in attenuation and frequency in the equations for the polarization vector of a system of oscillating atoms (6.7.1). These changes are due to the correlation of atoms' positions.

When correlations are taken into account, the coefficient of radiation friction of a separate atom is replaced by γ_{eff}, see (6.5.8, 9). The change in frequency in zero approximation with respect to the lagging of electromagnetic interaction can be determined using the "Lorentz correction." The effects due to the correlation of the particles' positions are most important near transition points. For example, near the point of liquid – gas transition, the effective coefficient of extinction increases abruptly. The change in frequency due to the Lorentz correction can lead to the appearance of a soft mode (see Sect. 6.7), which gives rise to a phase transition of the second kind.

Recall, further, that when the temperature approaches the critical point from above, the spectral density of fluctuations of a transverse electromagnetic field at low frequencies increases abruptly [as $(T - T_c)^{-1/2}$] (Sect. 6.8). This peculiarity is due to the increase in the function $\mathrm{Re}\{\varepsilon\}$ according to Curie's law. At temperatures below critical the parameter of order increases (Sect. 6.9).

Let us now analyze a similar phase transition in a system made up of two-level atoms and a field [13.1 – 4]. For the basic one we employ the set of equations (12.1.3 – 6), making the replacement

$$E \to E + 4\pi\beta P \equiv \mathscr{E},\tag{13.1.1}$$

which corresponds to the inclusion of the "Lorentz correction" (at $\beta = 1/3$).

Since on approaching the point of a phase transition the peculiarities are only observed in the behavior of the spectral density of fluctuations of transverse fields (Sect. 6.8), we shall employ the equation for the electromagnetic field strength (12.1.13).

Instead of using the set of equations (12.1.3 – 6), it is more convenient here to employ the set for the polarization vector P and the population difference D (12.1.11, 12). In these equations we also perform the substitution given in (13.1.1).

There is a correspondence between the classical and the quantum equations for the polarization vector [(6.7.1) and (12.1.11)]; it is

$$\gamma_{eff} \leftrightarrow 2\gamma_{ab}, \quad \omega_0 \leftrightarrow \omega_{ab}, \quad \frac{en}{\mu} \leftrightarrow \frac{2}{3} \frac{|d_{ab}|^2 n \omega_{ab} |D|}{\hbar}. \tag{13.1.2}$$

This allows us immediately to employ many results obtained in Sect. 6.7. For instance, from (6.7.4, 17) we find the expression for the real part of permittivity in the static limit,

$$\mathrm{Re}\{\varepsilon(0)\} = \frac{8\pi}{3} \frac{|d_{ab}|^2 n |D^0|}{\hbar \omega_{ab}} \bigg/ \left(1 - \frac{8\pi}{3} \beta \frac{|d_{ab}|^2 n |D^0|}{\hbar \omega_{ab}}\right). \tag{13.1.3}$$

Here we have taken into account that $D = D^0$ in a weak field at $\omega \to 0$. It is convenient to write this expression

$$\mathrm{Re}\{\varepsilon(0)\} = \frac{\lambda |D|}{1 - \beta\lambda |D|}, \quad \lambda = \frac{8\pi |d_{ab}e|^2}{\hbar \omega_{ab}} n. \tag{13.1.4}$$

In the second equation e is the unit vector along the field. If the vectors d_{ab} and e are statistically independent, then $|d_{ab}e|^2 = (|d_{ab}|^2)/3$. The equation which defines the critical temperature follows if the soft mode appears [the transition of the denominator in (13.1.4) to zero],

$$|D^0|_c \equiv \tanh\left(\frac{\hbar \omega_{ab}}{2 \mathscr{K} T_c}\right) = \frac{1}{\beta\lambda}. \tag{13.1.5}$$

Recall that for the Lorentz model $\beta = 1/3$.

As in classical theory (Sect. 6.8), a rapid increase in the function $\mathrm{Re}\{\varepsilon(\omega)\}$ approaching the critical point leads to an abrupt increase in the spectral density of field fluctuations in the low-frequency region.

To consider states below the critical point, when the parameter of order

$$\eta = \langle P \rangle_{F=0} \tag{13.1.6}$$

is nonzero, we again turn to the equations for the polarization vector and the difference in populations (12.1.11, 12). In these equations we perform the substitution shown in (13.1.1), shift the term with the "Lorentz correction" into the left-hand side of the equation for P, and in the term which includes γ_{ab} substitute

$$\gamma_{ab}(P) \to \gamma_{ab}(P - \eta) \,. \tag{13.1.7}$$

As a result we get the following set of equations:

$$\frac{\partial D}{\partial t} + \gamma(D - D^0) = \frac{2}{\hbar \omega_{ab} n} \left[\frac{\partial P}{\partial t} + \gamma_{ab}(P - \eta) \right] (E + 4\pi\beta P) \,, \tag{13.1.8}$$

$$\frac{\partial^2 P}{\partial t^2} + 2\gamma_{ab} \frac{\partial P}{\partial t} + \omega_{ab}^2 (1 + \lambda\beta D) P = -\frac{1}{4\pi} \lambda \omega_{ab}^2 D E \,. \tag{13.1.9}$$

Here we have employed the designation for λ given in (13.1.4). This set of equations has a special solution

$$D = D^0 \,, \quad P = \eta \,, \quad E = 0 \tag{13.1.10}$$

at the condition

$$1 + \lambda\beta D^0 = 0 \quad \text{or} \quad \frac{1}{\lambda\beta} = |D^0| \,. \tag{13.1.11}$$

For above-critical temperatures, when the parameter of order is equal to zero and

$$|D| = |D^0| = \tanh\left(\frac{\hbar\omega_{ab}}{2 \mathcal{H} T}\right) \,, \tag{13.1.12}$$

(13.1.11) is not satisfied since the left-hand side is greater than zero. At $T = T_c$ a soft mode appears (13.1.5). This gives rise to the problem of finding the function D^0 for below-critical temperatures. This can be done in the following manner.

At below-critical temperatures the atoms are influenced by a permanent field whose strength is determined by the parameter of order. This strength equals $4\pi\beta\eta$. In a constant field, provided that the diagonal matrix elements are zero ($d_{aa} = 0$, $d_{bb} = 0$), the quadratic Stark effect is observed. The corresponding frequency shift can be determined from (10.2.8, 10). By setting $\omega_0 = 0$ (static field) in (10.2.10) and taking only the transition between two levels ($n = a$, $m = b$) into consideration, the new frequency of transition $\bar{\omega}_{ab}$ can be attained with the expression

$$\bar{\omega}_{ab} = \omega_{ab} \left[1 + \frac{2}{3} \frac{|d_{ab}|^2}{\hbar^2 \omega_{ab}^2} (4\pi\beta\eta)^2 \right] = \omega_{ab}^2 \left(1 + \frac{1}{2} \frac{\beta^2 \lambda^2 \eta^2}{n^2 |d_{ab}e|^2} \right) \,. \tag{13.1.13}$$

This result corresponds to the perturbation theory approximation. In an arbitrary field [13.3]

$$\tilde{\omega}_{ab} = \omega_{ab} \sqrt{1 + \frac{\beta^2 \lambda^2 \eta^2}{n^2 |d_{ab}e|^2}} \,.$$

(13.1.14)

Thus, for below-critical temperatures, one has to substitute $\omega_{ab} \rightarrow \tilde{\omega}_{ab}$ in (13.1.11, 12). Consequently we get the desired expression for the difference in populations,

$$|D^0| = \tanh \left(\frac{\hbar \tilde{\omega}_{ab}}{2 \mathcal{H} T} \right), \quad T \ll T_c ,$$

(13.1.15)

and the corresponding equation

$$\frac{1}{\beta \bar{\lambda}} = \tanh \left(\frac{\hbar \tilde{\omega}_{ab}}{2 \mathcal{H} T} \right), \quad \bar{\lambda} = \frac{8 \pi |d_{ab}e|^2}{\hbar \tilde{\omega}_{ab}} n ,$$

(13.1.16)

which in this case in the Curie – Weiss equation, i.e., the equation determining the parameter of order η.

We have discussed a special solution of the set of equations (13.1.8, 9), which is true for the approximation which does not take the fluctuations of the parameter of order into account. Let us now analyze fluctuations with respect to this special solution. We shall see that this task is similar to the problem discussed in Sects. 6.7 – 9 regarding the classical system of atom oscillators and a field.

13.2 Fluctuations in the Polarization and the Field at Above-Critical Temperatures

In calculating these fluctuations we shall widely employ analogies with the results obtained in the classical theory of phase transitions in a system of atom oscillators and a field (Sect. 6.9).

We have seen that the value of the parameter of order is determined by the Curie – Weiss equation (13.1.16) in a system made up of two-level atoms and a field, in the approximation which does not account for fluctuations. Expanding this equation with respect to η^2, in zero approximation we get (13.1.5), which determines the critical temperature.

It is also possible to write the corresponding equation for the polarization vector P in the same approximation. Let us turn to (13.1.9). Since in calculating fluctuations in the critical region we are mainly concerned with slow fluctuations, we can drop the term with the second time derivative in (13.1.9). Using the designation

$$a = \frac{\omega_{ab}^2}{2 \gamma_{ab}} (1 - \beta \gamma |D^0|) ,$$

(13.2.1)

in linear approximation we get with respect to the polarization vector P the following equation:

$$\frac{\partial P}{\partial t} + aP = \frac{\lambda \omega_{ab}^2}{8 \pi \gamma_{ab}} |D^0| E + y . \tag{13.2.2}$$

It is similar to the classical equation (6.9.1) at $b = 0$.

In order to find the expression which gives the spectral density of the Langevin source, we employ the classical formula (6.9.2) and the relations given in (13.1.2). This results in the following expression:

$$(y \cdot y)_{\omega, k} = 2 \frac{|d_{ab}|^2}{\hbar \gamma_{ab}} n \omega_{ab} |D^0| \mathscr{H} T , \tag{13.2.3}$$

which holds for the low-frequency region when the inequality

$$\hbar \tilde{\omega}_{ab} \ll \mathscr{H} T \tag{13.2.4}$$

is satisfied. Otherwise one has to make a substitution in (13.2.3),

$$\mathscr{H} T \rightarrow \frac{\hbar \tilde{\omega}_{ab}}{2} \coth \frac{\hbar \tilde{\omega}_{ab}}{2 \mathscr{H} T} , \tag{13.2.5}$$

where, according to (13.2.1), the renormalized frequency is

$$\tilde{\omega}_{ab} = \omega_{ab} \sqrt{1 - \beta |D^0|} . \tag{13.2.6}$$

The renormalization is due to the Lorentz field.

In order to find the equation for the spectral density of fluctuations δP^{source} [see (6.8.4, 6 and 6.9.7)], we employ the quantum expression for the permittivity, which can be obtained from (13.2.2),

$$\varepsilon(\omega) = 1 + \frac{\lambda \omega_{ab}^2}{2 \gamma_{ab}} |D^0| \frac{1}{-i \omega + a} . \tag{13.2.7}$$

In the static case it coincides with (13.1.4).

The unknown spectral density can be found with the aid of the Langevin equation (13.2.2) (at $E = 0$); using (13.2.3, 5) it can be written

$$(\delta P \, \delta P)_{\omega, k}^{\text{source}} = \frac{3}{4 \pi} \frac{\text{Im}\{\varepsilon(\omega)\}}{\omega} \hbar \tilde{\omega}_{ab} \coth \frac{\hbar \tilde{\omega}_{ab}}{2 \mathscr{H} T} . \tag{13.2.8}$$

Although this equation formally coincides with the classical formula (6.8.4) if the inequality (13.2.4) is satisfied, the equation for the function $\text{Im}\{\varepsilon\}$ is, of course, different. It follows from (13.2.7). With the aid of (13.1.4) it can be presented as

$$\text{Im}\{\varepsilon(\omega)\} = \frac{4\pi |d_{ab}|^2 n \omega_{ab}}{3 \hbar \gamma_{ab}} \frac{\omega}{\omega^2 + a^2} |D^0|. \tag{13.2.9}$$

Finally, let us write the equation for the spatial spectral density of fluctuations δP^{source} [cf. the second equation of (6.9.8)],

$$(\delta P \, \delta P)_k^{\text{source}} = \text{Re}\{\alpha(0)\} 3 \mathcal{K} T, \quad \text{Re}\{\alpha(0)\} = \frac{|d_{ab}|^2}{3 \hbar \gamma_{ab}} n \omega_{ab} \frac{|D^0|}{a}. \tag{13.2.10}$$

Thus, the intensity of fluctuations increases approaching the critical point according to Curie's law. Since fluctuations $P^{\text{source}}(R, t)$ are δ correlated in space, their correlation radius is zero.

The spectral density of field fluctuations under the condition in (13.2.4) is given by the same equations (6.8.10, 12).

Thus, the Rayleigh – Jeans equation remains true. The only difference is that one now has to employ the expression for $\text{Re}\{\varepsilon(\omega)\}$, which follows from (13.2.7). Approaching the critical point, the level of field fluctuations at low frequencies becomes abnormally high.

13.3 Fluctuations in the Polarization and the Field in the Critical Region

13.3.1 The Equation for the Polarization Vector

Investigating fluctuations of polarization and field at temperatures below critical, as well as in the critical region, is similar to the classical analysis in Sect. 6.9. The quantum system, however, is much more complicated. This can even be deduced from the fact that we are now dealing with the basic system of two equations, for the polarization vector and for the difference in populations, instead of with a single Langevin equation – the equation for the polarization vector (6.9.1).

Consider the simplest case by assuming that only the fluctuations of polarization increase significantly on approaching the critical point, while the fluctuations in the difference in populations remain essentially on the same level. This restriction allows one to employ (13.1.15) again for the function D^0, with the difference that the Stark effect in (13.1.13, 14) is not determined by the square of the parameter of order η^2, but by the square of the polarization vector P^2. Instead of the special solution (13.1.10), we thus employ an approximate solution which corresponds to the quasi-static approximation,

$$D = D^0; \quad P, E = 0. \tag{13.3.1}$$

This is justified provided that the range of frequencies responsible for the phase transition is governed by the conditions, cf. (6.7.13),

$$\omega \sim |a| \ll \gamma_{ab}, \gamma . \tag{13.3.2}$$

Naturally, more complex situations are also possible.

At these conditions we can in (13.1.9) drop the second time derivative, set $D = D^0$, and employ (13.1.15) for D^0, at the same time replacing $\eta^2 \to P^2$ in the expression for the frequency $\bar{\omega}_{ab}$.

At temperatures close to critical, the function D^0 can be expanded in P^2. In zero-order approximation we arrive at (13.1.12), and the equation for the polarization vector coincides with (13.2.2).

Retaining in the left-hand side of (13.1.9) the first term of expansion in P^2 as well (for the right-hand side zero approximation is sufficient), we get the following equation for the polarization vector:

$$\frac{\partial P}{\partial t} + (a + bP^2)P = \frac{\lambda \omega_{ab}^2}{8\pi\gamma_{ab}}|D^0|_c E + y \equiv \frac{|d_{ab}|^2}{3\hbar\gamma_{ab}} n\omega_{ab}|D^0|E + y . \tag{13.3.3}$$

Here we have employed the designation introduced in (13.1.4) for λ.

Equation (13.3.3) corresponds to the classical equation (6.9.1), see (13.1.2). The strength of the random source in (13.3.3) is given by (13.2.3). The expression for the constant b is determined by the coefficient at P^2 in the expansion of the function $|D^0|$.

The correspondence between the quantum and the classical equations for the polarization vector enables one to apply the results of classical theory, developed in Sect. 6.9, in the quantum domain. Let us emphasize some of these results.

The parameter of order as well as its fluctuations are both determined by (6.9.3, 5). The equation for the function $\delta P_k(t)$ then is

$$\left[\frac{\partial}{\partial t} + (a + 3b\eta^2)\right]\delta P_k(t) = \frac{|d_{ab}|^2}{3\hbar\gamma_{ab}} n\omega_{ab}\,\delta E_k(t) + y_k(t) . \tag{13.3.4}$$

Hence, the function δP can again be presented in the form of (6.9.7).

The fluctuations δP^{source} are δ correlated. The spatial spectral density can again be presented in the form of (13.2.10). Now, however, the static susceptibility is given by a more general equation,

$$\text{Re}\{\alpha(0)\} = \frac{|d_{ab}|^2}{3\hbar\gamma_{ab}} n\omega_{ab} \frac{|D^0|_c}{a + 3b\eta^2} . \tag{13.3.5}$$

Thus at temperatures above and below critical the spatial spectral density (13.2.10) (at $\eta^2 = |a|/b$) obeys Curie's law. According to (6.9.11), the same behavior is displayed by the correlation times of fluctuations δP^{source} (6.9.12). Only at the critical point itself is the correlation time of fluctuations δP^{source} proportional to $V^{1/2}$ (6.9.13); in the thermodynamic limit it tends to infinity.

From all of this it follows that the correlations of fluctuations at below-critical temperatures cannot ensure that a coherent state exists in which all the particles in the system are correlated. As in classical theory, the coherent state is created thanks to the appearance of long-life induced fluctuations of polarization after crossing the critical point.

13.3.2 Induced Fluctuations of Polarization

In order to gain some knowledge about the behavior of fluctuations in the critical point itself, we perform the substitution shown in (6.9.14). This means that the coherent state is characterized by the correlator of induced fluctuations of polarization, averaged over volume, and not by the parameter of order (6.9.3), i.e., that is characterized by the function

$$\langle P_{k=0}^2(t) \rangle = \langle P^2(t) \rangle, \quad P(t) = \int P(R, t) \frac{dR}{V}. \tag{13.3.6}$$

The index "ind" is omitted.

The equation for the two-time correlator of induced fluctuations now is [cf. (6.9.15)]

$$\left[\frac{\partial}{\partial t} + (a + b\langle P^2 \rangle) \right] \langle PP \rangle_\tau = \frac{|d_{ab}|^2}{3\hbar\gamma_{ab}} n\omega_{ab} |D^0|_c \langle EP \rangle_\tau. \tag{13.3.7}$$

It must be supplemented by the relevant equation for the one-time correlator, the initial condition at $\tau = 0$ [cf. (6.9.20)],

$$(a + b\langle P^2 \rangle) \langle P^2 \rangle = \frac{|d_{ab}|^2}{3\hbar\gamma_{ab}} n\omega_{ab} |D^0|^2 \langle EP \rangle. \tag{13.3.8}$$

Instead of the set of equations (13.3.7, 8) one can employ the equation for P^{ind}. It is similar to (6.9.21), and leads to the expression for the polarizability,

$$\alpha(\omega) = \frac{|d_{ab}|^2}{3\hbar\gamma_{ab}} n\omega_{ab} |D^0|_c \frac{1}{-i\omega + a + b\langle P^2 \rangle}. \tag{13.3.9}$$

Consequently we again come to the expression for the correlation time of induced fluctuations (6.9.23). It implies, in particular, that the correlation time for induced fluctuations tends to infinity, increasing with the distance from the critical point, when the state with $\langle P^2 \rangle = \eta^2 = |a|/b$ is established; it does not decrease in accordance with Curie's law, as is the case of fluctuations δP^{source}.

13.3.3 Field Fluctuations

The calculation of the field fluctuations coincides entirely with that performed for a classical system (Sect. 6.9). The only difference is that one has to employ the relevant quantum expression for the dielectric permittivity.

Thus, we can again employ the expression (6.9.18) for the correlation radius of field fluctuations. From this equation it follows that for the main mode, when $k_{min} \sim V^{-1/3}$, the radius of correlation is

$$r_c(\omega_{min}) \propto V^{1/3} \left\{ \frac{\text{Re}\{\varepsilon\}}{\text{Im}\{\varepsilon\}} \right\}_{\omega=0}. \qquad (13.3.10)$$

The correlation radius of field fluctuations, which determine the induced fluctuations of polarization, thus exceeds the dimensions of the system. This makes possible the existence of induced fluctuations of polarization, which can ensure the creation of a coherent state. Naturally, the coherent state can only be established when the correlation time of the induced fluctuations of polarization in the thermodynamic limit tends to infinity.

Now let us turn back to the equation for the correlation time of the induced fluctuations of polarization (6.9.23). It was shown that this equation can be transformed into (6.9.26). Using these equations and depicting the correspondence between the classical and the quantum results (13.1.2), we get the following expression for the correlation time of the induced fluctuations of polarization:

$$\tau_{ind} \sim \frac{V}{2 \mathcal{H} T} \frac{3 \hbar \gamma_{ab}}{|d_{ab}|^2 |D^0|_c} \frac{\langle p^2 \rangle}{n \omega_{ab}} \propto V. \qquad (13.3.11)$$

Just like in classical theory, the correlation times τ_{source} and τ_{ind} are of the same order of magnitude at above-critical temperatures and at the critical point itself. However, departing from the critical point, when $\langle P^2 \rangle \to |a|/b$, the correlation time τ_{source} decreases obeying Curie's law, while the correlation time of the induced fluctuations $\tau_{ind} \propto V$ and in the thermodynamic limit therefore tends to infinity. This means that a phase transition is taking place − a transition into a state where the spectrum of the induced fluctuations of polarization in the thermodynamic limit is infinitesimally narrow ($\Delta \omega \propto 1/\tau_{ind} \propto V^{-1}$).

The phase transition in a system of two-level atoms and a field, as analyzed here, differs markedly from those described by the Ginzburg − Landau equation. Here there are no grounds for separating the "region of scale invariance", where the critical indices would differ significantly from those of Landau theory. As in the case of the classical system of oscillating atoms interacting via the field (Sect. 6.9), we deal here only with two regions instead of with three (Sects. 5.7, 8), namely with the region adjacent (with respect to the parameter $1/N$) to the critical point, and that where Landau theory is applicable. The transition to the thermodynamic limit having been performed, the first region is reduced to zero, and Landau theory becomes applicable up to the critical point itself.

The two classes of systems described here represent the two extreme cases. Of course, intermediate situations are also possible where, e.g., a region of scale invariance exists, however vague. An example is single-axis ferroelectrics, where dipole – dipole interaction is important [13.5, 6].

13.4 Laser-Radiation-Induced Phase Transitions in a System of Two-Level Atoms

In Chap. 12 we analyzed the processes in which coherent electromagnetic radiation is generated in a system of two-level atoms and a field, and in the preceding sections of this chapter we have shown that a phase transition of the second kind is possible in such a system. We have already pointed out the analogy between these two phenomena. However, in spite of their analogous character, these phenomena are in a sense antipodal.

In a quantum generator the coherent radiation is generated at frequencies which are close to the frequency of the atomic transition. The principal role is played by the active dissipative nonlinearity. At the same time, for a phase transition of the second kind the coherent state is observed in the frequency band approaching zero. In this case the principal role is played by nondissipative nonlinearity. These differences mean that crossing the generation threshold is necessarily a nonequilibrium process. Moreover, for the generation to be stationary, the system must be an open one since a constant inflow of energy from an external source is necessary. Phase transitions are also known to occur in closed systems at statistical equilibrium.

Today we know substances which are characterized by equally strong dissipative and nondissipative nonlinearities, such as ferroelectric crystals including optically active dope atoms [13.7], or liquid crystals with certain impurities [13.8].

It would be interesting to investigate the influence of the equilibrium and the nonequilibrium transitions on each other in such systems, that is, the influence of the phase transition on generation, and vice versa, the influence of coherent radiation on phase transition. The final sections of this book deal with exactly these two problems; we shall now deal with the second of them [13.9].

Recall that the basis of the theory of the nonequilibrium (Chap. 12) and equilibrium (Sects. 13.1 – 3) phase transitions discussed above is formed by the self-consistent equations for the elements of the density matrix and field. The very important difference is that a distinction is made between the Maxwell field and the acting field (the Lorentz correction) in the equations for the density matrix when describing an equilibrium phase transition.

In both cases we assumed that only the off-diagonal matrix elements of the atom's dipole moment vector were nonzero, i.e., $d_{aa} = 0$, $d_{bb} = 0$. Let us now consider a more complicated system, where the diagonal matrix elements for the chosen pair of levels are not equal to zero and moreover, not equal to one another,

$$d_{aa} - d_{bb} \equiv d \neq 0 . \tag{13.4.1}$$

This condition can be satisfied either in noncentrally symmetric systems, or in centrally symmetric systems with degenerate energy levels, permitting the existence of the linear Stark effect.

The field E_a acting on the atoms can be given as

$$E_a(R, t) = \mathscr{E}(R, t) + E_L(R, t) . \tag{13.4.2}$$

Here $E_L(R, t)$ is the field of laser radiation, which can be created either within the system (where the active atoms belong to the impurities) or by an external source. Here we assume this field to be set in advance; $\mathscr{E}(R, t)$ is the internal field which can be presented as a sum of two parts (13.1.1). The effects of the two field components in (13.4.2) differ substantially since, as is well-known, abnormal growth at a phase transition is only displayed by the low-frequency field.

At the conditions given in (13.4.1, 2) the set of equations (12.1.3 – 6) for the elements of the density matrix are

$$\frac{\partial D}{\partial t} + \gamma(D - D^0) = \frac{2i}{\hbar}(d_{ab} f_{ba} - f_{ab} d_{ba})(\mathscr{E}(R, t) + E_L(R, t)] , \tag{13.4.3}$$

$$\frac{\partial R}{\partial t} + \gamma(R - R^0) = 0 , \tag{13.4.4}$$

$$\left(\frac{\partial}{\partial t} + \gamma_{ab} + i\omega_{ab}\right)f_{ab} = -\frac{id_{ab}}{\hbar}D(\mathscr{E} + E_L) + \frac{i}{\hbar}df_{ab}(\mathscr{E} + E_L) , \tag{13.4.5}$$

$$f_{ba} = f_{ab}^* . \tag{13.4.6}$$

In order to close this set of equations one must, of course, include the field equations as well. At the condition (13.4.1), an additional term appears in the expression for the polarization vector (12.1.10) which is proportional to D. Thus,

$$P = n(d_{ba} f_{ab} + d_{ab} f_{ba}) + \frac{n}{2}dD \equiv P_{ab} + P_D . \tag{13.4.7}$$

Here we do not include the term proportional to $(d_{aa} + d_{bb})R$, since at $R = \text{const}$ (as here assumed) this term drops out after averaging over the whole ensemble.

Let us demonstrate that there exists a phase transition in this system at $E_L \neq 0$, which results in the establishment of a state characterized by the parameter of order,

$$\eta = \langle P_D \rangle . \tag{13.4.8}$$

In the designations accepted here, the parameter of order at a phase transition, discussed in Sects. 13.1 − 3, is defined as

$$\eta = \langle P_{ab} \rangle . \tag{13.4.9}$$

We shall see that the position of the critical point, where the transition occurs to the state with the parameter of order (13.4.8), is given by the values of the four parameters

$$T, n, a_s |E_L|^2, \quad \Delta = \omega_{ab} - \omega_L . \tag{13.4.10}$$

Here T is the temperature, n is the concentration of the dope atoms, a_s is the saturation parameter (12.2.9), and Δ is the mistuning between the frequency of transition ω_{ab} and the frequency of the laser radiation ω_L.

According to (13.4.8), to determine the parameter of order one has to employ the equation for the difference in populations D. This equation, however, involves an unknown quantity D^0. In the phase transition discussed above the quantity D^0 was given by (13.1.15), where $\bar\omega_{ab}$ is the frequency of the transition which accounts for the quadratic Stark effect in the field $4\pi\eta$. At $d \neq 0$ the linear Stark effect is possible. Then [cf. (10.9.15)]

$$\bar\omega_{ab} = \omega_{ab} \left(1 \mp \frac{d\eta}{\hbar\omega_{ab}} \right), \tag{13.4.11}$$

where the minus sign corresponds to d parallel to the field $4\pi\eta$, and the plus sign corresponds to d antiparallel to the field. With the linear Stark effect taken into account, the quantity D^0 is given by (13.1.15) with $\bar\omega_{ab}$ represented by (13.4.11).

Returning to (13.4.5), for the basic eigenfunctions we can choose the functions which account for the linear Stark effect. For the quickly changing part of the function f_{ab} we get

$$f_{ab} = - \frac{d_{ab}}{2\hbar} \frac{DE_L}{\omega_L - \bar\omega_{ab} + i\gamma_{ab}}; \quad (f_{ba} = f_{ab}^*), \tag{13.4.12}$$

where E_L is the complex field amplitude at the frequency ω_L (12.2.1). Substituting this solution into the right-hand side of (13.4.3) and singling out the slowly changing terms, for the stationary state we get the following expression for the difference in populations (designated D_L):

$$D_L = D^0 \frac{(\bar\omega_{ab} - \omega_L)^2 + \gamma_{ab}^2}{(\bar\omega_{ab} - \omega_L)^2 + \gamma_{ab}^2(1 + a_s |E_L|^2)} . \tag{13.4.13}$$

The dependence on the parameter of order η at $E_L \neq 0$ enters this expression via both multiplicands. At $E_L = 0$ the second multiplicand is equal to one; therefore $D_L = D^0$. At $E_L \neq 0$, the dependence on the parameter of order is stronger for the second multiplicand (at least for values of $a_s|E_L|^2$ which are not too small) than for the first. Indeed, at small η the expansion of the first multiplicand is performed with respect to $d\eta/\hbar\omega_{ab}$, while the second multiplicand is expanded in the parameter

$$\frac{d\eta}{\hbar(\omega_{ab}-\omega_L)} \sim \frac{d\eta}{\hbar\gamma_{ab}}, \qquad (13.4.14)$$

which is ω_L/γ_{ab} times larger. For that reason we shall substitute the frequency ω_{ab} into (13.1.15) for D^0 instead of $\bar{\omega}_{ab}$.

Let us now return to (13.4.3 for D, assuming the condition

$$\gamma_{ab} \gg \gamma \qquad (13.4.15)$$

to be fulfilled. The contribution from the fast components of the functions f_{ab} and f_{ba} is accounted for using (13.4.12), while the slow components of these functions

$$f_{ab}^0 \sim f_{ba}^0 \sim \frac{d_{ab}\mathscr{E}}{\hbar\omega_{ab}}D. \qquad (13.4.16)$$

The substitution of these expressions for D in the right-hand side of (13.4.3) leads to the appearance there of the term of the order

$$\frac{d_{ab}^2\mathscr{E}^2}{(\hbar\omega_{ab})^2}\omega_{ab}D \sim a_s\mathscr{E}^2\frac{\gamma_{ab}}{\omega_{ab}}\gamma D. \qquad (13.4.17)$$

Here we have employed (12.2.9) for the parameter of saturation a_s.

Thus, the term determined by the functions f_{ab}^0 and f_{ba}^0 is proportional to the product of two small parameters $a_s\mathscr{E}^2$ and γ_{ab}/ω_{ab}, and can therefore be omitted. Consequently the difference in population is

$$\frac{\partial D}{\partial t} + \gamma(D-D^0) = -\gamma a_s|E_L|^2\frac{\gamma_{ab}^2}{(\tilde{\omega}_{ab}-\omega_L)^2+\gamma_{ab}^2}D. \qquad (13.4.18)$$

Hence (13.4.13) follows for the stationary state at $D \equiv D_L$.

Using the definition (13.4.8), one can then obtain the equation for the parameter of order which is linked to D_L by the equality

$$\eta = \frac{n}{2}\langle \mp dD_L\rangle_{\mp}, \qquad (13.4.19)$$

where the angular brackets $\langle\rangle_{\mp}$ designate averaging over either of the two possible values $\pm d$. For the stationary state we therefore get the following equation for the parameter of order:

$$\eta = \frac{n}{2} \left\langle \mp dD^0 \frac{(\bar{\omega}_{ab} - \omega_L)^2 + \gamma_{ab}^2}{(\bar{\omega}_{ab} - \omega_L)^2 + \gamma_{ab}^2(1 + a_s|E_L|^2)} \right\rangle_{\mp},$$

$$\bar{\omega}_{ab} = \omega_{ab} \left(1 \mp \frac{d\eta}{\hbar \omega_{ab}} \right). \tag{13.4.20}$$

This equation corresponds to the Curie – Weiss equation (13.1.16). From this equation for the parameter of order, it follows that in the absence of the laser field ($E_L = 0$)

$$\eta = \frac{n}{2} \langle \mp dD^0 \rangle_{\mp} = 0. \tag{13.4.21}$$

Carrying out expansion in the right-hand side of (13.4.20) with respect to the parameter of order only holds the terms in odd powers of η. Retaining the first two nonvanishing terms, we obtain an equation, which can be written in the spirit of Landau's theory of phase transitions,

$$(a + b\eta^2)\eta = 0. \tag{13.4.22}$$

Here we have introduced the designations ($D^0 < 0$)

$$a = 1 + \frac{4\pi d^2 n}{\hbar \omega_{ab}} D^0 \frac{\Delta \gamma_{ab}^2 \omega_{ab}}{(\Delta^2 + \gamma_E^2)(\Delta^2 + \gamma_{ab}^2)} a_s |E_L|^2, \tag{13.4.23}$$

$$b = \beta \lambda \frac{d^2}{\hbar^2} D^0 \frac{2(\gamma_E^2 - \Delta^2)}{(\gamma_E^2 + \Delta^2)^2}, \quad \gamma_E^2 = \gamma_{ab}^2(1 + a_s|E_L|^2). \tag{13.4.24}$$

The usual expression for the parameter of order follows from (13.4.22) at $a < 0$, $b > 0$,

$$\eta = \sqrt{|a|/b}. \tag{13.4.25}$$

For the stationary state the condition $a < 0$ is only fulfilled (at $D^0 < 0$) at positive mistunings ($\Delta = \omega_{ab} - \omega_L > 0$). The condition $b > 0$ can be fulfilled only if $|\Delta| < \gamma_E$; otherwise the character of the phase transition will change.

The critical parameters T, n, $a_s |E_L|^2$, Δ are found, as usual, from the equation

$$a = 0. \tag{13.4.26}$$

The induced phase transition discussed here, though possible at an equilibrium occupation of atomic levels, is however nonequilibrium. It stands halfway between the nonequilibrium phase transition in lasers (Chap. 12) and the equilibrium phase transition, discussed in Sects. 13.1 – 3.

Indeed, for the induced phase transition the Curie – Weiss equation (13.4.20) only holds in the presence of external laser radiation. This equation follows from the relaxational nonreversible equation (13.4.18), and the transition is therefore dissipative since pumping by an external field has to compensate the loss which occurs in the relaxation toward the equilibrium state.

Here we have discussed the simplest example of an induced phase transition in a system of two-level atoms. Our model, naturally, is rather idealized. Calculations for more practical models for ferroelectrics have been carried out in [13.10 – 12].

13.5 The Influence of a Phase Transition on Generation

The joint influence of equilibrium and nonequilibrium phase transitions can, of course, give rise to a number of new phenomena. A simultaneous occurrence of both types of phase transitions (equilibrium and nonequilibrium) is possible, e.g., in ferroelectrics with optically active dope atoms (or ions), which could serve as an active medium for lasers [13.7, 13]. The active medium for a laser can also be provided by dope dye atoms in liquid crystals. For example, the dependence of the generation threshold of a laser on the proximity to the critical point in nematic liquid crystals is dealt with in [13.8, 14]. It was observed that approaching the critical point the generation threshold declines, together with the radiation wavelength. Here we shall discuss two processes, which are possibly responsible for the alteration in the characteristics of laser radiation on approaching the point of a phase transition in the "solvent."

13.5.1 In Ferroelectrics

Consider generation at operating levels a, b of dope atoms (or ions) in ferroelectrics, taking into account the influence of the phonon subsystem.

To calculate polarization at the operating levels a, b according to the theory developed in Chap. 12 the equations for the elements of the density matrix f_{ab}, f_{ba}, $D = f_a - f_b$ must be employed (12.1.3 – 6).

We can supplement (12.3.5) for f_{ab} to account for the resonant interaction with phonons [13.15], giving the function f_{ab} in the form

$$\left(\frac{\partial}{\partial t} + \gamma_{ab} + i\,\omega_{ab}\right)f_{ab} = -\frac{i}{\hbar}d_{ab}DE - \frac{i}{\hbar}CUf_{ab}. \tag{13.5.1}$$

The vector C characterizes the interaction of the dope atoms with the phonon subsystem, and U is the vector of the lattice displacement.

Besides the interaction accounted for in this equation, interaction of the form $iCUD/\hbar$ is also possible; it, however, does not give a resonant contribution at optical frequencies, and will therefore be neglected[1].

Consider for example the influence of the optical phonons on generation. We designate the eigenfrequency of the oscillations of ions in the lattice with ω_i, and the frequency of the soft mode (at $T > T_c$) with $\tilde{\omega}_i$. Then at the condition

$$\hbar\tilde{\omega}_i \ll \mathscr{K}T \tag{13.5.2}$$

we have the expressions for the first two moments of the function U,

$$\langle U \rangle = 0, \quad \langle (\delta U)^2 \rangle = \frac{3\,\mathscr{K}T}{M\tilde{\omega}_i^2}. \tag{13.5.3}$$

The equation for the fluctuation δU can be written in the form of the Langevin equation,

$$\frac{d^2\delta U}{dt^2} + \Gamma\frac{d\delta U}{dt} + \tilde{\omega}_i^2\delta U = y. \tag{13.5.4}$$

At low frequencies $\Omega \ll \Gamma$ [cf. (6.7.13)], the spectral density of the fluctuations δU is given by

$$(\delta U)^2_\Omega = \frac{2\tilde{\omega}_i^2}{\Gamma}\frac{\langle(\delta U)^2\rangle}{\Omega + (\tilde{\omega}_i^2/\Gamma)^2}. \tag{13.5.5}$$

At $\gamma_{ab} \sim 10^8 - 10^9\,\mathrm{s}^{-1}$ the following inequality is fulfilled for the dope atoms:

$$\tilde{\omega}_i^2/\Gamma \gg \gamma_{ab}. \tag{13.5.6}$$

At this condition the spectrum of fluctuations δU can be considered to be a "white-noise" spectrum; the correlator is therefore

$$\langle \delta U \delta U \rangle_\tau = 2D\delta(\tau); \quad D = (\delta U)^2_{\Omega=0} = \frac{\langle(\delta U)^2\rangle}{\tilde{\omega}_i^2/\Gamma}. \tag{13.5.7}$$

To analyze the influence of fluctuations in the phonon subsystem on laser generation, we shall calculate the polarization for the operating transition $a - b$ with the aid of (13.5.1). In [13.15] such a calculation is carried out using methods from perturbation theory with respect to the interaction of the dope atoms with phonons. Here we shall do it without employing perturbation theory, though

---

[1] However, this term is important in the so-called vibron theory of phase transitions in ferroelectrics [13.16, 17]. This theory is based on the assumption that the appearance of the soft mode in the phonon subsystem is due to the interaction between the latter and the two-level electron subsystem, which is described by equations similar to (13.5.1). A number of theories of phase transitions in ferroelectrics are presented in [13.18, 19].

with the condition in (13.5.6), i. e., when the spectrum of fluctuations can be considered as white noise [13.16].

The technique is similar to that employed in Sect. 5.2 in deriving the Fokker – Planck equation. The only difference is that the white noise in the Langevin equations (5.1.3) is an additive one, while it enters (13.5.1) parametrically.

Due to the fluctuation effect of the phonon subsystem, the function f_{ab} in (13.5.1) is a random function. Calculating the polarization determined by the function f_{ab} averaged over the ensemble, we get with (13.5.1)

$$\left(\frac{\partial}{\partial t} + \gamma_{ab} + i\,\omega_{ab}\right)\langle f_{ab}\rangle = -\frac{i\,d_{ab}}{\hbar}D^0\langle E\rangle - i\frac{C}{\hbar}\langle\delta U\,\delta f_{ab}\rangle. \qquad (13.5.8)$$

Since we are interested in the state close to the generation threshold, in the right-hand side of this equation the function $D = D^0$ $(D^0 > 0)$ and is therefore determined by the preset pumping.

Equation (13.5.8) is not closed since it involves the correlator $\langle\delta U\,\delta f_{ab}\rangle$, cf. (5.2.5). Therefore we shall write the equation for the function δf_{ab}. In the white-noise approximation this equation can be presented as [cf. (5.2.8)]

$$\frac{\partial\delta f_{ab}}{\partial t} = -i\frac{C}{\hbar}\delta U\langle f_{ab}\rangle,$$

$$\delta f_{ab} = -i\frac{C}{\hbar}\int_0^\infty \delta U(t-\tau)\langle f_{ab}\rangle_{t-\tau}d\tau. \qquad (13.5.9)$$

With the aid of this solution and taking the expression for the correlator of fluctuations δU (13.5.7) into account, we get the following expression for the second term in the right-hand side of (13.5.8):

$$-i\frac{C}{\hbar}\langle\delta U\,\delta f_{ab}\rangle = -\frac{C^2}{3\,\hbar^2}D^0\langle f_{ab}\rangle. \qquad (13.5.10)$$

The intensity of noise D is given by (13.5.7).

In this way we have succeeded in expressing the correlator $\langle\delta U\,\delta f_{ab}\rangle$ in terms of the function $\langle f_{ab}\rangle$ without employing perturbation theory with regard to the interaction of the dope atoms with the phonon subsystem. This allows us to write the equation for the function $\langle f_{ab}\rangle$ as (omitting the angular brackets $\langle\,\rangle$)

$$\left(\frac{\partial}{\partial t} + \tilde{\gamma}_{ab} + i\,\omega_{ab}\right)f_{ab} = -\frac{i\,d_{ab}}{\hbar}D^0 E, \qquad (13.5.11)$$

which differs from (12.1.5) (at $D = D^0$) in that the quantity γ_{ab}, which accounts for the interaction with phonons, is replaced by

$$\tilde{\gamma}_{ab} = \gamma_{ab} \left(1 + \frac{C^2 D}{3\hbar^2 \gamma_{ab}}\right) = \gamma_{ab} \left(1 + \frac{C^2 \Gamma}{\hbar^2 \gamma_{ab}} \frac{\varkappa T}{M\omega_i^4}\right). \tag{13.5.12}$$

Let us now estimate the role of the additional term, proportional to C^2. At temperatures far from the critical temperature T_c, the frequency $\tilde{\omega}_i = \omega_i$. Then, replacing the quantities in (13.5.12) by their numeric values ($C \sim 10^{-7}$ erg/cm, $\Gamma \sim 10^{11}$ s^{-1}, $\omega_i \approx 3 \cdot 10^{13}$ s^{-1}, $M \sim 10^{-22}$ g, $\gamma_{ab} \sim 10^9$ s^{-1}, $T \sim 10^2$ K) we find that the second term in parentheses is of the order 10^{-4}, and that therefore the interaction with phonons is of little significance.

Approaching the critical point the role of this interaction increases, and can become dominating. Indeed,

$$\omega_i^4 / \tilde{\omega}_i^4 \propto 1/(T - T_c)^2, \tag{13.5.13}$$

therefore at, e.g., $\omega_i \approx 3 \cdot 10^{11}$ s^{-1} (that is, at a relatively slight "softening" of the mode), the second term in parentheses in (13.5.12) is of the order 10^4, so the interaction with phonons is predominant.

To determine how the interaction between the dope atoms and the phonon subsystem influences the generation threshold, we employ the energy balance equation (12.2.14). For the imaginary part of permittivity (12.2.15) we use the corresponding expression which accounts for the interaction between the dope atoms and the phonon subsystem,

$$\text{Im}\{\varepsilon(\omega)\} = -\frac{4\pi |d_{ab}|^2}{3\hbar} \frac{\tilde{\gamma}_{ab}}{(\omega - \omega_{ab}) + \tilde{\gamma}_{ab}^2} nD^0, \quad D^0 > 0. \tag{13.5.14}$$

Substituting this expression into the equation for energy balance (12.2.14), we get

$$\frac{1}{Q} = \frac{4\pi |d_{ab}|^2}{3\hbar} nD^0 \frac{\tilde{\gamma}_{ab}}{(\omega - \omega_{ab})^2 + \tilde{\gamma}_{ab}^2}, \tag{13.5.15}$$

which yields the threshold values of the dope atom concentration and the difference in populations (the function nD^0), with the effect of the phonon subsystem taken into account.

The sign of the effect (either raising or lowering of the generation threshold on approaching the point of phase transition) depends on the relative values of mistuning $|\omega - \omega_{ab}|$ and the linewidth $\tilde{\gamma}_{ab}$. At zero mistuning, for example, (13.5.15) takes the form

$$\frac{1}{Q} = \frac{4\pi |d_{ab}|^2}{3\hbar \tilde{\gamma}_{ab}} nD^0, \tag{13.5.16}$$

and therefore the threshold value of nD^0 on approaching the critical point increases as

$$nD^0 \propto \frac{\omega_i^4}{\tilde{\omega}_i^4} \propto \frac{1}{(T-T_c)^2} .$$ (13.5.17)

The inverse effect (lowering the threshold value of nD^0) is only possible at large mistunings, when $|\omega - \omega_{ab}| > \gamma_{ab}$. Such a situation in quantum optical generators is not, as a rule, advantageous from the energetic viewpoint.

13.5.2 In Liquid Crystals

Let us now consider another mechanism by which fluctuations at a phase transition influence the characteristics of laser radiation, on the example of a solution of optically active molecules in a liquid crystal. The state of orientational ordering in nematic liquid crystals is characterized by the tensor [13.21 – 24]

$$S_{ij}(R, t) = \int \langle v_i v_j - \tfrac{1}{3}\delta_{ij} \rangle f(v, R, t) dv ,$$ (13.5.18)

where v is the unit vector along the longer axis of the molecule, and f is the corresponding distribution function.

In the crystal phase the state can be characterized by the parameter of order η_{ij}. At temperatures above critical the parameter of order equals zero.

In the theory of phase transitions the parameter of order is usually defined as a quasi-mean value of the tensor S_{ij} [cf. (5.8.10, 6.9.3)]. The average (over the ensemble) value of the tensor S_{ij} is zero, $\langle S_{ij} \rangle = 0$.

The phase transition can again [see (5.8.27, 6.9.14)] be characterized by the correlator of the tensor $S_{ij}(R, t)$,

$$\langle S_{ij}(t) S_{kl}(t) \rangle ; \quad S_{ij}(t) = \int S_{ij}(R, t) \frac{dR}{V} .$$ (13.5.19)

For the random function $S_{ij}(R, t)$ one can write the corresponding Ginzburg – Landau equation, cf. (5.8.1). For temperatures above critical, when the non-linear term is insignificant, this equation has the form [13.23]

$$\frac{\partial S_{ij}}{\partial t} + aS_{ij} - g\frac{\partial^2 S_{ij}}{\partial R^2} = y_{ij}(R, t), \quad a = a_0\frac{T-T_c}{T_c} .$$ (13.5.20)

The moments of the Langevin source are given by the expressions

$$\langle y_{ij} \rangle = 0, \quad \langle y_{ij}(R, t) y_{kl}(R', t') \rangle = 2\frac{D}{n} I_{ijkl}\delta(R-R') \delta(t-t') .$$ (13.5.21)

Here we have introduced the designation for the tensor

$$I_{ijkl} = \frac{1}{2}\left(\delta_{ik}\delta_{jl} + \delta_{il}\delta_{jk} - \frac{2}{3}\delta_{ij}\delta_{kl}\right),$$ (13.5.22)

where D is the corresponding diffusion coefficient which characterizes the molecular motion in the liquid crystal.

With the aid of (13.5.20) we find the expressions for the spectral densities of the tensor,

$$(S_{ij}S_{kl})_{\omega,k} = \frac{2DI_{ijkl}}{\omega^2 + (a + gk^2)^2},$$ (13.5.23)

$$(S_{ij}S_{kl})_k = \frac{DI_{ijkl}}{a + gk^2}.$$ (13.5.24)

Further we shall also need the expression for the correlator $\langle S_{ij}(R)S_{kl}(R)\rangle_\tau$, which can be obtained by dint of (13.5.23).

In calculating field fluctuations near the generation threshold it can be assumed that the width of the spectrum of fluctuations S_{ij} is much greater than the value of $\omega_{ab}(|\mathrm{Im}\{\varepsilon\}| - 1/Q)$. In this case one can employ the approximation

$$[S_{ij}(R)S_{kl}(R)]_\tau = (S_{ij}(R)S_{kl}(R))_{\omega=0}\delta(\tau).$$ (13.5.25)

The expression for the intensity follows from (13.5.23),

$$(S_{ij}(R)S_{kl}(R))_{\omega=0} = \frac{2DI_{ijkl}}{(2\pi)^3}\int\frac{dk}{(a + gk^2)^2}.$$ (13.5.26)

If a dimensionless variable $x = k\sqrt{g/a}$ is placed under the integral, it becomes clear that approaching the critical point the integral increases as $1/\sqrt{a}$, and that therefore the intensity (13.5.26) is

$$(S_{ij}(R)S_{kl}(R))_{\omega=0} \propto \frac{1}{\sqrt{T - T_c}}.$$ (13.5.27)

To see how the fluctuations of the tensor $S_{ij}(R,t)$ influence the generation threshold, we employ the equation for the amplitude of laser radiation. At zero mistuning ($\omega = \omega_{ab}$) it has the form

$$\frac{dE}{dt} + \frac{\omega_{ab}}{2}\left[\frac{1}{Q} + \mathrm{Im}\{\varepsilon_d(\omega_{ab})\}\right]E = 0.$$ (13.5.28)

Equation (12.2.14) then follows for the stationary regime.

The function $\mathrm{Im}\{\varepsilon_d\}$ in (13.5.28) is the imaginary part of permittivity for the dope molecules at frequency ω_{ab}. This function is given by (12.2.15), which we shall write (at $\omega = \omega_{ab}$) in a different form,

$$\mathrm{Im}\{\varepsilon_d(\omega_{ab})\} = -\frac{4\pi e_i(d_{ba})_i(d_{ab})_j}{\hbar\gamma_{ab}}nD^0, \quad D^0 > 0,$$ (13.5.29)

where $\hat{e}$ is the unit vector along the direction of the field.

With the fluctuations of the tensor of orientational ordering S_{ij} not taken into account, the dope system can be assumed to be isotropic, and the dipole moments tensor presented in the form of (12.1.14). Our further analysis will, however, be based on the assumption that with the ordering action from the side of the solvent being taken into account, the structure of the dipole moments tensor is

$$(d_{ba})_i (d_{ab})_j = \tfrac{1}{3} \delta_{ij} |d_{ab}|^2 + C S_{ij} . \tag{13.5.30}$$

The constant C defines the extent to which the orientational ordering of the liquid crystal is accepted by the dope molecules. Under this assumption, the imaginary part of permittivity is given by

$$\mathrm{Im}\{\varepsilon_d(\omega_{ab})\} = \mathrm{Im}\{\varepsilon_d^0\} + \mathrm{Im}\{\delta\varepsilon_d\} . \tag{13.5.31}$$

According to (13.5.30), the second term is proportional to the tensor S_{ij}.

Let us average (13.5.28). For the stationary state we get the following equation for energy balance:

$$\frac{1}{Q} = |\mathrm{Im}\{\varepsilon_d^0\}| - \frac{\langle \mathrm{Im}\{\delta\varepsilon_d \delta E\} \rangle}{\langle E \rangle} , \tag{13.5.32}$$

which, in contrast to (12.2.14), is not closed. Since the correlator $\langle \mathrm{Im}\{\delta\varepsilon_d \delta E\} \rangle$ can be calculated similar to obtaining (13.5.10), we can present the final result,

$$-\frac{\langle \mathrm{Im}\{\delta\varepsilon_d \delta E\} \rangle}{\langle E \rangle} = \frac{\omega_{ab}}{4} \left(\frac{4\pi C n D^0}{\hbar \gamma_{ab}} \right)^2 e_i e_j e_k e_l (S_{ij}(R) S_{kl}(R))_{\omega=0} , \tag{13.5.33}$$

which was obtained with the aid of (13.5.25 – 27). Equation (13.5.32) can now be written

$$\frac{1}{Q} = \frac{4\pi |d_{ab}|^2}{3\hbar \gamma_{ab}} n D^0 + \frac{\omega_{ab}}{4} \left(\frac{4\pi C n D^0}{\hbar \gamma_{ab}} \right)^2 e_i e_j e_k e_l (S_{ij}(R) S_{kl}(R))_{\omega=0} . \tag{13.5.34}$$

Hence, the value of the square of the modulus of the matrix elements of the dope molecules is effectively increased when the ordering action of the solvent is taken into account. This results in a lowering of the threshold value of $(nD^0)_{\mathrm{th}}$.

Far from the critical point [using (13.5.27)], the lowering of the threshold on approaching the critical point obeys the law

$$\Delta(nD^0)_{\mathrm{th}} \propto \frac{1}{\sqrt{T - T_c}} . \tag{13.5.35}$$

On further approaching the critical point the generation threshold decreases less rapidly, and in the region where the second term in the right-hand side of (13.5.34) dominates,

$$(nD^0)_{\text{th}} \sim \sqrt[4]{T - T_{\text{c}}} \, . \tag{13.5.36}$$

In this way we have determined how the fluctuations of the tensor S_{ij} influence the generation threshold of radiation from dope molecules in a liquid crystal. In experiments described in [13.8], the shortening of the generation wavelength was also observed on approaching the point of phase transition.

To describe this phenomenon we write the dispersion equation (12.2.13), which follows from the equation for the phase of the emitted wave, in the form

$$[\omega_{ab}^2 \, \text{Re}\{\varepsilon(\omega_{ab})\} - c^2 k^2]E = 0 \, . \tag{13.5.37}$$

This assumes the mistuning to be zero.

Due to fluctuations in molecular orientation in the liquid crystal, there is an additional term in the expression for the permittivity [cf. (13.5.31)]

$$\text{Re}\{\varepsilon\} = \text{Re}\{\varepsilon^0\} + \text{Re}\{\delta\varepsilon\} \, . \tag{13.5.38}$$

This equality, however, includes the total permittivity at the frequency of the transition ω_{ab} since the refraction index is determined by the summary polarization of the molecules of both the impurity and the solvent. In addition, since the function $\text{Re}\{\varepsilon_{\text{d}}\}$ is zero at the transition frequency, the polarizability of the solvent, i. e., of the liquid crystal, plays the principal role.

The permittivity tensor in a liquid crystal, as is well known [13.23, 24], can be presented as

$$\varepsilon_{ij} = \tfrac{1}{3}(\varepsilon_{\parallel} + 2\varepsilon_{\perp})\delta_{ij} + (\varepsilon_{\parallel} - \varepsilon_{\perp})S_{ij} \, , \tag{13.5.39}$$

where $\varepsilon_{\parallel}$ and $\varepsilon_{\perp}$ are the longitudinal and transverse (with respect to the direction of the vector n) permittivities, respectively. By comparing (13.5.38) and (13.5.39) we find

$$\text{Re}\{\varepsilon^0\} = \tfrac{1}{3}\text{Re}\{\varepsilon_{\parallel} + 2\varepsilon_{\perp}\}, \quad \text{Re}\{\delta\varepsilon\} = \text{Re}\{\varepsilon_{\parallel} - \varepsilon_{\perp}\}e_i S_{ij} e_j \, . \tag{13.5.40}$$

Averaging (13.5.37) we get an equation which, in contrast to (12.2.13), is not closed,

$$\omega_{ab}^2(\text{Re}\{\varepsilon^0\} + \langle\text{Re}\{\delta\varepsilon\delta E\}\rangle/\langle E\rangle) = c^2/k^2 \, . \tag{13.5.41}$$

The expression for the correlator entering this equation can be obtained in the same manner as the expression for the correlator (13.5.10). Consequently we get

$$\langle\text{Re}\{\delta\varepsilon\delta E\}\rangle/\langle E\rangle = \frac{\omega_{ab}}{4}\text{Re}\{\varepsilon_{\parallel} - \varepsilon_{\perp}\}\frac{4\pi CnD^0}{\hbar\gamma_{ab}}e_i e_j e_k e_l \langle S_{ij}(R)S_{kl}(R)\rangle_{\omega=0} \, . \tag{13.5.42}$$

We substitute this expression into (13.5.41), giving the value of the wave vector in the form

$$k = \frac{\omega_0 \sqrt{\mathrm{Re}\{\varepsilon^0\}}}{c} + \Delta k \equiv k_0 + \Delta k . \tag{13.5.43}$$

For Δk in linear approximation we get

$$c\Delta k = \frac{\omega_{ab}^3}{2ck_0} \mathrm{Re}\{\varepsilon_| - \varepsilon_\perp\} \frac{\pi C n D^0}{\hbar \gamma_{ab}} e_i e_j e_k e_l (S_{ij}(R) S_{kl}(R))_{\omega=0} . \tag{13.5.44}$$

We see that the sign of Δk depends on the combination of the signs of two quantities, $\varepsilon_\| - \varepsilon_\perp$ and C. If C in (13.5.30) has the same sign as $\varepsilon_\| - \varepsilon_\perp$ (which is expected), then Δk is positive, and the alteration of the radiation wavelength ($\Delta\lambda = -\Delta k/k_0^2$) is therefore negative.

We conclude that approaching the critical point the wavelength of radiation decreases provided that $(\varepsilon_\perp - \varepsilon_\|)C > 0$. Far from the critical point, when the dependence of the threshold value $(nD^0)_{th}$ on temperature in (13.5.43) is not yet pronounced,

$$-\Delta\lambda \propto 1/\sqrt{T - T_c} . \tag{13.5.45}$$

Approaching the critical point this dependence weakens, and at (13.5.36) $-\Delta\lambda \propto 1/\sqrt[4]{T - T_c}$.

Thus, the two examples discussed here have enabled us to clarify the manner in which a phase transition in ferroelectrics and in liquid crystals can influence the characteristics of laser radiation created in dope atoms and molecules. Since our results have been obtained under certain simplifying assumptions, our understanding of the physical processes which determine the phenomena discussed here is far from complete.

14. Conclusion

It goes without saying that we have only been able to discuss a few of the problems of the kinetic theory of electromagnetic processes. As pointed out in Chap. 1, from the great variety of processes determined by electromagnetic interaction we have chosen only those dominated either by interactions of atoms and free charged particles which slowly decrease with distance, or by interactions with an electromagnetic field. As we have shown, the polarization approximation proved to be effective in such cases.

Even under these considerable restrictions it was possible to describe a wide variety of phenomena, although only to a limited extent, of course. Our efforts were directed toward a greater uniformity in describing apparently different phenomena.

Let us point out certain problems which the method developed in this book might be advantageously applied to.

1) First of all we must admit that the class of phenomena determined by strong pair interactions of particles was not treated. The problem of pair interactions is excellently treated in the theory of collisions. However, in such a complex object as described by the somewhat evasive term "partially ionized plama", taking both short-range strong interactions and collective interactions into consideration simultaneously is very significant. This was already mentioned in Sect. 10.13 in connection with the problem of spectral line broadening (see also [14.1, 2]). Simultaneously taking both strong and weak (collective) interactions into consideration is also necessary in approaching many other problems, including the problem of the ionization equilibrium in nonideal plasma, and the kinetics of nonideal plasma.

2) Almost all the calculations of small-scale fluctuations, which determine the dissipative processes in the kinetic equations, were performed without taking the effects of the mean field into account. For that reason the collision integrals obtained here do not explicitly depend on the field. Today we know many cases where the fields are so strong that even small-scale fluctuations ought to be calculated taking the mean field into account. Then the collision integrals, and therefore all the kinetic coefficients, will be explicit functions of field.

Calculations of this kind have been carried out, e. g., for completely ionized plasma ([Ref. 14.3, Chap. 10] and [Ref. 14.4, Chap. 8]). In particular, research has been carried out on the influence of the field on the conductivity of plasma. Naturally, such calculations of nonequilibrium fluctuations in partially ionized plasma with the field taken into account prove much more complicated, but the problems are far more fascinating.

3) Far from the equilibrium, e.g., due to the action of external fields, the polarization approximation can become insufficient for calculating fluctuations. Calculations then have to incorporate nonlinear fluctuation processes. A great number of problems fall within this class (see, e.g., the review [14.5]).

Naturally, nonlinear fluctuation processes are most interesting when we deal with large-scale fluctuations. We encountered these problems in the analysis of fluctuations in classical and quantum self-oscillatory systems, as well as in the study of fluctuations in the critical region at phase transitions of the second kind. In other words, nonlinear fluctuation processes can be said to be equally important both at equilibrium and nonequilibrium phase transitions.

Both these types of phase transitions can be attributed to the branch of physics called *nonlinear thermodynamics of irreversible processes*. This field is now developing rapidly, and the results obtained are of great importance not only for physics, but for chemistry and biology as well.

These problems are attracting great interest, as shown by the fact that the last Solvay Congress, which took place in November 1978 in Brussels, was wholly dedicated to the problem of "Order and Fluctuations in Equilibrium and Nonequilibrium Statistical Mechanics."

Important results in nonlinear thermodynamics (fluctuation dissipation theorem for nonlinear systems, generalized Onsager relations) were obtained in recent works by G. F. Yefremov and R. L. Stratonovich. These results were further developed in the articles [14.6, 7], where the main works on this subject are also listed.

One of the principal tasks of the kinetic theory of electromagnetic processes lies in further investigating nonlinear fluctuational processes on both the thermodynamic and kinetic levels. One of the most interesting subjects here is, of course, the problem of self-organization in irreversible processes. The first book dealing with this problem was that by *Prigogine* and *Nicolis* [14.8]. To emphasize the special role of collective or cooperative interactions in the process of self-organization, *Haken* has coined the term "synergetics," and published the first book [14.9] giving a systematic account of the whole scope of questions related to this new branch of science [14.1, 2, 10 – 15].

I conclude this book with the hope that it will contribute to whatever extent to solving these most fascinating problems.

References*

Chapter 1

1.1 S. Chapman, T. G. Cowling: *The Mathematical Theory of Nonuniform Gases* (Cambridge University Press, 1939)

1.2. N. N. Bogolyubov: *Problemy dinemicheskoi teorii v statisticheskoi fizike* (Problems of the Dynamic Theory in Statistical Physics) (Gostekhizdat, Moscow 1946)

1.3 G. E. Uhlenbeck, G. W. Ford: *Lectures in Statistical Mechanics* (Am. Math. Soc., Providence 1963)

1.4 I. Prigogine: *Non-Equilibrium Statistical Mechanics* (Wiley, New York 1962)

1.5 Yu. L. Klimontovich: *The Statistical Theory of Non-Equilibrium Processes in Plasma* (Pergamon, Oxford 1967)

1.6 K. P. Gurov: *Osnovaniya kineticheskoi teorii* (Foundations of Kinetic Theory) (Nauka, Moscow 1966)

1.7 M. N. Kogan: *Dinamika razrezhennogo gaza* (Nauka, Moscow 1967) [English transl.: *Rarefied Gas Dynamics* (Plenum, New York 1969)]

1.8 R. Balescu: *Statistical Mechanics of Charged Particles* (Wiley, London 1962)

1.9 V. P. Silin: *Vvedeniye v kineticheskuyu teoriyu gazov* (Introduction to Kinetic Theory of Gases) (Nauka, Moscow 1971)

1.10 C. Cercigniani: *Mathematical Methods in Kinetic Theory* (Plenum, New York 1969)

1.11 Yu. L. Klimontovich: *Kinetic Theory of Non-Ideal Gases and Non-Ideal Plasma* (Pergamon, Oxford 1982)

1.12 J. H. Ferziger, H. G. Kaper: *Mathematical Theory of Transport Processes in Gases* (North-Holland, Amsterdam 1972)

1.13 S. Ichimaru: *Basic Principles of Plasma Physics* (Benjamin, New York 1973)

1.14 G. Ecker: *Theory of Fully Ionized Plasmas* (Academic, New York 1972)

1.15 B. B. Kadomtsev: *Kollektivnye yavleniya v plazme* (Collective Phenomena in Plasma) (Nauka, Moscow 1976)

1.16 V. N. Tsytovich: *Teoriya turbulentnoi plazmy* (Atomizdat, Moscow 1971) [English transl.: *Theory of Turbulent Plasma* (Plenum, New York 1974)]

1.17 A. G. Sitenko: *Fluktuatsii i nelineinoye vzaimodeistviye voln v plazme* (Fluctuations and Nonlinear Interaction of Waves in a Plasma) (Naukova Dumka, Kiev 1977)

1.18 S. A. Akhmanov, R. V. Khokhlov: *Problemy nelineinoi optiki* (Academy of Sciences of the USSR, Moscow 1964) [English transl.: *Nonlinear Optics* (Gordon and Breach, New York 1972)]

1.19 V. S. Letokhov, V. P. Chebotayev: *Printsipy nelineinoi lazernoi spektroskopii* (Nauka, Moscow 1975) [English transl.: *Nonlinear Laser Spectroscopy,* Springer Series in Optical Sciences, Vol. 4 (Springer, Berlin, Heidelberg, New York 1977)]

1.20 V. M. Fain: *Fotony i nelineinye sredy* (Photons and Nonlinear Media) (Sovetskoye Radio, Moscow 1972)

1.21 D. N. Zubarev (ed.): *Nervanovesnaya statisticheskaya termodinamika* (Nauka, Moscow 1971) [English transl.: *Nonequilibrium Statistical Thermodynamics* (Plenum, New York 1974)]

1.22 M. Lax: *Fluctuation and Coherence Phenomena in Classical and Quantum Physics* (Gordon and Breach, London 1968)

---

* Please note that articles in Soviet journals are in Russian. The titles have been translated here for the reader's convenience.

1.23 F. S. Dyson, E. W. Montroll, M. Kac: In *Statistical Physics, Phase Transitions and Super-fluidity*, ed. by M. Chretien (Gordon and Breach, New York 1968);
M. E. Fisher: "The Theory of Critical Point Singularities", Enrico Fermi Summer School of Critical Phenomena, 1970

1.24 V. L. Ginzburg: *Rasprostraneniye elektromagnitnykh voln v plazme* (Nauka, Moscow 1967) [English transl.: *Propagation of Electrodynamic Waves in Plasmas*, 2nd ed. (Pergamon, Oxford 1971)]

1.25 A. A. Sokolov, I. M. Ternov: *Relyativistkii elektron* (The Relativistic Electron) (Nauka, Moscow 1974)

1.26 V. B. Baranov, K. V. Krasnobayev: *Gidrodinamicheskaya teoriya kosmicheskoi plazmy* (Hydrodynamic Theory of Cosmic Plasma) (Nauka, Moscow 1977)

1.27 C. V. Heer: *Statistical Mechanics, Kinetic Theory and Stochastic Processes* (Academic, New York 1972)

1.28 M. Born, H. S. Green: *A General Kinetic Theory of Liquids* (Cambridge University Press, Cambridge 1949)

1.29 J. G. Kirkwood: The statistical mechanical theory of transport processes. J. Chem. Phys. **14**, 180, 347 (1946); **15**, 72 (1947)

1.30 I. Yvon: *La théorie statistique des fluides et l'équation d'état* (Hermann, Paris 1935)

1.31 H. Haken: *Synergetics. An Introduction*, 2nd ed., Springer Series in Synergetics, Vol. 1 (Springer, Berlin, Heidelberg, New York 1978)

Chapter 2

2.1   Yu. L. Klimontovich: *The Statistical Theory of Non-Equilibrium Processes in Plasma* (Pergamon, Oxford 1967)

2.2   Yu. L. Klimontovich: *Kinetic Theory of Non-Ideal Gases and Non-Ideal Plasmas* (Pergamon, Oxford 1982)

2.3   S. Ichimaru: *Basic Principles of Plasma Physics* (Benjamin, New York 1973)

2.4   G. Ecker: *Theory of Fully Ionized Plasma* (Academic, New York 1972)

2.5   W. Heitler: *Quantum Theory of Radiation* (Clarendon, Oxford 1954)

2.6   V. L. Ginzburg: *Teoriticheskaya fizika i astrofizika* (Theoretical Physics and Astrophysics) (Nauka, Moscow 1975)

2.7   V. M. Fain: *Fotony i nelineinye sredy* (Photons and Nonlinear Media) (Sovetskoye Radio, Moscow 1972)

2.8   W. E. Brittin: Statistical mechanical theory of transport phenomena in a fully ionized plasma. Phys. Rev. **106**, 843 (1957)

2.9   Yu. L. Klimontovich: Energy loss of charged particles for the excitation of oscillations in a plasma. Zh. Eksp. Teor. Fiz. **36**, 1405 (1959)

2.10 Yu. L. Klimontovich: Kinetic equations for relativistic plasma. Zh. Eksp. Teor. Fiz. **37**, 735 (1959)

2.11 V. V. Bely, Yu. L. Klimontovich: Low-frequency non-equilibrium fluctuations in electron–phonon systems. Physica **73**, 327 (1974)

Chapter 3

3.1   P. Mazur: Adv. Chem. Phys. **1**, 309 (1958)

3.2   S. R. De Groot: *The Maxwell Equations* (North-Holland, Amsterdam 1969)

3.3   L. D. Landau, E. M. Lifshitz: *Course of Theoretical Physics*, Vol. 2. The Classical Theory of Fields, 4th ed. (Pergamon, New York 1976)

3.4   Yu. L. Klimontovich: *Kinetic Theory of Non-Ideal Gases and Non-Ideal Plasmas* (Pergamon, Oxford 1982)

Chapter 4

4.1   Yu. L. Klimontovich: *Kinetic Theory of Non-Ideal Gases and Non-Ideal Plasmas* (Pergamon, Oxford 1982)

4.2   V. P. Silin: *Vvedeniye v kineticheskuyu teoriyu gazov* (Introduction to Kinetic Theory of Gases) (Nauka, Moscow 1971)

4.3   S. Ichimaru: *Basic Principles of Plasma Physics* (Benjamin, New York 1973)

4.4   L. D. Landau, E. M. Lifshitz: *Quantum Physics* (Pergamon, Oxford 1962)

4.5   A. A. Vlasov: *Teoriya mnogikh chastits* (Gostekhizdat, Moscow 1950) [English transl.: *Many-Particle Theory and Its Application to Plasma* (Gordon and Breach, New York 1961)]

4.6   R. Balescu: *Statistical Mechanics of Charged Particles* (Wiley, London 1962)

4.7   G. Ecker: *Theory of Fully Ionized Plasmas* (Academic, New York 1972)

4.8   M. L. Levin, S. M. Rytov: *Teoriya ravnovesnogo elektromagnitogo izlucheniya* (Theory of Equilibrium Electromagnetic Radiation (Nauka, Moscow 1967)

4.9   L. D. Landau, E. M. Lifshitz: *Course of Theoretical Physics*, Vol. 5. Statistical Physics, 2nd ed. (Pergamon, New York 1969)

4.10   L. D. Landau: The kinetic equation for the case of Coulomb interaction. Zh. Eksp. Teor. Fiz. **7**, 203 (1937)

4.11   Yu. L. Klimontovich: *The Statistical Theory of Non-Equilibrium Processes in Plasma* (Pergamon, Oxford 1967)

4.12   Yu. L. Klimontovich: Determination of the eigenvalues of physical quantities with the quantum distribution function. Dokl. Akad. Nauk SSSR **108**, 1033 (1956)

4.13   D. A. Kirzhnitz: On statistical many particle theory. Tr. Fiz. Inst. Akad. Nauk SSSR **16**, 3 (1961)

4.14   P. Batnagar, E. Gross, M. Krook: A model for collision processes in gases. Phys. Rev. **94**, 511 (1954); **102**, 593 (1956)

4.15   V. L. Ginzburg, A. A. Rukhadze: *Volny v magnitoaktivnoi plazme* (Waves in Magnetoactive Plasma) (Nauka, Moscow 1975)

4.16   A. I. Akhiezer, I. A. Akhiezer, R. V. Polovin, A. G. Sitenko, K. N. Stepanov: *Elektrodinamika plazmy* (Nauka, Moscow 1974) [English transl.: *Plasma Electrodynamics* (Pergamon, Oxford 1975)]

4.17   A. F. Aleksandrov, L. S. Bogdankevich, A. A. Rukhadze: *Osnovy elektrodinamiki plazmy* (Basic Principles of Plasma Electrodynamics) (Nauka, Moscow 1974)

4.18   B. B. Kadomtsev: On fluctuations in a gas. Zh. Eksp. Teor. Fiz. **32**, 943 (1957)

Chapter 5

5.1   S. Chandrasekhar: *Radiation Transfer* (Oxford University Press, London 1960)

5.2   L. D. Landau, E. M. Lifshitz: On hydrodynamic fluctuations. Zh. Eksp. Teor. Fiz. **32**, 618 (1957)

5.3   V. I. Klyatzkin: *Statisticheskoye opisaniye dinamicheskikh sistem s fluktuiruyushchimi para-metrami* (Statistical Description of Dynamic Systems with Fluctuating Parameters) (Nauka, Moscow 1975)

5.4   Yu. L. Klimontovich: *Kinetic Theory of Non-Ideal Gases and Non-Ideal Plasmas* (Pergamon, Oxford 1982)

5.5   R. F. Fox, G. E. Uhlenbeck: Theory of hydrodynamical fluctuation. Phys. Fluids **13**, 1893 (1970)

5.6   R. L. Stratonovich: *Izbrannye voprosy teorii fluktuatsii v radiotekhnike* (Selected Questions in the Theory of Fluctuations in Radio Engineering) (Sovetskoye Radio, Moscow 1961)

5.7   S. M. Rytov: *Vvedeniye v statisticheskuyu radiofiziky. Sluchainye protsessy* (An Introduction to Statistical Radio Physics. Random Processes) (Nauka, Moscow 1976)

5.8   S. E. Pitovranov, V. M. Chetverikov: Corrections to the diffusion approximation in stochastic differential equations. Teor. Mat. Fiz. **35**, 211 (1978)

5.9   L. D. Landau, E. M. Lifshitz: *Mechanics of Continuous Media* (Pergamon, Oxford 1960)

5.10   Yu. A. Krutkov: "Issledovaniya po teorii braunovskogo dvizheniya" (Investigations of the Theory of Brownian Motion), in *Braunovskoye dvizheniye* (Brownian Motion) (ONTI, Moscow 1934)

5.11   N. N. Bogolyubov: "Kvazisredniye v zadachakh statisticheskoi mekhaniki" (The Quasi-Means in Problems of Statistical Mechanics), in *Izbrannye trudy* (Selected Works) (Naukova Dumka, Kiev 1971) Vol. 3, p. 174 (in Russian)

5.12 L. D. Landau, E. M. Lifshitz: *Course of Theoretical Physics*, Vol. 5. Statistical Physics, 2nd ed. (Pergamon, New York 1969)

5.13 L. P. Kadanoff: Scaling laws for Ising models near T_c. Physics **2**, 263 (1961)

5.14 A. Z. Patashinsky, V. L. Pokrovsky: *Fluktuatsionnaya teoriya fazovykh perekhodov* (Fluctuational Theory of Phase Transitions) (Nauka, Moscow 1975)

5.15a K. G. Wilson, G. Kogut: The renormalization group and the ε expansion. Phys. Rep. **2**, 75 – 109 (1974)

5.15b M. E. Fysher: "The Theory of Critical Point Singularities", Enrico Fermi Summer School of Critical Phenomena, 1970

5.16 A. P. Levanyuk: On the theory of scattering of light near the points of a phase transition of the second kind. Zh. Eksp. Teor. Fiz. **36**, 810 (1959)

5.17 V. L. Ginzburg: Some notes on phase transitions of the second kind and microscopic theory of ferroelectrics. Fiz. Tverd. Tela **2**, 2031 (1960)

5.18 R. Balescu: *Equilibrium and Nonequilibrium Statistical Mechanics* (Wiley, New York 1975)

5.19 Yu. L. Klimontovich: On fluctuations in the critical region. Zh. Eksp. Teor. Fiz. **76**, 1632 (1979)

5.20 L. D. Landau, I. M. Khalatnikov: On anomalous sound absorption near the points of phase transitions of the second kind. Dokl. Akad. Nauk SSSR **96**, 469 (1954)

5.21 E. M. Lifshitz, L. P. Pitayevsky: *Fizicheskaya kinetika* (Physical Kinetics) (Nauka, Moscow 1979)

5.22 A. N. Malakhov: *Fluktuatsii v avtokolebatel'nykh sistemakh* (Fluctuations in Self-Oscillatory Systems) (Nauka, Moscow 1968)

5.23 Yu. M. Romanovsky, N. V. Stepanova, D. S. Chernavsky: *Matematicheskoye modelirovaniye v biofizike* (Mathematical Modeling in Biophysics) (Nauka, Moscow 1975)

5.24 W. Ebeling: *Strukturbildung bei irreversiblen Prozessen* (Teubner, Leipzig 1976)

5.25 G. Nicolis, I. Prigogine: *Self-Organization in Non-Equilibrium Systems* (Wiley, New York 1977)

5.26 H. Haken: *Synergetics. An Introduction*, 2nd ed., Springer Series in Synergetics, Vol. 1 (Springer, Berlin, Heidelberg, New York 1978)

5.27 S. P. Zeiger, Yu. L. Klimontovich, P. S. Landa, E. G. Lariontsev, E. E. Fradkin: In *Volnovye i fluktuatsionnye protsessy v lazerakh* (Wave and Fluctuational Processes in Lasers), ed. by Yu. Klimontovich (Nauka, Moscow 1974)

5.28 Yu. L. Klimontovich: On spectrum width of nearly harmonic self-oscillatory systems. Pis'ma Zh. Eksp. Teor. Fiz. **26**, 180 (1977)

5.29 I. L. Bershtein: Fluctuations in a self-oscillatory system and determination of natural sharpless-ness of frequency of a vacuum-tube generator. Zh. Eksp. Teor. Fiz. **11**, 305 (1941)

5.30 V. V. Belyi, Yu. L. Klimontovich: Kinetic theory of fluctuations at Brownian motion. Teor. Mat. Fiz. **34**, 233 (1978)

5.31 Yu. L. Klimontovich: Residual time correlations and the $1/\omega$ spectrum in Brownian movement. Zh. Eksp. Teor. Fiz. **80**, 2243 (1981)

5.32 Yu. L. Klimontovich: Nonequilibrium sources of hydrodynamic fluctuations. Kinetic coefficients considering the effects of hydrodynamic motion and turbulence pulsations. Pis'ma Zh. Eksp. Teor. Fiz. **7**, 1181 (1981)

5.33 Yu. Klimontovich: Calculation of averaged turbulent motion of liquid in a tube and in boundary layer. Pis'ma Zh. Tekh. Fiz. **8**, 311 (1982)

Chapter 6

6.1 Yu. L. Klimontovich: *Kinetic Theory of Non-Ideal Gas and Non-Ideal Plasmas* (Pergamon, Oxford 1981)

6.2 L. D. Landau, E. M. Lifshitz: *Course of Theoretical Physics*, Vol. 8. Electrodynamics of Continuous Media (Pergamon, New York 1960)

6.3 Yu. L. Klimontovich: Questions of the statistical theory of the interaction of atoms with radiation. Usp. Fiz. Nauk **101**, 578 (1970)

6.4 L. D. Landau, E. M. Lifshitz: *Course of Theoretical Physics*, Vol. 2. The Classical Theory of Fields, 4th ed. (Pergamon, New York 1976)

6.5   Yu. L. Klimontovich, V. S. Fursov: The influence of the interaction between molecules on the radiational retardation term in the classical theory of light scattering. Zh. Eksp. Teor. Fiz. **19**, 819 (1949)

6.6   L. I. Mandelshtam: "K teorii dispersii" (On the Theory of Disperion) in *Polnoye sobraniye trudov* (Complete Works) (Academy of Sciences of the USSR, Moscow 1948) Vol. 1, p. 125

6.7   V. A. Alekseyev, A. V. Vinogradov, I. I. Sobelman: On the microscopic approach to the effects of radiational interaction of atoms and molecules. Usp. Fiz. Nauk **102**, 43 (1970)

6.8   V. G. Vaks: *Vvedeniye v mikroskopicheskuyu teoriyu segnetoelektrikov* (Introduction to the Microscopic Theory of Ferroelectrics) (Nauka, Moscow 1973)

6.9   R. Blinc, B. Zeks: *Soft Modes in Ferroelectrics and Antiferroelectrics* (American Elsevier, New York 1974)

6.10 N. N. Kristofel, P. I. Konsin: Vibron theory of ferroelectricity. Usp. Fiz. Nauk **120**, 507 (1976)

6.11 E. M. Lifshitz, L. P. Pitayevsky: In *Course of Theoretical Physics*, Vol. 5. Statistical Physics, 2nd ed., by L. D. Landau, E. M. Lifshitz (Pergamon, New York 1969) Part 2

6.12 N. N. Bogolyubov, D. N. Zubarev, Yu. A. Tserkovnikov: Asymptotically exact solution for the model Hamiltonian in the superconductivity theory. Zh. Eksp. Teor. Fiz. **39**, 120 (1960)

6.13 N. N. Bogolyubov: "K voprosy o model'nom gamil'toniana v teorii sverkhprovodimosti" (On the Question of Model Hamiltonian in the Superconductivity Theory), in *Izbrannye trudy* (Selected Works) (Naukova Dumka, Kiev 1971) Vol. 3, p. 110

6.14 N. N. Bogolyubov, Jr.: *Metod issledovaniya model'nykh gamil'tonianov* (Nauka, Moscow 1974) [English transl.: *A Method for Studying Model Hamiltonians* (Pergamon, Oxford 1972)]

6.15 A. I. Larkin, D. E. Khmelnitsky: Phase transition in single-axis ferroelectrics. Zh. Eksp. Teor. Fiz. **56**, 2087 (1969)

Chapter 7

7.1   L. D. Landau, E. M. Lifshitz: *Course of Theoretical Physics*, Vol. 3. Quantum Mechanics, 3rd ed. (Pergamon, New York 1977)

7.2   N. N. Bogolyubov: "Lektsii po kvantovoi statistike" (Lectures on Quantum Statistics), in *Izbrannye trudy* (Selected Works) (Naukova Dumka, Kiev 1971) Vol. 2, p. 287 [English transl.: *Lectures on Quantum Statistics* (Gordon and Breach, New York 1967)]

7.3   Yu. L. Klimontovich: *Kinetic Theory of Non-Ideal Gases and Non-Ideal Plasmas* (Pergamon, Oxford 1982)

7.4   Yu. L. Klimontovich: Statistical theory of inelastic processes in plasma. 1. Kinetic equations for a Coulomb plasma with inelastic processes taken into account. Zh. Eksp. Teor. Fiz. **52**, 1233 (1967)

7.5   Yu. L. Klimontovich: Statistical theory of inelastic processes in plasma. 2. Processes determined by a transverse electromagnetic field. Zh. Eksp. Teor. Fiz. **54**, 136 (1968)

Chapter 8

8.1   Yu. L. Klimontovich: *Kinetic Theory of Non-Ideal Gases and Non-Ideal Plasma* (Pergamon, Oxford 1982)

8.2   Yu. L. Klimontovich: Statistical theory of inelastic processes in plasma. Zh. Eksp. Teor. Fiz. **52**, 1233 (1967)

8.3   R. Balescu: *Statistical Mechanics of Charged Particles* (Wiley, New York 1962)

8.4   V. P. Silin: *Vvedeniye v kineticheskuyu teoriyu gazov* (Introduction to Kinetic Theory of Gases) (Nauka, Moscow 1971)

8.5   V. L. Ginzburg, A. A. Rukhadze: *Volny v magnitoaktivnoi plazme* (Waves in Magnetoactive Plasma) (Nauka, Moscow 1975)

8.6   A. I. Akhiezer, I. A. Akhiezer, R. V. Polovin, A. G. Sitenko, K. N. Stepanov: *Elektrodinamika plazmy* (Nauka, Moscow 1974) [English transl.: *Plasma Electrodynamics* (Pergamon, Oxford 1975)]

8.7   A. B. Mikhailovsky: *Teoriya plazmennykh noustoichivostei* (Theory of Plasma Instabilities) (Atomizdat, Moscow 1975) Vols. 1, 2

8.8   A. F. Aleksandrov, L. S. Bogdankevich, A. A. Rukhadze: *Osnovy elektrodinamiki plazmy* (Basic Principles of Plasma Electrodynamics) (Vysshaya Shkola, Moscow 1978)

8.9   A. I. Larkin: Thermodynamic functions of nonisothermic plasma. Zh. Eksp. Teor. Fiz. **38**, 1896 (1969)

8.10 W. Ebeling, W. D. Kraeft, D. Kremp: *Theory of Bound States and Ionization Equilibrium in Plasmas and Solids* (Akademie, Berlin 1976)

8.11 L. P. Kudrin: *Statisticheskaya fizika plazmy* (Statistical Physics of Plasma) (Atomizdat, Moscow 1974)

8.12 L. D. Landau, E. M. Lifshitz: *Course of Theoretical Physics*, Vol. 3. Quantum Mechanics, 3rd ed. (Pergamon, New York 1977)

8.13 Ya. B. Zel'dovich, Yu. P. Raizer: *Fizika udarnykh voln i vysokotemperaturnye gidrodinamicheskiye yavleniya* (Physics of Shock Waves and High-Temperature Hydrodynamic Phenomena) (Nauka, Moscow 1966)

8.14 Yu. L. Klimontovich: "Kinetic Theory of Fluctuations in Gases and Plasma", in Fundamental Problems in Statistical Mechanics, IV ed. by E. G. D. Cohen, W. Fiszdon (Ossolinskich, Warszawa 1978) (in Polish)

8.15 V. V. Belyi, Yu. L. Klimontovich: Kinetic fluctuations in partially ionized plasma and chemically reacting gases. Zh. Eksp. Teor. Fiz. **74**, 1660 (1978)

8.16 G. Nicolis, L. Prigogine: *Self-Organization in Non-Equilibrium Systems* (Wiley, New York 1977)

8.17 N. G. Van Kampen: The expansion of the master equation. Adv. Chem. Phys. **34**, 245 (1976)

Chapter 9

9.1   Yu. L. Klimontovich: *Kinetic Theory of Non-Ideal Gases and Non-Ideal Plasmas* (Pergamon, Oxford 1982)

9.2   V. L. Ginzburg: Some questions on the theory of radiation from hyperrelativistic motion in a medium. Usp. Fiz. Nauk **69**, 537 (1959)

9.3   V. L. Ginzburg: *Teoreticheskaya fizika i astrofizika* (Theoretical Physics and Astrophysics) (Nauka, Moscow 1975)

9.4   L. D. Landau, E. M. Lifshitz: *Course of Theoretical Physics*, Vol. 3. Quantum Mechanics, 3rd ed. (Pergamon, New York 1977)

9.5   L. D. Zel'dovich, E. M. Lifshitz: *Fizika udarnykh voln i vysokotemperaturnye gidrodinamicheskiye yavleniya* (Physics of Shock Waves and High-Temperature Hydrodynamic Phenomena) (Nauka, Moscow 1966)

9.6   L. A. Vainshtein, I. I. Sobel'man, E. A. Yukov: *Vozbuzhdeniye atomov i ushireniye spektral'nykh linii* (Excitation of Atoms and Spectral Line Broadening) (Nauka, Moscow 1979)

9.7   T. W. Hänsch, A. L. Shavlow: Cooling of gases by laser radiation. Opt. Commun. **13**, 68 (1975)

9.8   A. P. Kasantsev: Resonant light pressure. Usp. Fiz. Nauk **124**, 113 (1978)

9.9   V. S. Letokhov, V. G. Minogin, B. D. Pavlik: Cooling and capture of atoms and molecules by resonant light field. Zh. Eksp. Teor. Fiz. **72**, 1328 (1977)

9.10 I. V. Krasnov, N. Ya. Shaparev: Separation of gases by resonant electromagnetic field. Pis'ma Zh. Tekh. Fiz. **1**, 875 (1975)

9.11 Yu. L. Klimontovich, S. N. Luzgin: Kinetic theory of cooling monoatomic gases by resonant electromagnetic radiation. Zh. Tekh. Fiz. **48**, 2217 (1978)

9.12 Yu. L. Klimontovich, S. N. Luzgin: On the possibility of simultaneous self-focusing of atomic and light beams. Pis'ma Zh. Eksp. Teor. Fiz. **30**, 654 (1979)

Chapter 10

10.1 R. G. Breene: *The shift and Shape of Spectral Lines* (Pergamon, New York 1961)

10.2 I. I. Sobel'man: *Vvedniye v teoriyu atomnykh spektrov* (Fizmatgiz, Moscow 1963) [English transl.: *Introduction to the Theory of Atomic Spectra* (Pergamon, Oxford 1972)]

10.3 L. A. Vainshtein, I. I. Sobel'man, E. A. Yukov: *Vozbuzhdeniye atomov i ushireniye spektral'nykh linii* (Excitation of Atoms and Spectral Line Broadening) (Nauka, Moscow 1979)

10.4 G. Griem: *Spectral Line Broadening By Plasmas* (Academic, New York 1974)

10.5 V. S. Lisitsa: Stark broadening of hydrogen lines in plasma. Usp. Fiz. Nauk **122**, 369 (1977)

10.6 Yu. L. Klimontovich: Questions on the statistical theory of the interaction of atoms with radiation. Usp. Fiz. Nauk *101*, 578 (1970)

10.7 E. A. Asmarian, Yu. L. Klimontovich: On the theory of spectral line broadening in a nonequilibrium partially ionized gas. Vestn. Mosk. Univ. No. 3, 272 (1974)

10.8 Yu. L. Klimontovich: *Kinetic Theory of Non-Ideal Gases and Non-Ideal Plasmas* (Pergamon, Oxford 1982)

10.9 L. D. Landau, E. M. Lifshitz: *Course of Theoretical Physics*, Vol. 3 Quantum Mechanics, 3rd ed. (Pergamon, New York 1977)

10.10 A. A. Vlasov, V. S. Fursov: Theory of the width of spectral lines in a homogenous gas. Zh. Eksp. Teor. Fiz. **6**, 378 (1936)

10.11 A. P. Kazantsev: Kinetic equations for a gas of excited atoms. Zh. Eksp. Teor. Fiz. **51**, 1751 (1966)

10.12 P. Berman, W. Lamb: Influence of resonance and foreign gas collision on line shapes. Phys. Rev. **187**, 221 (1969)

10.13 Yu. A. Vdovin, V. M. Galitsky: Dielectric permittivity of a gas of resonant atoms. Zh. Eksp. Teor. Fiz. **52**, 1345 (1967)

10.14 S. G. Rautian: Some questions on the theory of gas quantum generators. Tr. Fiz. Inst. Akad. Nauk SSSR **43**, 3 (1968)

10.15 E. G. Pestov, S. G. Rautian: On the kinetic equation for the density matrix. Zh. Eksp. Teor. Fiz. **56**, 901 (1969)

10.16 S. G. Rautian, I. I. Sobel'man: The influence of collisions on the Doppler broadening of spectral lines. Usp. Fiz. Nauk **90**, 209 (1966)

10.17 Yu. L. Klimontovich, S. A. Sukhin: Broadening of spectral lines considered as a problem of fluctuation theory in nonequilibrium plasma. Vestn. Mosk. Univ. No. 5 (1980)

10.18 L. M. Biberman: On the diffusion of resonant radiation. Dokl. Akad. Nauk SSSR **27**, 220 (1940)

10.19 L. M. Biberman: On the theory of diffusion of resonant radiation. Zh. Eksp. Teor. Fiz. **17**, 416 (1947)

10.20 T. Holstein: Imprisonment of resonance radiation in gases I, II. Phys. Rev. **72**, 1212 (1947); **83**, 1169 (1951)

10.21 M. I. D'yakonov, V. I. Perel': Coherent relaxation at the diffusion of resonant radiation. Zh. Eksp. Teor. Fiz. **47**, 1483 (1964)

10.22 M. I. D'yakonov, V. I. Perel': The influence of the capture of resonant radiation upon the characteristics of a gas laser. Zh. Eksp. Teor. Fiz. **58**, 1090 (1970)

10.23 Yu. A. Vdovin, V. M. Yermachenko: Propagation of radiation in a resonant media. Zh. Eksp. Teor. Fiz. **54**, 148 (1968)

10.24 I. M. Beterev, Yu. A. Mityugin, S. G. Rautian, V. P. Chebotayev: On capture of resonant radiation in gas lasers. Zh. Eksp. Teor. Fiz. **58**, 1243 (1970)

10.25 P. V. Yelutin, V. D. Krivchenkov: *Kvantovaya mekhanika* (Quantum Mechanics) (Nauka, Moscow 1976)

10.26 A. A. Sokolov, Yu. M. Loskutov, I. M. Ternov: *Kvantovaya mekhanika* (Quantum Mechanics) (Prosvescheniye, Moscow 1962)

10.27 S. Chandrasekhar: *Radiation Transfer* (Oxford University Press, Oxford 1960)

10.28 G. Ecker: "Raspredeleniye mikropolei v plazme" (Distribution of Microfields in a Plasma), in *Problemy teorii plazmy* (Problems of Plasma Theory) (Naukova Dumka, Kiev 1976)

10.29 G. Traving: "Ushireniye i sdvig spektral'nykh linii" (Broadening and Shift of Spectral Lines), in *Metody issledovaniya plazmy* (Methods of Investigating Plasmas) (Mir, Moscow 1971)

10.30 D. Voslamber: Influence of time ordering on unified line profiles. Z. Naturforsch. **27**, 1783 (1972)

10.31 H. Capes, D. Voslamber: Electron correlations in the unified model for Stark broadening. Phys. Rev. A**5**, 2528 (1972)

10.32 V. S. Lisitsa, S. I. Yakovenko: Optical and radiational collisions. Zh. Eksp. Teor. Fiz. **66**, 1550 (1974); Nonlinear theory of broadening. Zh. Eksp. Teor. Fiz. **68**, 479 (1975)

10.33 V. S. Lisitsa: Misshaping of hydrogen spectral line in a plasma by a strong electromagnetic field. Zh. Eksp. Teor. Fiz. **69**, 195 (1975)

10.34 A. V. Andreyev, Yu. A. Ilyinsky, R. V. Khokhlov: Narrowing of γ-resonance line by a continuous radio-frequency field. Zh. Eksp. Teor. Fiz. **67**, 1647 (1974)

10.35 V. I. Kogan (ed.): In *Fizika plazmy i problemy upravlyayemogo termoyadernogo sinteza* (Plasma Physics and Problems of Controlled Thermonuclear Fusion), by M. A. Leontovich (Academy of Sciences of the USSR, Moscow 1958) Vol. 4 [English transl.: *Reviews of Plasma Physics*, Vol. 4 (Plenum, New York)]

10.36 H. Capes, D. Voslamber: Spectral line profiles in weakly turbulent plasmas. Phys. Rev. A 15, 1751 (1977)

10.37 E. A. Oks, G. V. Sholin: On Stark line profiles of the hydrogen spectrum in a plasma with Langmuir turbulence. Zh. Eksp. Teor. Fiz. 68, 974 (1975)

10.38 A. I. Zhuzhanashvili, E. A. Oks: The technique of optical polarization measurements of the Langmuir turbulence spectrum of a plasma. Zh. Eksp. Teor. Fiz. 73, 2142 (1977)

Additional References

V. A. Alekseyev, T. L. Andreyeva, I. I. Sobel'man: A method of quantum kinetic equation for atoms and molecules, and its application for calculating the optical characteristics of gases. Zh. Eksp. Teor. Fiz. 62, 614 (1972)

A. V. Vinogradov, V. P. Shevel'ko: On the frequency of inelastic collisions in a plasma. Zh. Eksp. Teor. Fiz. 71, 1037 (1976)

Chapter 11

11.1 V. L. Ginzburg: *Pasprostraneniye elektromagnitnykh voln v plazme* (Propagation of Electromagnetic Waves in a Plasma) (Nauka, Moscow 1957)

11.2 B. B. Kadomtsev: Acting field in a plasma. Zh. Eksp. Teor. Fiz. 33, 151 (1957)

11.3 Yu. L. Klimontovich: *Kinetic Theory of Non-Ideal Gases and Non-Ideal Plasmas* (Pergamon, Oxford 1982)

11.4 Yu. L. Klimontovich, W. Ebeling: Quantum kinetic equations for non-ideal gases and nonideal plasmas. Zh. Eksp. Teor. Fiz. 63, 905 (1972)

11.5 L. D. Landau, E. M. Lifshitz: *Course of Theoretical Physics*, Vol. 5. Statistical Physics (Pergamon, New York 1969)

11.6 H. B. Callen, T. A. Welton: Irreversibility and generalized noise. Phys. Rev. 83, 34 (1951)

11.7 R. Kubo: Statistical-mechanical theory of irreversible processes. J. Phys. Soc. Jpn. 12, 570 (1957)

11.8 D. N. Zubarev: *Neravnovesnaya statisticheskaya termodinamika* (Nauka, Moscow 1967) [English transl.: *Nonequilibrium Statistical Thermodynamics* (Plenum, New York 1974)]

11.9 M. L. Levin, S. M. Rytov: *Teoriya ravnovesnogo elektromagnitnogo izlucheniya* (Theory of Equilibrium Electromagnetic Radiation) (Nauka, Moscow 1967)

11.10 R. Balescu: *Equilibrium and Non-Equilibrium Statistical Mechanics* (Wiley, New York 1975)

11.11 V. P. Silin, A. A. Rukhadze: *Elektromagnitnye svoistva plazmy i plazmopodobnykh sred* (Electromagnetic Properties of Plasma and Plasma-Like Media) (Atomizdat, Moscow 1961)

11.12 S. P. Zeiger, Yu. L. Klimontovich, P. S. Landa, E. G. Lariontsev, E. E. Fradkin: In *Volnovye i fluktuatsionnye protsessy v lazerakh* (Wave and Fluctuational Processes in Lasers), ed. by Yu. Klimontovich (Nauka, Moscow 1974)

11.13 M. Lax: *Fluctuation and Coherence Phenomena in Classical and Quantum Physics* (Gordon and Breach, London 1968)

11.14 A. G. Sitenko: *Fluktuatsii i nelineinye bzaimodeistviya voln v plazme* (Fluctuations and Nonlinear Interactions of Waves in a Plasma) (Naukova Domka, Kiev 1977)

11.15 V. P. Silin: *Vvedeniye v kineticheskuyu teoriyu gazov* (Introduction to the Kinetic Theory of Gases) (Nauka, Moscow 1971)

11.16 I. Prigogine: *Non-Equilibrium Statistical Mechanics* (Wiley, New York 1962)

11.17 Yu. A. Kukharenko: Kinetic equation for a many-particle system with strong interaction and the entropy increase law. Teor. Mat. Fiz. 31, 133 (1977)

11.18 Yu. L. Klimontovich: Callen-Welton and Kubo formulae for non-equilibrium states. Zh. Eksp. Teor. Fiz. 75, 361 (1978)

11.19 J. W. Gibbs: *Basic Principles of Statistical Mechanics* (Longmans, Green, New York 1931)

11.20 Yu. L. Klimontovich, V. S. Fursov: The influence of interaction between molecules upon the term for radiational retardation in the classic theory of light dispersion. Zh. Eksp. Teor. Fiz. 19, 819 (1949)

11.21 L. D. Landau, E. M. Lifshitz: *Course of Theoretical Physics*, Vol. 8. Electrodynamics of Continuous Media (Pergamon, New York 1960)
11.22 Yu. L. Klimontovich: Statistical Physics (Nauka, Moscow 1982)
11.23 Yu. L. Klimontovich: The measure of incompletness of statistical description and irreversibility. Fluctuation-dissipation relatinos for many-particle distribution function. Pis'ma Zh. Eksp. Teor. Fiz. **36**, 35 (1982)

Chapter 12

12.1 Yu. L. Klimontovich: *Kinetic Theory of Non-Ideal Gases and Non-Ideal Plasmas* (Pergamon, Oxford 1982)
12.2 A. N. Orayevsky: *Molekulyarnye generatory* (Molecular Generators) (Nauka, Moscow 1964)
12.3 S. P. Zeiger, Yu. L. Klimontovich, P. S. Landa, E. G. Lariontsev, E. E. Fradkin: In *Volnovye i fluktuatsionnye protsessy v lazerakh* (Wave and Fluctuational Processes in Lasers), ed. by Yu. L. Klimontovich (Nauka, Moscow 1974)
12.4 M. Lax: *Fluctuation and Coherence Phenomena in Classical and Quantum Physics* (Gordon and Breach, London 1968)
12.5 H. Haken: Theory of intensity and phase fluctuations of a homogeneously broadened laser. Z. Phys. **190**, 327 (1966)
12.6 Yu. L. Klimontovich, A. S. Kovalyov, P. S. Landa: Natural fluctuations in lasers. Usp. Fiz. Nauk **106**, 279 (1972)
12.7 A. L. Schawlow, C. H. Townes: Infrared and optical masers. Phys. Rev. **112**, 1940 (1958)
12.8 W. E. Lamb: The theory of an optical maser. Phys. Rev. **134**, A1429 (1964)
12.9 A. P. Kasantsev, G. I. Surdutovich: On the quantum theory of lasers. Zh. Eksp. Teor. Fiz. **56**, 200 (1969); **58**, 245 (1970)
12.10 I. L. Bershtein, I. A. Andronova, Yu. I. Zaitsev: Fluctuations of strength and frequency of radiation in quantum optical generators. Izv. Vyssh. Uchebn. Zaved. Radiofiz. **10**, 59 (1967)
12.11 Yu. I. Zaitsev, D. P. Stepanov: Fluctuations of frequency of a gas laser and determination of natural width of its spectral line. Zh. Eksp. Teor. Fiz. **55**, 1645 (1968); **56**, 718 (1969)
12.12 F. T. Arecchi, V. Digiorgio: *Measurement of the Statistical Properties of Optical Fields*. Laser Handbook. (North-Holland, Amsterdam 1972) Vol. 1, p. 191 – 264
12.13 S. A. Akhmanov, R. V. Khokhlov: *Problemy nilineinoi optiki* (Academy of Sciences of the USSR, Moscow 1964) [English transl.: *Nonlinear Optics* (Gordon and Breach, New York 1972)]
12.14 S. A. Akhmanov, A. S. Chirkin: *Statisticheskiye yavleniya v nelineinoi optike* (Statistical Phenomena in Nonlinear Optics) (Moscow State University, Moscow 1971)
12.15 S. A. Akhmanov, V. B. Pakhalov, A. S. Chirkin: Formation of spatial coherence of laser radiation on overstepping the generation threshold. Pis'ma Zh. Eksp. Teor. Fiz. **23**, 391 (1976)
12.16 R. L. Stratonovich: Phase transitions in three-dimensional two-level systems. Kvantovaya Elektron. **4**, 2141 (1977)
12.17 V. B. Pakhalov, A. S. Chirkin: Phase transition on formation of spatially coherent beams in a laser. Kvantovaya Elektron. **4**, 1268 (1977)

Chapter 13

13.1 K. Hepp, E. H. Lieb: On superradiance phase transition for molecules in a quantized radiation field: the Dicke model. Ann. Phys. **86**, 360 (1973)
13.2 Y. K. Wang, F. T. Hioe: Phase transition in the Dicke model of superradiance. Phys. Rev. A**7**, 831 (1973)
13.3 V. I. Yemelyanov, Yu. L. Klimontovich: The 4th Vavilov Conference on linear optics, Novosibirsk, 1975. Kvantovaya Elektron. **3**, 848 (1976)
13.4 V. F. Elesin, Yu. V. Kopayev: On the electromagnetic theory of ferroelectricity. Pis'ma Zh. Eksp. Teor. Fiz. **24**, 78 (1976)
13.5 A. P. Levanyuk, K. A. Minayeva, B. A. Strukov: On anomalous sound absorption near the Curie point of a single-axis ferroelectric. Fiz. Tverd. Tela **10**, 2443 (1968)

13.6   A. I. Larkin, D. E. Khmelnitsky: Phase transition in single-axis ferroelectrics. Zh. Eksp. Teor. Fiz. **56**, 2087 (1969)

13.7   A. A. Kaminsky, V. A. Kotsik, Yu. A. Maskayev, I. I. Naumov, L. N. Rashkovich, S. E. Sarkisov: Stimulated emission of ions in ferroelectric crystal barium-sodium niobate (BSN). Pis'ma Zh. Eksp. Teor. Fiz. **1**, 439 (1975)

13.8   S. Kuroda, K. Kubota: Dye laser action in a liquid crystal. Appl. Phys. Lett. **29**, 737 (1976)

13.9   V. I. Yemelyanov, Yu. L. Klimontovich: A phase transition in a system of two-level atoms, induced by a laser field. Pis'ma Zh. Eksp. Teor. Fiz. **27**, 7 (1978)

13.10 M. A. Osipov: The influence of external monochromatic field upon phase transition in ferroelectrics of the Seignettesalz type. Fiz. Tverd. Tela **20**, 3142 (1978)

13.11 M. A. Osipov: A phase transition in ferroelectrics with two sublattices induced by an external field. Fiz. Tverd. Tela **21**, 1475 (1979)

13.12 V. I. Yemelyanov, M. V. Indenbom: Structural phase transition in centrally nonsymmetric and symmetric media, induced by laser radiation. Fiz. Tverd. Tela **21**, 688 (1979)

13.13 A. A. Kaminsky: *Lazernye kristally* (Laser Crystals) (Nauka, Moscow 1975)

13.14 S. A. Akopyan, G. A. Vardanyan, G. A. Lyakhov, Yu. S. Chilingaryan: A retunable dye laser near the point of a phase transition of the second kind. Pis'ma Zh. Tekh. Fiz. **5**, 531 (1979)

13.15 V. I. Yemelyanov, Yu. L. Klimontovich: The influence of a phase transition upon the characteristics of stimulated radiation of dope atoms in ferroelectrics. Pis'ma Zh. Tekh. Fiz. **4**, 180 (1978)

13.16 Yu. L. Klimontovich: Effect of fluctuations in the critical region during a phase transition on the second kind on nonequilibrium phase transition. Zh. Eksp. Teor. Fiz. **78**, 2394 (1982)

13.17 I. B. Bersuker, B. G. Vekhter, I. Ya. Ogurtsov: Tunnel effects in many-atom systems with electronic degeneration and pseudo-degeneration. Usp. Fiz. Nauk **116**, 605 (1975)

13.18 N. N. Kristofel, P. I. Konsin: Vibron theory of ferroelectricity. Usp. Fiz. Nauk **120**, 507 (1976)

13.19 V. G. Vaks: *Vvedeniye v mikroskopicheskuyu teoriyu segnetoelektrikov* (Introduction to the Microscopic Theory of Ferroelectrics) (Nauka, Moscow 1973)

13.20 R. Blinc, B. Zeks: *Soft Modes in Ferroelectrics and Antiferroelectrics* (North-Holland, Amsterdam 1974)

13.21 P. G. De Genns: *The Physics of Liquid Crystals* (Clarendon, Oxford 1974)

13.22 L. D. Landau, E. M. Lifshitz: *Course of Theoretical Physics*, Vol. 5. Statistical Physics (Pergamon, New York 1969)

13.23 R. L. Stratonovich: On fluctuations in liquid crystals. Zh. Eksp. Teor. Fiz. **70**, 1290 (1976)

13.24 L. M. Blinov: *Elektro- i magnitooptika zhidkikh kristallov* (Electric and Magnetic Optics of Liquid Crystals) (Nauka, Moscow 1978)

Chapter 14

14.1   Yu. L. Klimontovich: The measure of incompletness of statistical description and irreversibility. Fluctuation-dissipation relations for many-particle distribution functions. Pis'ma Zh. Eksp. Teor. Fiz. **36**, 35 (1982)

14.2   Yu. L. Klimontovich: *Statistical Physics* (Nauka, Moscow 1982)

14.3   V. P. Silin: *Vvedeniye v kineticheskuyu teoriyu gazov* (Introduction to the Kinetic Theory of Gases) (Nauka, Moscow 1971)

14.4   Yu. L. Klimontovich: *Kinetic Theory of Non-Ideal Gas and Non-Ideal Plasmas* (Pergamon, Oxford 1982)

14.5   S. V. Gantsevich, V. L. Gurevich, R. Katilis: Theory of fluctuations in non-equilibrium electron gas. Riv. Nuovo Cimento **2**, 1 (1979)

14.6   G. N. Bochkov, Yu. E. Kuzovlev: On the general theory of thermal fluctuations in nonlinear systems. Zh. Eksp. Teor. Fiz. **72**, 238 (1979)

14.7   G. N. Bochkov, Yu. E. Kuzovlev: Fluctuations dissipation relationships for nonequilibrium processes in open systems. Zh. Eksp. Teor. Fiz. **76**, 1071 (1979)

14.8   G. Nicolis, I. Prigogine: *Self-Organization in Non-Equilibrium Systems* (Wiley, New York 1977)

14.9   H. Haken: *Synergetics. An Introduction*, 2nd ed., Springer Series in Synergetics, Vol. 1 (Springer, Berlin, Heidelberg, New York 1978)

14.10 W. Ebeling: *Strukturbildung bei irreversiblen Prozessen* (Teubner, Leipzig 1976)
14.11 A. M. Zhabotinsky: *Kontsentratsionnye avtokolebaniya* (Concentrational Autooscillations) (Nauka, Moscow 1974)
14.12 Yu. M. Romanovsky, N. V. Stepanova, D. S. Chernavsky: *Matematicheskoye modelirovaniye v biofizike* (Mathematical Modeling in Biophysics) (Nauka, Moscow 1975)
14.13 Yu. M. Svirezhev, D. O. Logofet: *Ustoichivost' biologicheskikh soobshchestv* (Stability of Biological Communities) (Nauka, Moscow 1978)
14.14 L. A. Artsimovich, R. Z. Sagdeyev: *Fizika plazmy dlya fizikov* (Plasma Physics for Physicists) (Atomizdat, Moscow 1979)
14.15 V. V. Belyi, Yu. L. Klimontovich, V. P. Nalivaiko: On kinetic theory of anomalous conductivity of collisional turbulent plasma. Fiz. Plazmy (in press)

Subject Index

Atom oscillator 48, 50

Balescu – Lenard kinetic equation 69
BBGKY chain of equations 81
Bogolyubov principle of damping
 initial correlations 81
Boltzmann equation 70

Callen – Welton formula 7, 283
Capture radiation 259
Charge density, microscopic 9, 36,
 37
Coherent radiation 6
Collision integral
 Balescu – Lenard 69
 Fokker – Planck 72, 76
 Landau 69
 large-scale (long-range)
  fluctuations of phase densities
   field oscillators 83, 84
   particles 83
  partially ionized plasma 207,
   208, 210, 223, 229, 247
Collision frequencies 60, 258
Collision integrals, properties 210,
 211, 212, 225, 226
Cooling atoms by resonant field 235
Correlation
 functions 66
 length 67
 radius 67, 68, 115
 time 81
Coulomb logarithm 72
Crystal liquid 6, 342
Critical
 indice 5, 117
 point 5, 113, 118
 temperature 113

Curie law 109
Current density, microscopic 9, 36,
 39

Damping coefficient 60
 radiative friction 155
Debye
 correlation function 272
 potential 272
 radius 56, 251
Density matrix operator
 equation 182
 equations 184
 pairs of particles 184
 particles 182
Dielectric permittivity
 atom-oscillator system 149
 fully ionized plasma 67, 70, 89,
  90
 partially ionized plasma 199, 220
 system of particles with strong
  interactions 293
Diffusion
 coefficient 94, 96
 equation 98
Dimensionality index 117
Dipole approximation 33
Dispersional equation 168
Dissipation-fluctuation theorem 7,
 282
Distribution functions
 classical 65, 95
 quantum 194
 two-time 107

Effective
 frequency 166
 Lorentz field 163, 277

Effective
    polarizability 167
    radiation friction 164
Einstein coefficients 228, 300
Einstein formula 304
Entropy 298
Equation
    Balescu – Lenard 69
    Boltzmann 70
    Einstein – Smoluchowsky 98
    Fokker – Planck 95, 98, 128
    Ginzburg – Landau 111
    Vlasov 63
Extinction coefficient 303

Field strengths
    mean 62
    microscopical 2
First moments approximation for
    field oscillators 63
    particles 62

Gibbs
    distribution function 280, 297
    theorem 297
Ginzburg parameter 117
    renormalized 115, 118, 119

Haken – Lax formula 316
Heating atoms by resonant field
    232, 235
Holzmark function 267
H-theorem 296

Induced fluctuations
    phase density 67, 148
    polarization 170, 331
Ionization coefficients
    photoionization 231
    shock ionization 214
Irreversible process 348
Isothermal compressibility, coefficient
    of 5, 164

Kadanoff – Patashinsky – Pokrovsky
    hypothesis of scale invariance 117
Klimontovich – Fursov formula 164

Kubo formula 283

Landau theory of phase transition
    113
Langevin formula 143
Langevin sources
    amplitude equation 133
    Brownian motion 94, 98, 99
    density matrix equation 288
    dissipative nonlinear equation
        125
    field equations 78, 91
    fluctuations
        order parameter 106
        phase density of particles 75,
            85, 86
        phase density of field oscil-
            lators 75, 85, 86
        polarization 170, 175, 330
    Fokker – Planck equation 144,
        145
    Ginzburg – Landau equation 111
    Liouville equation 288
    phase equation 133
    Van der Pol equation 127
Liouville equation 286
Lorentz equations 9, 40
Lorentz – Lorenz formula 168

Maxwell distribution function 96
Maxwell microscopic field equations
    9, 40
Mean (quasi) value 105
Microscopic phase density
    atoms 50
    free charged particles 2
    pairs of charged particles 3, 34
Moments of phase density
    field oscillators 63, 74, 75
    free charged particles 62, 63, 67,
        68

Nonequilibrium phase transition
    classical self-oscillating systems
        127 – 142
    induced by laser radiation 333

influence of equilibrium phase
  transitions
    in ferroelectrics 338
    in liquid crystals 342
  quantum distributed systems
  318, 320
  quantum self-oscillating systems
  (lasers) 308 – 318
Nonlinear thermodynamics 348
Nyquist formula
  classical 284
  quantum 285

Order and fluctuations 348
Order parameter
  classical systems 105, 173
  quantum systems 325, 330

Partially ionized plasma 1
  Coulomb plasma 194
  electromagnetic interactions 220
  spectral emission line broadening
  338
Phase densities, microscopic
  electromagnetic oscillators 20
    equations 20
  free charged particles 1, 9, 14
    equations 2, 9
  pairs of charged particles 3, 34
    equations 3, 34, 35
  structureless atoms 50
    equations 50
Phase density operators in Wigner
(mixed) presentation
  structureless particles 182
    equations 183
  atoms 191
    equations 192, 193
Photoionization coefficient 231
Photorecombination coefficients
  induced 231
  spontaneous 231
Physically infinitesimally small
  length 57, 58, 60, 159
  time interval 159
  volume 159
Planck – Larkin formula 203

Plasma parameter 56
Polarization approximation
  classical atom-field system 148
  fully ionized plasma 73
  partially ionized plasma 195
Principle of partial attenuation of
initial correlations 81

Quasi-mean value 105

Rayleigh formula for extinction
  coefficient 304
Recombination coefficients
  induced 231
  spontaneous 231
  triple collision 214
Relaxation time
  in gases 60
  phase transitions
    equilibrium 123, 124, 173
    nonequilibrium 136, 315, 316
  radiation 60
    influence of correlation of
      particles 164
Retardation 11, 158, 159, 276

Saha formula 203
Schawlow – Townes formula 316
Self-organization 348
Solvay Congress 348
Sources in equations for correlators
  of fluctuations phase densities of
  free particles 86, 87
    Brownian particles 88
    field oscillators 88
Sources of fluctuations
  field oscillators 77
  field strengths 89, 91
  phase densities of free charged
    particles 67, 75, 85
  phase density of pairs of charged
    particles 148
Spectral densities of fluctuations
  density matrix of macroscopical
    systems 297
  density of Brownian motion 145

Spectral densities of fluctuations
  electrical field   67, 70, 205, 222,
    293, 294
  field oscillators   76
  internal parameters   283, 285
  Langevin sources   80
  matrix densities of pairs of
    particles   195
  order parameter   107
  parameters of oscillating system
      amplitude   134
      energy   132, 136
  phase densities of particles and
    field   67
Spectral densities of sources of
  fluctuations
  electrical field   68, 91, 92
  polarization vector   79, 80, 149,
    293
Spectral line broadening
  influence of

collective long-range interactions   271
correlations of free charged
  particles   273
dynamic polarization   273
strong short-range inter-
  actions   271
turbulence   273
Structureless atoms   1
Synergetics   348

Transparency region   225, 300

Vlasov equations   63
Vlasov – Fursov formula   254

Weisskopf radius   60
Wigner operator distribution function
  182, 191
Wilson parameter   121

Exciton Dynamics in Molecular Crystals and Aggregates

With contributions by **V. M. Kenkre, P. Reineker**

1982. 37 figures. IX, 226 pages. (Springer Tracts in Modern Physics, Volume 94). ISBN 3-540-11318-5

Contents: *V. M. Kenkre:* The Master Equation Approach: Coherence, Energy Transfer, Annihilation, and Relaxation. – *P. Reineker:* Stochastic Liouville Equation Approach: Coupled Coherent and Incoherent Motion, Optical Line Shapes, Magnetic Resonance Phenomena.

Solitons

Editors: **R. K. Bullough, P. J. Caudrey**

1980. 20 figures. XVIII, 389 pages. (Topics in Current Physics, Volume 17). ISBN 3-540-09962-X

Contents: *R. K. Bullough, P. J. Caudrey:* The Soliton and Its History. – *G. L. Lamb Jr., D. W. McLaughlin:* Aspects of Soliton Physics. – *R. K. Bullough, P. J. Caudrey, H. M. Gibbs:* The Double Sine-Gordon Equations: A Physically Applicable System of Equations. – *M. Toda:* On a Nonlinear Lattice (The Toda Lattice). – *R. Hirota:* Direct Methods in Soliton Theory. – *A. C. Newell:* The Inverse Scattering Transform. – *V. E. Zakharov:* The Inverse Scattering Method. – *M. Wadati:* Generalized Matrix Form of the Inverse Scattering Method. – *F. Calogero, A. Degasperis:* Nonlinear Evolution Equations Solvable by the Inverse Spectral Transform Associated with the Matrix Schrödinger Equation. – *S. P. Novikov:* A Method of Solving the Periodic Problem for the KdV Equation and Its Generalization. – *L. D. Faddeev:* A Hamiltonian Interpretation of the Inverse Scattering Method. – *A. Luther:* Quantum Solitons in Statistical Physics. – Futher Remarks on John Scott Russel and on the Early History of His Solitary Wave. – Note Added in Proof. – Additional References with Titles. – Subject Index.

Real-Space Renormalization

Editors: **T. W. Burkhardt, J. M. J. van Leeuwen**

1982. 60 figures. XIII, 214 pages. (Topics in Current Physics, Volume 30). ISBN 3-540-11459-9

Contents: *T. W. Burkhardt, J. M. J. van Leeuwen:* Progress and Problems in Real-Space Renormalization. – *T. W. Burkhardt:* Bond-Moving and Variational Methods in Real-Space Renormalization. – *R. H. Swendsen:* Monte Carlo Renormalization. – *G. F. Mazenko, O. T. Valls:* The Real-Space Dynamic Renormalization Group. – *P. Pfeuty, R. Jullien, K. A. Penson:* Renormalization for Quantum Systems. – *M. Schick:* Application of the Real-Space Renormalization to Absorbed Systems. – *H. E. Stanley, P. J. Reynolds, S. Redner, F. Family:* Position-Space Renormalization Group for Models of Linear Polymers, Branched Polymers, and Gels.

Springer-Verlag
Berlin
Heidelberg
New York

H. Grabert
Projection Operator Techniques in Nonequilibrium Statistical Mechanics

1982. 4 figures. Approx. 220 pages. (Springer Tracts in Modern Physics, Volume 95)
ISBN 3-540-11635-4
In preparation

Contents: Introduction and Survey. - The Projection Operator Technique. Statistical Thermodynamics. - The Fokker-Planck Equation Approach. The Master Equation Approach. Response Theory. - Damped Nonlinear Oscillator. - Simple Fluids. - Spin Relaxation. - References. - Subject Index.

Monte Carlo Methods in Statistical Physics

Editor: **K. Binder**
1979. 91 figures, 10 tables. XV, 376 pages. (Topics in Current Physics, Volume 7)
ISBN 3-540-09018-5

Contents: *K. Binder:* Introduction: Theory and "Technical" Aspects of Monte Carlo Simulations. - *D. Levesque, J. J. Weis, J. P. Hansen:* Simulation of Classical Fluids.. - *D. P. Landau:* Phase Diagrams of Mixtures and Magnetic Systems. - *D. M. Ceperley, M. H. Kalos:* Quantum Many-Body Problems. - *H. Müller-Krumbhaar:* Simulation of Small Systems. - *K. Binder, M. H. Kalos:* Monte Carlo Studies of Relaxation Phenomena: Kinetics of Phase Changes and Critical Slowing Down. - *H. Müller-Krumbhaar:* Monte Carlo Simulation of Crystal Growth. - *K. Binder, D. Stauffer:* Monte Carlo Studies of Systems with Disorders. - *D. P. Landau:* Applications in Surface Physics.

M. Toda, R. Kubo, N. Saito
Statistical Physics I

Equilibrium Statistical Mechanics
1982. 90 figures. Approx. 250 pages. (Springer Series in Solid-State Sciences, Volume 30)
ISBN 3-540-11460-2
In preparation

The fundamentals of equilibrium statistical mechanics are discussed in this text, focusing on basic physical aspects. It assumes no previous knowledge of thermodynamics or of the molecular theory of gases. Topics are drawn from simple materials and photon systems in order to elucidate basic ideas and methods involved.

The first chapter lays the ground for a discussion of probability and kinetics. Chapter 2 is concerned with general principles of statistical mechanics, the starting point of the theory of microscopic systems. Chapte 3 and 4 are devoted to fundamental applications, including quantum statistics and classical statistical mechanics as a limiting case. Imperfect gases and electrolytes are also covered. The basic ideas of treating phase change are described, with attention paid to the relation between exact and approximate theories. Recent theories of critical phenomena are presented in some detail. In the last chapter the ergodic problem is considered in an effort to explain the mechanical basis of statistical mechanics.

R. Kubo, M. Toda, N. Hashitsume
Statistical Physics II

Non-Equilibrium Statistical Mechanics
1982. (Springer Series in Solid-State Sciences, Volume 31). ISBN 3-540-11461-0
In preparation

The textbook starts out with the basic concepts and phenomenological theories of simple non-equilibrium processes as described by stochastic processes, emphasizing the viewpoint that a physical process in generated by an underlying, more microscopic process. The physical and mathematical meaning of coarse graining is thus illustrated by simple examples of Brownian motion and its generalization. Their general features of relaxation and dissipative processes are discussed in detail. The linear-response theory is introduced to link non-equilibrium processes to fluctuations in equilibrium characterized by relevant sorts of correlation of Green's functions. The field-theoretical methods are developed for microscopic calculations of Green's functions taking examples of plasma physics.

Springer-Verlag
Berlin
Heidelberg
New York